Ecology of Marine Zooplankton

Ecology of Marine Zooplankton

Editors

Marco Uttieri
Ylenia Carotenuto
Iole Di Capua
Vittoria Roncalli

Basel • Beijing • Wuhan • Barcelona • Belgrade • Novi Sad • Cluj • Manchester

Editors
Marco Uttieri
Stazione Zoologica Anton Dohrn
Naples, Italy

Ylenia Carotenuto
Stazione Zoologica Anton Dohrn
Naples, Italy

Iole Di Capua
Stazione Zoologica Anton Dohrn
Naples, Italy

Vittoria Roncalli
Stazione Zoologica Anton Dohrn
Naples, Italy

Editorial Office
MDPI
St. Alban-Anlage 66
4052 Basel, Switzerland

This is a reprint of articles from the Special Issue published online in the open access journal *Journal of Marine Science and Engineering* (ISSN 2077-1312) (available at: https://www.mdpi.com/journal/jmse/special_issues/Tian_Ecology_Marine_Zooplankton).

For citation purposes, cite each article independently as indicated on the article page online and as indicated below:

Lastname, A.A.; Lastname, B.B. Article Title. *Journal Name* **Year**, *Volume Number*, Page Range.

ISBN 978-3-0365-9879-6 (Hbk)
ISBN 978-3-0365-9880-2 (PDF)
doi.org/10.3390/books978-3-0365-9880-2

Cover image courtesy of Iole Di Capua

Contents

About the Editors

Marco Uttieri

Marco Uttieri (PhD) is a researcher at the Department of Integrative Marine Ecology at the Stazione Zoologica Anton Dohrn (Naples, Italy). His primary research focuses on the biology and ecology of zooplanktonic organisms from both marine and freshwater systems, using an integrated multidisciplinary approach. His research addresses the analysis of individual behavior, the implementation of numerical models, and the interpretation of the ecological significance of behavioral adaptations.

Ylenia Carotenuto

Ylenia Carotenuto (PhD) is a researcher at the Department of Integrative Marine Ecology at the Stazione Zoologica Anton Dohrn (Naples, Italy). Her scientific interests are in the field of physiological ecology, chemical ecology, and the molecular ecology of zooplankton copepods. She investigates prey–predator interactions at gene, organism and population levels, and their impact on copepod reproductive fitness and secondary production at sea. Her integrated multidisciplinary approach involves the assessment of the reproductive and developmental output of species and individual gene expression using RT-qPCR and transcriptomic techniques; this is performed in laboratory-controlled set-up and in situ environments over various spatial and temporal scales.

Iole Di Capua

Iole Di Capua (MSc) is a technologist at the Marine Organism Taxonomy Facility of the Department of Research Infrastructures for marine biological resources at the Stazione Zoologica Anton Dohrn (Naples, Italy). Her primary research interests are the diversity, phylogeny and ecology of marine metazoans (especially zooplankton). She uses the strengths of different techniques and integrated approaches (morphological and molecular) to identify the temporal and spatial occurrences of species, assemblages and communities, and to integrate environmental DNA analyses into marine biodiversity and ecosystem functioning research.

Vittoria Roncalli

Vittoria Roncalli (PhD) is a researcher at the Department of Integrative Marine Ecology at the Stazione Zoologica Anton Dohrn (Naples, Italy). Her main research focuses on understanding how zooplankters thrive and survive in response to environmental cues using a multidisciplinary approach that integrates laboratory "in vivo" experiments with molecular techniques (-omics). Investigating the physiological state of organisms provides insights on species-specific interactions, niche partitioning, and can help to understand the ecology of marine zooplankters and to predict changes at the community level.

Preface

As Guest Editors, we are very pleased to present this book based on the Special Issue "Ecology of Marine Zooplankton", published in the *Journal of Marine Science and Engineering*. Zooplanktonic organisms preside over a crucial position within marine food webs, providing the link between primary producers and higher trophic levels, but also sustaining the vertical exchanges of matter and energy within deep-sea and benthic communities. The contributions collected in this volume provide novel insights into the biology and ecology of zooplanktonic organisms with regard to multiple lines of research, scales and environments, and will surely stimulate discussion in the reference scientific community.

Marco Uttieri, Ylenia Carotenuto, Iole Di Capua, and Vittoria Roncalli
Editors

Journal of
Marine Science and Engineering

MDPI

Editorial

Ecology of Marine Zooplankton

Marco Uttieri [1,2,*], Ylenia Carotenuto [1,*], Iole Di Capua [1,2,*] and Vittoria Roncalli [1,2,*]

1 Stazione Zoologica Anton Dohrn, Villa Comunale, 80121 Naples, Italy
2 National Biodiversity Future Center (NBFC), Piazza Marina 61, 90133 Palermo, Italy
* Correspondence: marco.uttieri@szn.it (M.U.); ylenia.carotenuto@szn.it (Y.C.); iole.dicapua@szn.it (I.D.C.); vittoria.roncalli@szn.it (V.R.)

1. Overview

Marine ecosystems, from coastal areas to open waters, teem with a multitude of heterotrophic and mixotrophic organisms collectively forming the zooplankton, the animal component of the plankton. Zooplankton is an extremely variegated group, with an outstanding phylogenetic, taxonomic and functional diversity [1], a biological richness that captivated even Charles Darwin during his voyage aboard the HMS Beagle, as described in [2]. Almost all phyla are represented in marine zooplankton, although crustaceans represent the dominant component [3].

Dimensionally speaking, the size of these organisms ranges between 2.0 μm (nanozooplaknton) and 20 m (megazooplankton) [4], covering an exceptionally wide gamut of size fractions. A typical distinction is made between holoplanktonic and meroplanktonic species, with the former spending their entire life cycle in a pelagic form and the latter spending only a transitory planktonic stage [5]. Dietary speaking, zooplankton as a whole includes bacterivores, herbivores, carnivores, omnivores, and detritivors, as well as parasitic forms [5]. From an ecological perspective, zooplanktonic organisms provide the linchpin between different trophic levels; they contribute to the biological carbon pump, regulate the biomass stock of other planktonic groups, affect ecosystem dynamics, are excellent beacons of climate change, and are crucial in providing ecosystem services, as recently reviewed in [6,7]. As such, improving our understanding of the ecological role of zooplankton implies improving our knowledge of the functioning of marine ecosystems as a whole [7].

The overarching goal of this Special Issue, themed "Ecology of Marine Zooplankton", is to present novel research on the biology and ecology of zooplanktonic organisms. The collection includes nine articles, one opinion paper, and one review. The subjects cover multiple themes, from host–parasite interactions to seasonal variability, over a wide range of scales—from the molecular to the population one scale—and systems investigated—from lagoons to hydrothermal vents. The result is a cross-cutting, strongly interdisciplinary volume that may attract the interest of researchers from different fields.

Citation: Uttieri, M.; Carotenuto, Y.; Di Capua, I.; Roncalli, V. Ecology of Marine Zooplankton. *J. Mar. Sci. Eng.* **2023**, *11*, 1875. https://doi.org/10.3390/jmse11101875

Received: 26 July 2023
Accepted: 10 September 2023
Published: 27 September 2023

2. Contributions

In their paper, Litvinyuk et al. [8] perform a study to assess the non-consumptive mortality rate of zooplanktonic organisms, mainly copepods, and the decomposition and sedimentation rates of carcasses in Sevastopol Bay. Their work reveals a high variability in these parameters, suggesting a reduced sedimentation rate of copepod carcasses in turbulent conditions, and a comparable rate of sedimentation and microbial decomposition, confirming the important role of copepod carcasses in coastal waters.

Köster and Paffenhöfer [9] investigate the role of the predation by the doliolid *Dolioletta gegenbauri* on the abundance of the small neritic copepod *Paracalanus quasimodo*. Their laboratory experiments show that *D. gegenbauri* can ingest *P. quasimodo* eggs at a rate similar to that with which the doliodid preys upon phytoplankton cells. Conversely, the predation on copepod nauplii is significantly lower, likely due to the ability of motile nauplii to detect *D. gegenbauri* feeding currents. Based on these outcomes, the authors speculate about the effect of doliolid predation on copepod community composition.

The effects of several coccolithophore species, differing in cell size, and carbon and calcite content, on copepod grazing (ingestion and egestion rates) are investigated in the copepods *Temora longicornis* and *Acartia clausi* by Toullec et al. [10]. The authors find that the cellular volume and calcite content of the species strongly affect the copepod foraging capability and production of faecal pellets. In particular, contrary to the optimal foraging theory, copepod ingestion rates increase exponentially with food availability, likely due to food quality (calcite content). A decoupling between ingestion and egestion rates is also associated with a possible obstruction of the copepod gut related to calcite itself. Their study has important implications for the production and sedimentary flux of copepod faecal pellets into deeper waters.

Martinelli Filho et al. [11] report for the second time the infection of paracalanid copepods by the alveolate parasite *Ellobiopsis chattoni* Caullery, 1910 in South Atlantic, in subtropical coastal areas in the south-east of Brazil. *E. chattoni* is mostly found attached to the cephalosome appendages of *Paracalanus* spp. And *Parvolacanus crossirostris*, and is rarely found in the copepod taxa (59) identified in the same samples. However, parasitized copepods are mainly females rather than males and juveniles, and the highest percentage of infected copepods is observed in the winter and summer seasons of different years. This study shows that this infection by the alveolate has a negative impact on the growth and fitness of future copepod populations.

Zooplankton communities are investigated in studies by Gubanova et al. [12] and Chaigneau et al. [13], respectively, in response to climate change and seasonal variations in the Black Sea and a lagoon in West Africa, areas where little is known about the zooplankton diversity. Gubanova et al. [12] assess the response of the mesozooplankton community in Sevastopol Bay, a semi-enclosed estuarine-type bay, to the most persistent and intense marine heat wave recorded in the Black Sea (summer 2010). Using long-term routine observations (2003–2014), the study reports seasonal variations in zooplankton composition, abundance, and structure; warm water and non-native species (e.g., *Oithona davisae* and *Acartia tonsa*) showed the maximum seasonal density, suggesting their greater flexibility to adapt in response to environmental changes. *O. davisae* is suggested as an indicator of the environmental conditions associated with the warming of the Black Sea and the whole Mediterranean basin.

Chaigneau et al. [13] investigate zooplankton diversity and abundance in the Nokoué Lagoon in southern Benin (West Africa). In response to the high seasonal variations of salinity, the authors report differences in the zooplanktonic assemblages: during high water periods (fresh water), zooplanktonic diversity and abundances are quite high, mostly dominated by rotifers, compared with brackish water periods, when diversity is minimal and abundance decreases slightly. However, in some areas of the lagoon, changes in zooplankton abundances are independent of salinity levels, suggesting other factors (e.g., riverine inputs, fish traps) as potential drivers.

The spatial and temporal variability of plankton depends on environmental parameters. In their contribution, Prakash et al. [14] investigate the role of salinity gradients on bacterioplankton, phytoplankton, and zooplankton abundance and diversity in the highly productive Hooghly River Estuary in West Bengal, India. They find zooplankton distribution strongly affected by water circulation, bacteria, and Chl *a* content, with higher abundances of rotifers and cladocerans in lower salinity stations and copepod dominance in downstream stations with higher salinity. Their results confirm the importance of foraging strategies (bacterivory, herbivory, and omnivory) in shaping plankton communities, which could have implications for the production of commercially valuable fish and shrimp species in the estuary.

The structure of the mesozooplankton community in relation to water mass conditions in the Southeast China Sea is also studied by Wang et al. [15]. The authors find significant changes in the mesozooplankton community structure and copepod assemblages in the upwelling cold dome region, formed by the Kuroshio Current intrusion in the Southeast China Sea during the southwest monsoon. Copepod species indicators of low temperature and nutrient-rich water masses characterize the cold dome with respect to the area sampled the following season.

Hydrothermal vents represent perfect natural laboratories to study the ocean biota in future climate scenarios. In their opinion paper, Dahms et al. [16] inquire into the appropriateness of these systems for zooplankton studies, reviewing the available literature on the topic. The authors conclude that shallow water vents can offer a unique possibility to understand the possible effects of global change on the resident and allochthonous zooplankton assemblage, and propose leading questions to be addressed in future studies.

The iron metabolism in copepods is investigated in Roncalli et al. [17]. Attention is focused on identifying transcripts encoding ferritin, a highly conserved and ubiquitous multimeric iron storage protein required for the maintenance of iron homeostasis. Using an in silico workflow on 27 publicly available copepod transcriptomes, the authors describe the diversity of these proteins and infer their functions using gene expression data in three target species exposed to stressors and across development. Results point to species-specific differences suggesting ferritins as potential copepod biomarkers of multiple processes, such as development, stress response, and iron storage.

An updated review of the distribution of the non-indigenous calanoid copepod *Pseudodiaptomus marinus* in European and neighboring waters is given in Uttieri et al. [18]. Starting from a previous survey, the authors summarize published literature (from fall 2019 to date) and present original evidence showing the continuous expansion of this species. The data presented provide a real-time snapshot of the occurrence of *P. marinus* and are used to hypothesize future distribution scenarios.

3. Conclusions

The contributions included in this Special Issue cast fresh light on the complexity of zooplankton ecology, and further our current knowledge on the mechanisms regulating processes and dynamics taking place at different spatial and temporal scales. Such comprehension, however, is still far from being exhaustive: much has been undertaken over the last decades, but more is yet to come. As guest editors, we gratefully acknowledge the dedication of all contributing authors, and the time devoted by the reviewers to assess the quality and merit of the submitted works. We are confident that the reference scientific community will be deeply inspired by the papers included in this topical collection, which will surely stimulate new and productive research ideas.

Author Contributions: Conceptualization, all authors; writing—original draft preparation, all authors; writing—review and editing, all authors. All authors have read and agreed to the published version of the manuscript.

Funding: I.D.C., V.R. and M.U. acknowledge the support of NBFC to Stazione Zoologica Anton Dohrn, funded by the Italian Ministry of University and Research, PNRR, Missione 4 Componente 2, "Dalla ricerca all'impresa", Investimento 1.4, Project CN00000033.

Acknowledgments: The guest editors would like to thank all of the contributing authors and reviewers who devoted much of their time to realize this Special Issue.

Conflicts of Interest: The authors declare no conflict of interest.

References

1. Bucklin, A.; Peijnenburg, K.T.C.A.; Kosobokova, K.N.; O'Brien, T.D.; Blanco-Bercial, L.; Cornils, A.; Falkenhaug, T.; Hopcroft, R.R.; Hosia, A.; Laakmann, S.; et al. Toward a global reference database of COI barcodes for marine zooplankton. *Mar. Biol.* **2021**, *168*, 78. [CrossRef]
2. Richardson, A.J.; Uribe-Palomino, J.; Slotwinski, A.; Coman, F.; Miskiewicz, A.G.; Rothlisberg, P.C.; Young, J.W.; Suthers, I.M. Coastal and Marine Zooplaknton: Identification, Biology and Ecology. In *Plankton: A Guide to Their Ecology and Monitoring for Water Quality*, 2nd ed.; Suthers, I.M., Rissik, D., Richardson, A.J., Eds.; CSIRO Publishing: Clayton South, Australia, 2019; pp. 141–208.
3. Suthers, I.; Dawson, M.; Pitt, K.; Miskiewicz, A.G. Coastal and Marine Zooplankton: Diversity and Biology. In *Plankton: A Guide to Their Ecology and Monitoring for Water Quality*; Suthers, I.M., Rissik, D., Eds.; CSIRO Publishing: Collingwood, Australia, 2009; pp. 181–222.
4. Sieburth, J.M.; Smetacek, V.; Lenz, J. Pelagic ecosystem structure: Heterotrophic compartments of the plankton and their relationship to plankton size fractions. *Limnol. Oceanogr.* **1978**, *23*, 1256–1263. [CrossRef]
5. Alcaraz, M.; Calbet, A. Zooplankton ecology. In *Marine Ecology*; Duarte, C.M., Lot Helgueras, A., Eds.; Encyclopedia of Life Support Systems (EOLSS), Developed under the Auspices of the UNESCO; EOLSS Publishers: Paris, France, 2009; Volume I, pp. 295–318.
6. Ratnarajah, L.; Abu-Alhaija, R.; Atkinson, A.; Batten, S.; Bax, N.J.; Bernard, K.S.; Canonico, G.; Cornils, A.; Everett, J.D.; Grigoratou, M.; et al. Monitoring and modelling marine zooplankton in a changing climate. *Nat. Commun.* **2023**, *14*, 564. [CrossRef]
7. Lomartire, S.; Marques, J.C.; Gonçalves, A.M.M. The key role of zooplankton in ecosystem services: A perspective of interaction between zooplankton and fish recruitment. *Ecol. Indic.* **2021**, *129*, 107867. [CrossRef]
8. Litvinyuk, D.; Mukhanov, V.; Evstigneev, V. The Black Sea zooplankton mortality, decomposition, and sedimentation measurements using vital dye and short-term sediment traps. *J. Mar. Sci. Eng.* **2022**, *10*, 1031. [CrossRef]
9. Köster, M.; Paffenhöfer, G.A. On the predation of doliolids (Tunicata, Thaliacea) on calanoid copepods. *J. Mar. Sci. Eng.* **2022**, *10*, 1293. [CrossRef]
10. Toullec, J.; Delegrange, A.; Perruchon, A.; Duong, G.; Cornille, V.; Brutier, L.; Hermoso, M. Copepod feeding responses to changes in coccolithophore size and carbon content. *J. Mar. Sci. Eng.* **2022**, *10*, 1807. [CrossRef]
11. Martinelli Filho, J.E.; Gusmão, F.; Alves-Júnior, F.A.; Lopes, R.M. The infection of paracalanid copepods by the alveolate parasite *Ellobiopsis chattoni* Caullery, 1910 in a subtropical coastal area. *J. Mar. Sci. Eng.* **2022**, *10*, 1816. [CrossRef]
12. Gubanova, A.; Goubanova, K.; Krivenko, O.; Stefanova, K.; Garbazey, O.; Belokopytov, V.; Liashko, T.; Stefanova, E. Response of the Black Sea zooplankton to the marine heat wave 2010: Case of the Sevastopol Bay. *J. Mar. Sci. Eng.* **2022**, *10*, 1933. [CrossRef]
13. Chaigneau, A.; Ouinsou, F.T.; Akodogbo, H.H.; Dobigny, G.; Avocegan, T.T.; Dossou-Sognon, F.U.; Okpeitcha, V.O.; Djihouessi, M.B.; Azémar, F. Physicochemical drivers of zooplankton seasonal variability in a west African lagoon (Nokoué Lagoon, Benin). *J. Mar. Sci. Eng.* **2023**, *11*, 556. [CrossRef]
14. Prakash, D.; Tiwary, C.B.; Kumar, R. Ecosystem variability along the estuarine salinity gradient: A case study of Hooghly River Estuary, West Bengal, India. *J. Mar. Sci. Eng.* **2023**, *11*, 88. [CrossRef]
15. Wang, Y.-G.; Tseng, L.-C.; Chen, X.-Y.; Sun, R.-X.; Xiang, P.; Xing, B.-P.; Wang, C.-G.; Hwang, J.-S. Cold dome affects mesozooplankton communities during the southwest monsoon period in the southeast East China Sea. *J. Mar. Sci. Eng.* **2023**, *11*, 508. [CrossRef]
16. Dahms, H.-U.; Thirunavukkarasu, S.; Hwang, J.-S. Can marine hydrothermal vents be used as natural laboratories to study global change effects on zooplankton in a future ocean? *J. Mar. Sci. Eng.* **2023**, *11*, 163. [CrossRef]
17. Roncalli, V.; Uttieri, M.; Carotenuto, Y. The distribution of ferritins in marine copepods. *J. Mar. Sci. Eng.* **2023**, *11*, 1187. [CrossRef]
18. Uttieri, M.; Anadoli, O.; Banchi, E.; Battuello, M.; Beşiktepe, Ş.; Carotenuto, Y.; Cotrim Marques, S.; de Olazabal, A.; Di Capua, I.; Engell-Sørensen, K.; et al. The distribution of *Pseudodiaptomus marinus* in European and neighbouring waters-A rolling review. *J. Mar. Sci. Eng.* **2023**, *11*, 1238. [CrossRef]

Journal of
Marine Science and Engineering

Article

The Black Sea Zooplankton Mortality, Decomposition, and Sedimentation Measurements Using Vital Dye and Short-Term Sediment Traps

Daria Litvinyuk [1,2,*], Vladimir Mukhanov [1,2] and Vladislav Evstigneev [2]

[1] A.O. Kovalevsky Institute of Biology of the Southern Seas, Russian Academy of Sciences, 299011 Sevastopol, Russia; v.s.mukhanov@gmail.com
[2] Laboratory of Marine Ecosystems, Institute for Advanced Research, Sevastopol State University, 299053 Sevastopol, Russia; vald_e@rambler.ru
* Correspondence: d.litvinyuk@ibss-ras.ru

Abstract: The principal objectives of this research are to measure the non-consumptive mortality rate of marine copepod zooplankton and the sedimentation rate of copepod carcasses, using short-term sediment traps, and to reveal a correlation between the rates of the two competitive processes—sedimentation and degradation of the carcasses under turbulent mixing conditions. The traps were moored in Sevastopol Bay and adjacent coastal waters (the Black Sea) during summer and autumn seasons. A simulation model was developed to describe a wide range of processes in the trap and the water column above it and to interpret the results obtained with the sediment traps. Significant changes in the abundance of copepod carcasses (from 280 to 12,443 ind. m^{-3}) and their fraction in the total zooplankton abundance (53 to 81%) were observed in the waters over short time periods, indicating a high variability of zooplankton mortality, sedimentation, and decomposition rates. Despite the high concentrations of copepod carcasses in the water column, the rates of their accumulation in the traps proved to be extremely low, which could be due to intense turbulent mixing of the waters. The carcass sedimentation rate and the flow of swimmers (motile copepods) into the traps were significantly higher in waters subjected to weaker turbulent mixing. The obtained estimates of the sedimentation rate of copepod carcasses (0.012 to 0.39 d^{-1}) were comparable in value with the rate of their microbial decomposition (0.13 and 0.05 d^{-1} in the bay and adjacent waters, respectively). This confirmed the hypothesis on microbial decomposition as one of the key controls of the fraction of live zooplankton organisms in zooplankton.

Keywords: mesozooplankton; copepod; mortality; carcasses; decomposition; sedimentation; sediment trap; fluorescein diacetate (FDA); Sevastopol Bay; Black Sea

Citation: Litvinyuk, D.; Mukhanov, V.; Evstigneev, V. The Black Sea Zooplankton Mortality, Decomposition, and Sedimentation Measurements Using Vital Dye and Short-Term Sediment Traps. *J. Mar. Sci. Eng.* **2022**, *10*, 1031. https://doi.org/10.3390/jmse10081031

Academic Editors: Marco Uttieri, Ylenia Carotenuto, Iole Di Capua and Vittoria Roncalli

Received: 20 June 2022
Accepted: 19 July 2022
Published: 27 July 2022

1. Introduction

Zooplankton are essential components of the marine food web, mediating the flow of primary production upwards to higher trophic levels [1], and directly affecting pelagic fish populations and the biological pump of carbon into the deep ocean [2]. Marine ecosystems were shown to be quite sensitive to zooplankton mortality which can modify elemental fluxes into the ocean abyss and alter the balance of pelagic assemblages [2,3]. So, it is crucial to improve zooplankton viability assays, develop the methods for reliable measurement of mortality rate, and have a good understanding of the processes associated with zooplankton mortality and carcass decomposition in the water column.

Dead plankton organisms, including copepod carcasses, have long been the object of hydrobiological research. Various reasons for the plankton mortality, from starvation and disease to algal bloom and environmental pollution, were also of great interest [4–8]. Nevertheless, there are hardly any studies on the linkage of such important phenomenon as non-consumptive (non-predatory) mortality of zooplankton to pollution and trophic status of marine waters. The rate of this process is very difficult to measure in situ due to a number of methodological complications [9]. This is the reason why researchers focus mainly on indirect indicators which include the fraction of live organisms (FLO) in the community. They identify and enumerate dead (or live) organisms, using simple and, to a great extent, subjective methods (light microscopy and visual identification) for calculating the live/dead organisms ratio [6,9–11]. The main innovations are aimed at automated methods to study abundance and taxonomic composition of plankton, applying such instruments as Zooscan, ZooCAM, and FlowCAM [12,13]. But the potential of these technologies to assess FLO and study the plankton mortality has not been fulfilled yet. In our research, we used a novel fluorescent marker, diacetate fluorescein (FDA), and an original method of semi-automated (i.e., excepting any subjectivity) sorting of live organisms in zooplankton samples [14,15].

Our previous evaluations of the mortality of dominant crustacean plankton in Sevastopol Bay and adjacent waters [14] have revealed that the mean annual FLO was higher in more polluted waters of the bay corner. On the contrary, in more clean waters outside the bay, FLO was low. In order to explain the contradiction between these results and the already well-established (and seemingly obvious) idea that pollution leads to the death of organisms and, accordingly, to a decrease in FLO [16], we put forward and experimentally confirmed the hypothesis of a more intense bacterial decomposition of dead organisms in the polluted and eutrophicated waters of the bay and, as a possible consequence, an increase in FLO [15]. Indeed, a pool of dead organisms is formed as a result of natural (non-consumptive) mortality of zooplankton, while carcasses are removed from the water column due to the two main processes: bacterial degradation [6] and sedimentation [4,17].

The sedimentation rate of dead organisms is measured in situ based the rate of their accumulation in sediment traps installed near the bottom in different water areas [5,9,18,19]. However, the majority of such investigations were carried out in calm fresh waters where sedimentation is the main way of removal of the carcasses from the water column. Consequently, other contributing factors, such as carcass degradation, were neglected, supposing that their sedimentation occurs faster than their degradation [8,18,19]. But a series of other studies demonstrated that under turbulence and stratification, dead zooplankton may get suspended in the water column and act as an additional source of matter and energy for bacteria [20,21].

Turbulent mixing of water, utilization of carcasses by detritophages, and their decomposition by bacteria are among the processes which are commonly ignored by the researchers of zooplankton mortality making the methods they use unreliable. In this study, we just tried to overcome some of these difficulties by developing an appropriate simulation model describing the processes in the sediment trap more precise. The principal objectives of our research are to measure the non-consumptive mortality rate of the copepod zooplankton and the sedimentation rate of copepod carcasses, and to reveal a correlation between the rates of the two competitive processes—sedimentation and degradation of the carcasses, which determine the FLO dynamics under turbulent mixing conditions.

2. Materials and Methods

2.1. Study Sites and Experimental Design

A total of three series of experiments were carried out in the coastal waters of the SW Crimea (the Black Sea). The experiment BAY11 was run at Station B in Sevastopol Bay on 28–30 November 2017. The experiments SEA05 and SEA11 were conducted at Station *S* in the adjacent waters (a mile off the entrance to the bay, with the depth down to 40 m) on 30–31 May 2017 and 16–23 November 2017, respectively (Figure 1).

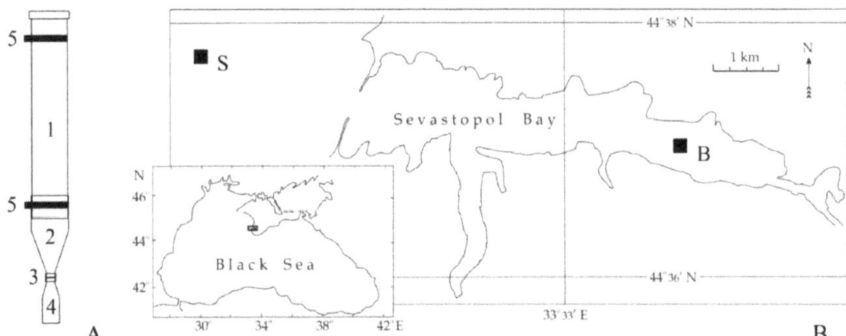

Figure 1. (**A**)—Design of a unit of the sedimentation trap according to [22]: 1—plastic tube; 2—funnel; 3—screwed joint connector; 4—collection cup; 5—connecting clamp; (**B**)—sites of the trap deployment.

The areas chosen for the research differ considerably in pollution level and trophic status. Sevastopol Bay is located at the south-western part of the Crimean Peninsula (the Black Sea). It is a semi-enclosed estuary-type body of water with a reduced water renewal rate (because of a mole at the entrance) and exposure to chronic industrial and anthropogenic stress. From the mouth of the Chernaya river in the corner of the bay (St. B) to the open water outside the bay (St. S) there were observed a gradual increase in water salinity, and a decrease in the levels of pollution and trophicity. Waters of the bay are characterized by high concentrations of nutrients, several times exceeding their background readings in the open sea: up to 290 μmol L^{-1} of nitrite nitrogen, up to 6.4 μmol L^{-1} of nitrate nitrogen, up to 41 μmol L^{-1} of ammonium nitrogen, up to 5.2 μmol L^{-1} of phosphates, up to 98 μmol L^{-1} of silicates [23]. The eutrophication E-TRIX index was shown to change on average from 5.05 in the bay at St. B to 4.70 in the open sea at St. S, characterizing the level of trophicity of Sevastopol Bay as a transition from medium to high [24]. Chronic oil pollution level increases from the open waters outside the bay to its central part. In particular, the total amount of chloroform extractable organic compounds ranges from 0.9 to 26.8 mg g^{-1} of air-dried bottom sediments. The highest oil pollution levels are revealed in the central part of the bay, with a maximum concentration of 13.4 mg g^{-1} [25].

Temperature and wind condition at the stations were also different during the trap deployment. At St.S, water surface temperature was 19.1 °C and remained unchanged through the exposition time of the trap during the experiment SEA05. Wind speed did not exceed 2.3 m s^{-1}, its direction changed from eastern to north-western. In November (SEA11 experiment), north-eastern winds prevailed with gusts up to 5–7 m s^{-1}, water surface temperature decreased from 13.4 °C to 11.9 °C during the 7-day trap exposition. At St. B in the bay (BAY11 experiment), average wind speed was 3 m s^{-1}, and temperature increased from 11.0 °C to 12.4 °C.

2.2. Design and Exposure Conditions of Sediment Traps

According to [22], the trap tube unit was custom-made of a 120-cm long plastic tube with entry-hole diameter of 110 mm (Figure 1A). The funnel was a 2-L plastic bottle, mounted on a pipe and by means of a screwed joint connector attached to a collection cup (0.5-L plastic bottle). The sediment trap in full assembly included four fastened tube units.

The traps were moored as close to the bottom as possible, anchoring the device with a 30-kg load at the depth of 8 m at St. B in the bay (BAY11 experiment) and 16–36 m at St. S in the open sea (SEA05 and SEA11 experiments). In order to keep the trap vertical, a submerged buoy was used. The trap location was marked with a signal buoy on the water surface.

Since few zooplankton carcasses were found in the traps after the 24-h experiment (SEA05), the trap exposition had to be prolonged from 2 (BAY11) to 7 (SEA11) days. Thus, the sediment traps were deployed for 1, 2, and 7 days during the experiments SEA05, BAY11, and SEA11, respectively.

In the experiments SEA11 and BAY11, the collection cups of two units (poisoned units) were filled with 40% formaldehyde (fin. conc. 2%). The other two units of the trap had no preservatives (non-poisoned units). First, the preservative prevented microbial decomposition of copepods in the trap and, second, it killed accidental "swimmers" getting in the trap, thus providing their accumulation in the collection cup. Consequently, a comparison of the data from the poisoned and non-poisoned units allowed us to estimate the two processes, the carcass decomposition rate and the swimming rate of alive copepods into the traps.

The contents of the sediment cups and the trap tubes in the poisoned and non-poisoned units were analyzed independently. Live organisms were identified in the non-poisoned units, using a vital stain (see below). The zooplankton found in the sediment cup of the poisoned unit were considered dead and, hence, were not stained for further viability assay. The contribution of the swimmers to the total abundance of copepods in the poisoned unit was estimated, using a simulation model.

2.3. Evaluation of Total Abundance of Zooplankton and Fraction of Live Organisms (FLO)

To study zooplankton species composition, abundance, and FLO, samples of zooplankton were taken at the trap location with a Juday net (entry-hole diameter 37 cm, 150-μm mesh, filtering cod end) at the beginning of the experiment, immediately after the trap deployment, and after its exposition.

Zooplankton samples from the water column above the trap and all the trap units were studied under a light microscope according to [26,27]. The abundance of live and dead organisms was evaluated after staining the samples with fluorescein diacetate (FDA), following the original protocol [14,28]. The FDA solution was prepared in dimethyl sulfoxide (DMSO) (5 mg mL^{-1}) and stored at +4 °C. For FDA staining, 1 μL of FDA solution was added to 1 mL of sample material according to the method widely used in marine phytoplankton research [29]. The sample was stained for 40 min in the dark. Earlier, this fluorescent stain was first used in field studies of marine zooplankton as viability marker in a series of our studies [14,28,30].

Organisms were microphotographed in a Bogorov's glass chamber under an inverted microscope (Nikon Eclipse TS100-F) (×4, ×10) equipped with a photo- and video camera (Ikegami ICD-848P) in the fluorescent mode (blue excitation filter set). The obtained images were processed with ImageRegionColor (IRC) software for semi-automated estimation of the proportions of the live and dead zooplankton. The updated version IRC 2.0.2. developed specifically for our tasks includes discriminant analysis, and allows statistically significant differentiation between dead and live organisms, depending on intensity of their staining.

2.4. Estimation of Abundance and Physiological Activity of Bacteria

Flow cytometry was used for measuring the total abundance of bacteria in the water column and collection cups of the trap at the beginning and after the trap exposition. Bacterial cells were counted with a Beckman Coulter flow cytometer (Cytomics FC 500) equipped with blue laser (15 mW, 488 nm). Aliquots (1 mL) of water samples previously preserved in formaldehyde (fin. conc. 2%) were stained with SYBR Green I (Molecular Probes Inc.) following the procedures described in [31,32]. Fluorescence of SYBR-Green I in the FL1 green light (525 nm) was assumed to be proportional to the content of intracellular nucleic acids and was interpreted as a measure of specific metabolic activity of bacterial cells, according to [33]. The nonparametric Mann–Whitney test was used to compare seasons and locations in terms of zoo- and bacterioplankton abundances. The significance level for all tests was set at <0.05.

2.5. Simulation Model of Live and Dead Zooplankton Dynamics in Sediment Traps and Water Column above Them

A simulation model was developed to comprehend and interpret the results of the field experiments, and estimate the matter flows through the community, including consumptive and non-consumptive mortalities, sedimentation, and decomposition. The model included three spatially homogeneous sub-models for describing live and dead copepod zooplankton dynamics (i) in the water column above the trap (Figure 2, I), (ii) in the non-poisoned units (Figure 2, II), and (iii) in the poisoned units (Figure 2, III). Table 1 provides description and units for all the variables measured in the field experiments, as well as the model parameters.

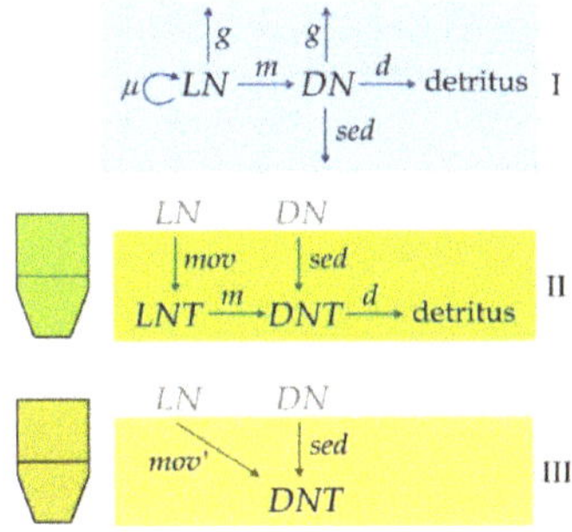

Figure 2. Simulation model describing the live and dead zooplankton dynamics in the water column above the sediment trap (sub-model I), in the non-poisoned trap unit (sub-model II), and in the poisoned trap unit (sub-model III). Description of the parameters is in Table 1.

Table 1. List of the measurable variables (*) and the model parameters.

Symbol	Description	Units
N_0	Initial zooplankton abundance in the water column *	ind. M^{-3}
N_t	Final zooplankton abundance in the water column *	ind. M^{-3}
FLO	Fraction of live organisms *	%
LN	Abundance of live organisms in the water column above the trap	ind. M^{-3}
LN_0	Initial abundance of live copepods in the water column above the trap *	ind. M^{-3}
LN_t	Final abundance oflivecopepodsin the watercolumnabovethetrap *	ind. M^{-3}
r_{live}	Apparent specific rate of growth/loss of live copepods *	d^{-1}
r_{dead}	Apparent specific rate of production/loss of dead copepods *	d^{-1}
DN	Abundance of dead copepods in the water column above the trap	ind. M^{-3}
DN_0	Initial abundance of dead copepods in the water column above the trap *	ind. M^{-3}
DN_t	Final abundance of dead copepods in the water column above the trap *	ind. M^{-3}
LNT	Abundance of live organisms (swimmers) in the trap *	ind.
DNT	Abundance of carcasses in the trap *	ind.
M	Non–consumptive mortality rate	d^{-1}
g	Consumptive mortality rate	d^{-1}
d	Carcass decomposition rate	d^{-1}
μ	Specific growth rate	d^{-1}
sed	Sedimentation rate	d^{-1}
mov	Net flow of swimmers into the non–poisoned trap unit	d^{-1}
mov'	Net flow of swimmers into the poisoned trap unit	d^{-1}
T	Duration of the experiment	d

The first sub-model (Figure 2, I) described dynamics of live (LN) and dead (DN) copepods in the water column above the trap and included specific growth rate of their populations (μ), non-consumptive (m), and consumptive mortality (g), as well as sedimentation (sed) and decomposition (d) of dead organisms:

$$\frac{dLN}{dt} = (\mu - g - m)\ LN \tag{1}$$

$$\frac{dDN}{dt} = m\ LN - (d + g + sed)\ DN \tag{2}$$

To simplify the model, a few assumptions were made, including: non-selective consumption of dead and live organisms by predators; equality of the non-consumptive mortality rate (m) in the non-poisoned unit and the above water column; equality of the decomposition rate (d) in the non-poisoned unit and the above water column. The value of d was calculated from previously obtained data (0.13 and 0.05 d^{-1} at the St. B and St. S, respectively, at 22.0 °C [15]), and the temperature coefficient Q_{10} = 2.4 [34].

The in situ sedimentation experiments allowed measuring the initial and final abundances of live and dead copepods in the water column above the trap (LN_0, DN_0, LN_t, and DN_t), and consequently, calculating the apparent growth rate of live (r_{live}) and dead (r_{dead}) copepods over the duration of the experiment (T):

$$r_{live} = \frac{(ln(LN_t) - ln(LN_0))}{T} \tag{3}$$

$$r_{dead} = \frac{(ln(DN_t) - ln(DN_0))}{T} \tag{4}$$

The empirical coefficients r_{live} and r_{dead} reflected the entire set of processes occurring in the water column and controlling the population of zooplankton: growth, mortality, sedimentation, and decomposition of organisms. The abundance of dead and live copepods in the water column (LN and DN, respectively) were external parameters for the sub-models II and III, describing processes in the traps. So, it was convenient to describe their dynamics with the equations:

$$\frac{dLN}{dt} = r_{live}\ LN \tag{5}$$

$$\frac{dDN}{dt} = r_{dead}\ DN \tag{6}$$

The sub-model II simulated the processes in the non-poisoned units of the trap (Figure 2, II). Apart from the copepod carcasses (DNT) sinking into the trap from the water column (sed) and suffering from bacterial decomposition there (d), the model took into account the accumulation of the live swimmers in the trap (LNT), their non-consumptive mortality (m), and decomposition of the carcasses (d) in the trap:

$$\frac{dDN}{dt} = m\ LNT + sed\ DN - d\ DNT \tag{7}$$

$$\frac{dLNT}{dt} = mov\ LN - m\ LNT \tag{8}$$

As it was stated above, the "external" variables LN and DN were defined from the Equations (5) and (6), using the empirical coefficients r_{live} and r_{dead} (the Equations (3) and (4)). Since the model did not account for the swimmers getting from the trap back to the water column, the coefficient mov presented the balance between inflow and outflow of the swimmers in/out the trap. Consequently, it appeared to differ significantly in the poisoned and non-poisoned units, and depend on behavior of the zooplankton. For example, organisms might have been attracted into the trap by excess of food or avoided the toxic content of the trap. On the contrary, the rates of sedimentation and accumulation of carcasses in both the units were supposedly the same.

The third sub-model (Figure 2, III) described dynamics of dead organisms (*DNT*) in the poisoned unit as a function of live (*LN*) and dead (*DN*) copepods abundance in the water column above the trap, net flow of the swimmers in/out the trap (*mov′*) and sedimentation of copepod carcasses from the water column (*sed*):

$$\frac{dDNT}{dt} = mov'LN + sed\ DN \tag{9}$$

Live organisms were absent in the poisoned unit according to the assumption that all the swimmers were immediately killed by formaldehyde ($mov' \times LN$ in the Equation (9)). For the same reason, copepod carcasses were not decomposed by bacteria whose activity was depressed by the fixator ($d \times DNT$ is absent in the Equation (9)). Same as in the sub-model II, the "external" variables *LN* and *DN* were defined from Equations (5) and (6).

Numerical experiments were carried out sequentially with each of the sub-model from III to I in such a way as to determine the ranges of the coefficients at which the dynamics of zooplankton in the water column and the traps would correspond well to the empirical data obtained during the experiments. For the same estimates of sed obtained in the sub-models II and III, the flow of the swimmers into the poisoned (*mov′*) and non-poisoned (*mov*) trap units were calculated and compared. Next, mortality (*m*) and sedimentation (*sed*) variability and interrelations were studied in the sub-model II. The values and nature of the relationship between these coefficients were used later in the sub-model I in order to calculate the specific growth rate (μ) and consumptive mortality (*g*) of zooplankton.

3. Results

3.1. Species Composition and Dynamics of Zooplankton in the Water Column above the Sediment Trap

In the samples taken in May in the open sea (the experiment SEA05), the heterotrophic dinoflagellate *Noctiluca scintillans* dominated the community (Table 2). Copepods were rare and presented by the eurythermal species *Acartia clausi*, the cold-water *Oithona similis*, and *Pseudocalanus elongatus*. In November, all the water samples collected during both the experiments (BAY11 and SEA11) were dominated by copepods. In particular, the invasive cyclopoid copepod *Oithona davisae* was highly abundant, that is a common species in coastal waters of the Black Sea in autumn [35]. Additionally, *Paracalanus parvus*, *P. elongatus*, *A. clausi*, and nauplii were found at both the stations. Meroplankton were represented by Bivalvia, Polychaeta, Cirripedia, Gastropoda, and Decapoda. These taxa were more abundant in the bay samples (Table 2).

In November, the total abundance of copepods in the water column above the trap was significantly lower at St. S (below 10^4 ind. m^{-3}) than at St. B (about 5×10^4 ind. m^{-3}) (Table 2). Over the two-day experiment BAY11 at St. B, the total abundance of copepods (and some other zooplankton groups) was decreasing more than three-fold, from 5.0 to 1.4×10^4 ind. m^{-3}, mostly due to the dominant species *O. davisae*. Such great changes in the zooplankton numbers could have been related to their abnormally high mortality rates, which was supposed to be verified using the model. But a transfer of the plankton with water masses also could not be excluded as a reason for this phenomenon.

The FLO values obtained for particular groups of zooplankton varied within a wide range from 4 to 100%, and the minima being registered in a few species of copepods, including the abundant *P. parvus* (Table 2). This species was found in November at both the stations, while its live individuals made up only 4% in the samples from the bay. On the contrary, FLO reached as high as 84 to 91% in *O. davisae* predominant in autumn samples. No carcasses were found among Cirripedia and copepod nauplii.

Table 2. Initial (N_0) and final (N_t) total abundances of zooplankton and the fraction of live organisms (FLO_0 and FLO_t, resapectively) in the water column, and the numbers of live (LNT) and dead (DNT) organisms found in the poisoned and non-poisoned trap units after their exposition.

Taxon	Water Column		Water Column		Non-Poisoned Unit		Poisoned Unit	
	N_0, ind. m^{-3}	FLO_0, %	N_t, ind. m^{-3}	FLO_t, %	LNT, ind.	DNT, ind.	LNT, ind.	DNT, ind.
Experiment SEA05 (St. S; depth: 36 m; time of exposition: 1 day)								
Total Copepoda	Nd *	53	849	67	12	0	–	–
Acartia clausi	nd	58	554	65	2	0	–	–
Pseudocalanus elongatus	nd	52	36	71	4	0	–	–
Oithona similis	nd	49	177	60	5	0	–	–
Copepoda nauplii	nd	88	144	94	10	1	–	–
Pleopis polyphemoides	nd	nd	29	nd	1	0	–	–
Noctiluca scintillans	nd	nd	7806	nd	5	0	–	–
Cirripedia nauplii	nd	nd	188	nd	3	1	–	–
Bivalvia larvae	nd	nd	87	nd	2	0	–	–
Experiment SEA11 (St. S; depth: 16 m; time of exposition: 7 d)								
Total Copepoda	9597	81	6181	75	43	28	0	178
Acartia clausi	801	87	1527	94	1	0	0	3
Paracalanus parvus	1858	67	1433	43	5	1	0	91
Oithona similis	445	88	203	47	0	0	0	7
Oithona davisae	2435	86	2859	86	10	11	0	32
Harpacticoida	0	nd	2	nd	27	4	0	19
Copepoda nauplii	861	100	797	95	10	7	0	21
Oikopleura dioica	0	85	0	nd	0	0	0	6
Bivalvia larvae	56	nd	135	nd	0	0	0	15
Experiment BAY11 (St. B; depth: 8 m; time of exposition: 11 d)								
Total Copepoda	49,691	75	13,981	72	372	125	0	1054
Acartia clausi	237	87	309	64	2	1	0	3
Paracalanus parvus	1707	32	459	4	0	1	0	11
Pseudocalanus elongatus	926	71	60	0	0	3	0	54
Pseudodiaptomus marinus	0	0	0	0	64	2	0	30
Oithona similis	250	89	75	nd	0	0	0	3
Oithona davisae	46,312	84	13,012	91	285	93	0	854
Harpacticoida	1,25	nd	0	nd	21	3	0	25
Copepoda nauplii	40	nd	89	nd	13	0	0	43
Cirripedia nauplii	584	nd	350	nd	8	2	0	12
Oikopleura dioica	0	nd	112	nd	0	0	0	19
Bivalvia larvae	1229	nd	131	nd	0	0	0	22
Polychaeta larvae	40	nd	62	nd	1	0	0	39
Gastropoda larvae	90	nd	44	nd	0	0	0	5

* nd—no data.

3.2. Accumulation of Zooplankton in the Sediment Trap

Results from the May pilot project (SEA05) showed that very few zooplankton (including 12 copepods and 10 copepod nauplii) were in the sediment trap after a 24-h exposure. The majority of the organisms in the trap were live (excepting 1 Copepoda nauplius and 1 Cirripedia nauplius) (Table 2). Despite the presence of a significant number of dead copepods in the water column (about 400 ind. m^{-3} of copepods), their absence in the traps indicated an important role of the processes hindering sedimentation of dead zooplankton, such as water mass movement and turbulent mixing. The latter seemed to not affect the ability of actively moving zooplankton to swim in and out of the trap.

When the experiments were extended to 2–7 days in November, it permitted to increase considerably the abundance of carcasses in the traps, especially during mass development of *O. davisae* (experiment BAY11), up to tens of individuals in the non-poisoned units (Table 2). However, the prolonged time of the trap exposition complicated the processes going on inside it. In particular, the abundance of live swimmers getting in the traps by accident grew considerably. At the same time, the chances of their death inside the trap also increased, which, in its turn, could lead to false mortality and sinking estimates. Thus, both the factors—turbulence and active swimmers—could be the source of miscalculations.

In the November experiments, the abundance of dead copepods in the non-poisoned trap units increased significantly (28 ind. in SEA11, and 125 ind. in BAY11). However, the number of the swimmers remained two to three times higher (43 and 372 ind., respectively). In the poisoned traps, the number of copepods more than doubled (178 ind. in SEA11, and 1054 ind. in BAY11), indicating a considerable proportion of the swimmers and their ability not only to swim into the trap, but also to leave it easily.

Apart from numerous copepods, the non-poisoned units contained live copepod nauplii, as well as single individuals of Cirripedia, larvae of Bivalvia and Polychaeta, while their carcasses were hardly present. The abundance of these organisms (and other taxons like Gastropoda larvae, Decapoda larvae, *Oikopleura dioica*) in the poisoned units was significantly higher, which was associated with their ability to swim in the trap, same as copepods (Table 2).

An interesting finding was a new and still rare in the Black Sea invasive copepod *Pseudodiaptomus marinus* [36], a few individuals of which were found in the near-bottom traps in Sevastopol Bay (experiment BAY11). This species is capable of active vertical migrations, while staying in the near-bottom layer during the day, that might explain its occurrence in the traps. Moreover, dozens of harpacticoid copepods were found in the traps exposed close to the bottom as they prefer to live in the near-bottom layer and are associated with seaweeds.

3.3. Dynamics of Bacterioplankton in the Water Column and the Traps

At St. S, bacterioplankton abundances differed insignificantly in May (1.34×10^6 cells mL^{-1}) and November (1.57×10^6 cells mL^{-1}). In the bay, bacteria were more abundant, up to 3.42×10^6 cells mL^{-1}. During the two-day experiment BAY11, the abundance of bacterioplankton decreased down to 1.38×10^6 cells mL^{-1}, thus, changing as considerably as the abundance of the copepod zooplankton (Table 2). This also indicated a complete change in plankton structure as a result of water mass movement over the experiment (Figure 3).

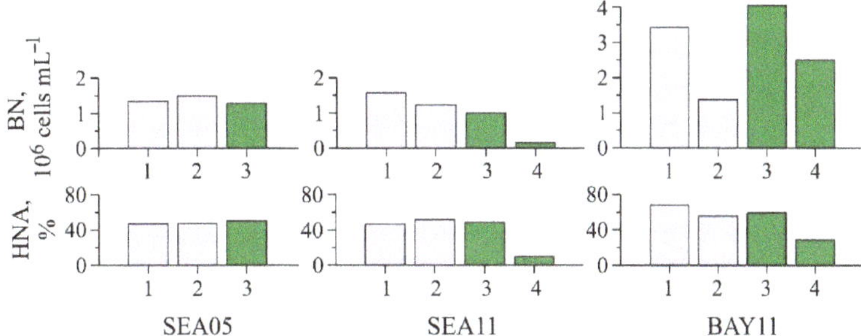

Figure 3. The abundance of bacterioplankton (BN) and the proportion of physiologically active bacteria (HNA) in May (SEA05) and November experiments (SEA11, BAY11): 1—initial values in the water column, 2—final values in the water column, 3—final values in the non-poisoned units, 4—final values in the poisoned units.

During the experiments SEA05 and SEA11, there was no significant bacterial growth in the non-poisoned units. On the contrary, in the bay (BAY11), where the abundances of bacteria and zooplankton were high, the bacterial numbers in the trap increased up to 4×10^6 cells mL^{-1}. In the presence of the fixative, the total number of bacteria and the proportion of physiologically active bacteria dropped significantly, but the complete death of microorganisms did not occur, probably due to the constant dilution of the fixative during the exposure (Figure 3).

3.4. Results of Numerical Experiments

Application of the simulation model has allowed us to study dynamics and major functional characteristics of zooplankton community, including the predominant species—invasive copepod *O. davisae*. The results of simulation of the copepod dynamics in the traps and in the water column above them at stations in the bay (BAY11) and in the open sea (SEA11) are presented in Figure 4. In both the numerical experiments, the abundance of live copepods dropped significantly, especially in the bay, while FLO did not decrease much, in the range between 80% and 70% (Figure 4A,C). Inside the non-poisoned trap units, the decrease in FLO was more pronounced due to the high rate of accumulation of carcasses (Figure 4B,D).

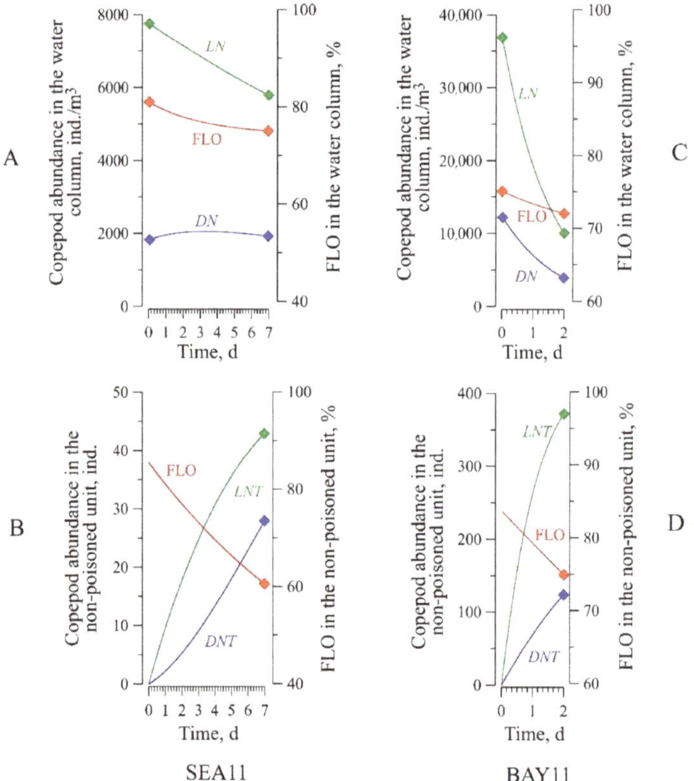

Figure 4. Simulation of dynamics of the copepod abundance and fraction of live organisms (FLO) in the water column (**A**,**C**) and the non-poisoned trap units (**B**,**D**) during the SEA11 (**A**,**B**) and BAY11 (**C**,**D**) experiments. *LN* and *LNT* are the abundances of live copepods in the water column and the trap, respectively; *DN* and *DNT* are the abundances of dead copepods in the water column and the trap, respectively. Symbols denote empirical data.

Figure 5 represents the ranges of values of the model parameters at which the simulated dynamics of copepods in the water column and traps corresponded well to the empirical data obtained in the experiments SEA11 (Figure 5A–C) and BAY11 (Figure 5D–F). In the sub-model III, dependences were obtained (straight line 1 in Figure 5A,D) between the rates of sedimentation of carcasses (*sed*) and flow of swimmers (*mov′*), which determined the accumulation of dead copepods in the poisoned trap unit by the end of the experiments SEA11 (DNT_t = 178 ind.) and BAY11 (DNT_t = 1054 ind.) (Table 2).

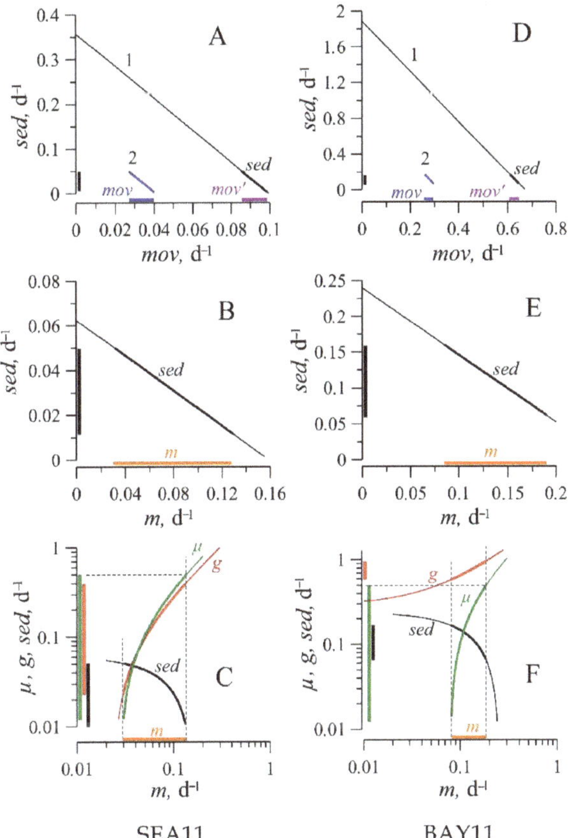

Figure 5. Ranges of the coefficients (see their description in Table 1) which characterize dynamics of the copepod community, and provide the best fit of the simulation model to the results of the SEA11 (**A–C**) and BAY11 (**D–F**) experiments. The ranges are marked with a bold line on the graphs, and are also represented by projections on the axes.

The sub-model II (non-poisoned unit) included two additional coefficients—decomposition of carcasses (*d*) and non-consumptive mortality of the swimmers (*m*). The values of *d* were set taking into account the water temperature and the coefficient Q_{10} and kept unchanged in each of the numerical experiments. For a wide range of values of non-consumptive mortality (*m*), we studied the dependences (straight line 2 in Figure 5A,D) between the rates of sedimentation of carcasses (*sed*) and swimming (*mov*), which determined the accumulation of carcasses and live copepods in the non-poisoned trap unit and ensured compliance model to the empirical data obtained by the end of the experiments SEA11 (LNT_t = 43 ind., DNT_t = 28 ind.) and BAY11 (LNT_t = 372 ind., DNT_t = 125 ind.). Such a correspondence was achievable at much lower values of the swimming rate (*mov* < *mov′*), because in the absence of the fixative, *mov* was the result of two opposite processes—the swimming of organisms into the trap and their swimming out of it. In addition, a relationship between *sed* and m was obtained in the sub-model II (Figure 5B,E), which was later used in the sub-model III, based on the assumption that the values of non-consumptive mortality of organisms in the non-poisoned unit and in the water column are the same.

In the sub-model I (water column), for the entire range of pairs of *m* and *sed* values (which were described in the sub-model II), the values of μ and g were determined, at which the model provided the best fit to the empirical data obtained in the experiment and presented in Table 2. The specific growth rate of copepods never exceeded 0.5 d^{-1} in accordance with the maximum values reported by other authors [37–39]. The ranges of values of the main coefficients (μ, g, *sed*, *m*), which adequately describe the dynamics of dead and living copepods in the water column, are shown in Figure 5C,E, in the form of projections on the axes and are marked with a thick line on the graphs. Each value of *m* (abscissa axis in Figure 5C,E) can be correlated with a corresponding set of values of other coefficients, which together ensure that the model corresponds to the experimental results.

Similar calculations were also made for the invasive copepod *O. davisae* (Figure 6), whose abundance was exceptionally high during the autumn experiments, especially in the bay (BAY11): more than 4 × 10^4 ind. m^{-3} (Table 2). Contribution of this species to the total abundance of the community exceeded 90%. The results obtained in both the experiments (SEA11 and BAY11) for two components of the zooplankton community, all Copepoda and the species *O. davisae*, are summarized in Table 3.

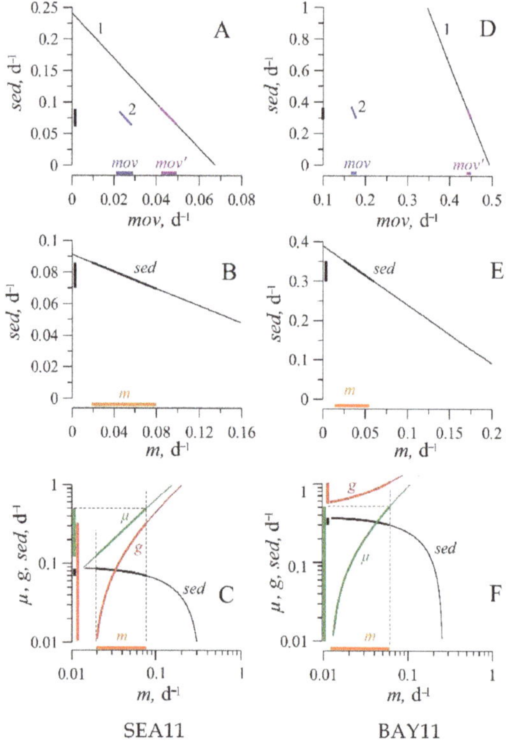

Figure 6. Ranges of the coefficients (see their description in Table 1) which characterize dynamics of the invasive copepod *Oithona davisae*, and provide the best fit of the simulation model to the results of the SEA11 (**A–C**) and BAY11 (**D–F**) experiments. The ranges are marked with a bold line on the graphs, and are also represented by projections on the axes.

Table 3. Ranges of the coefficients which describe dynamics of the copepod community and the invasive copepod *Oithona davisae*, and enable the simulation model to best fit the results of the experiments BAY11 (Sevastopol Bay) and SEA11 (adjacent waters). Description and units of the coefficients are in Table 1. The ranges of *sed* are descending as the *sed* maxima correspond to minima of the other constants, and vice versa.

	Experiment	*d*	*mov*	*mov′*	*sed*	*m*	*g*	*μ*
Copepoda	SEA11	0.02	0.02–0.04	0.08–0.10	0.05–0.01	0.03–0.13	0.00–0.40	0.00–0.50
	BAY11	0.05	0.24–0.29	0.59–0.64	0.16–0.06	0.08–0.19	0.60–0.97	0.00–0.50
O. davisae	SEA11	0.02	0.02–0.03	0.04–0.05	0.08–0.07	0.02–0.08	0.00–0.32	0.12–0.50
	BAY11	0.05	0.17–0.18	0.44–0.45	0.35–0.30	0.01–0.06	0.60–1.00	0.00–0.50

The estimates of *mov* and *mov′* obtained in the numerical experiments (up to 0.64 d^{-1}, Table 3) indicated a significant contribution of the swimmers to the accumulation of organisms in the trap, which can be comparable to and even exceed the sedimentation of carcasses. In the bay (BAY11), the values of *mov* and *mov′* were almost an order of magnitude higher than in the open sea (SEA11), which was difficult to explain by such a large difference in the motility of organisms. Since the ratio between the sedimentation rates *sed* at St. B and St. S was similar (Table 3) and could be due to turbulent mixing, a similar explanation may also be applicable to copepods swimming into the traps.

Model estimates were confirmed by the quite expected and explainable ratio between the values *mov′* and *mov* (Table 3): the net flow of swimmers into the poisoned trap unit (*mov′*) was the highest, since the live organisms getting inside could no longer leave it; lower values of *mov* were due to outflow of the swimmers" from the trap. The difference between these values (*mov′* minus *mov*) served as a measure of the outflow of the swimmers from the trap. Thus, the obtained results indicate that, first, live copepods did not avoid the poisoned trap units and actively swam in them, and second, in the absence of the poison, zooplankton left the trap freely.

As it was noted earlier, sedimentation had greater effect on zooplankton dynamics in the bay waters (0.16 d^{-1} versus 0.05 d^{-1} in open waters, Table 3), which was apparently associated with less intense water mixing in the semi-closed bay. For *O. davisae*, the same regularity was obtained, but higher estimates of *sed* (0.35 d^{-1} at St. B versus 0.08 d^{-1} at St. S, Table 3). The sedimentation rate (*sed*) was the only parameter that decreased with an increase in all other coefficients (Figures 5 and 6).

Non-consumptive mortality (*m*) of copepods in the waters of the bay (0.08–0.19 d^{-1}) was generally higher than in the adjacent waters (0.03–0.13 d^{-1}) (Figure 5, Table 3). The ranges of m obtained for *O. davisae* were equally wide, and their upper limit, on the contrary, was somewhat higher in the sea (0.08 d^{-1} at St. S vs. 0.06 d^{-1} at St. B). The minimum non-consumptive mortality of the species was observed in the bay (0.01–0.06 d^{-1}, Table 3).

The numerical experiments have allowed an alternative explanation of the significant decrease in the total abundance of copepods and the dominant species *O. davisae* in the waters of the bay (BAY11 experiment)—exceptionally high rates of predation on zooplankton (g = 0.60–0.97 d^{-1} for all copepods; 0.60–1.00 d^{-1} for *O. davisae*) (Table 3). For comparison, the same values were noticeably lower in the SEA11 experiment: 0.00–0.40 d^{-1} and 0.00–0.32 d^{-1}, respectively (Table 3).

4. Discussion

4.1. Rates of the Processes Controlling Copepod Carcasses Dynamics in the Water Column

The ranges of specific growth rate (μ) put into our model corresponded well to the estimates obtained by other authors for calanoid and cyclopoid copepods [40], as well as *Oithona* spp. [41]. However, it proved impossible to calculate this parameter precisely, since, in accordance with the simulation results, it could take values in the entire possible range from zero to the established maximum (Table 3). Only in the autumn experiment SEA11,

μ was not lower than 0.12 d^{-1} (Table 3), but this circumstance did not provide any useful information for understanding and interpreting the data obtained.

The results of the numerical experiments suggested that the sharp decrease in the abundance of copepods during the BAY11 experiment was associated, first of all, with their exceptionally high mortality due to predation (g), the specific rate of which reached 0.97 d^{-1} in the copepod community and 1.0 d^{-1} in *O. davisae* (Table 3). We had no information about the presence of predators that would be able to eat copepods so actively, since such a task was not set in this study. The same non-predatory mortality rates were earlier reported only for early developmental stages of planktonic copepods [20].

Another possible reason for such a sharp change in zooplankton abundance could be water mass movement. It is quite possible that at the end of the experiment we were dealing with a completely different community, which was brought to the exposition area of the trap with a stream of water. In such circumstances, none of the existing methods, including the one presented in our work, would make it possible to correctly estimate the mortality of zooplankton and the rates of other processes that characterize community dynamics. The possibility of water mass transfer in the BAY11 experiment could also be indicated by a significant decrease in the number of bacterioplankton in the water column above the trap (Figure 3); however, all the parameters involved in the model remained within acceptable limits, i.e., no negative or abnormally high values were obtained for them. This, in turn, gave no reason to doubt the results obtained using the simulation model.

The most important conclusion that can be drawn from the results of our study is the comparability of the rates of copepod mortality, sedimentation, and decomposition of carcasses in the water column. According to our results, bacteria were more abundant in the bay, that could serve an explanation of similar differences in the rate of bacterial decay of copepods carcasses in the bay and adjacent waters [12]. Thus, our earlier hypothesis about the significant effect of carcass decomposition on the dynamics of FLO in zooplankton of coastal waters [15] has received more confirmation.

It should also be noted that our estimates of *sed* proved to be significantly lower than the published values. It is known that the rate of sinking of carcasses depends on a number of internal (the degree of decomposition, the size and shape of the body) and external (salinity, temperature, water density) factors [42–46]. Crustacean zooplankton have well-developed organs for hovering—antennules, and some species have significant reserves of fat that prevent passive sinking even after the death of the organism [42]. The rate of sinking of a crustacean is also influenced by its position, whether it descends with its head or ventral side down. As the carcass decomposes, its buoyancy may remain negative and even become positive due, for example, to the release of gas bubbles and their accumulation under the carapace. In addition, with an increase in salinity and water temperature, the rate of sinking of dead organisms slows down, regardless of the stage of decomposition [43]. Water stratification and hydrology also contribute to a decrease in the velocity of sinking of crustacean carcasses [42].

The sinking velocity of copepod carcasses, measured by different authors, changed in a wide range, according to some estimates, from 36 m d^{-1} (for small *Paracalanus parvus*) to 294 m d^{-1} (for *Calanus euxinus*) [42], according to others, from 242 to 10,835 m d^{-1}, i.e., 0.3 to 12.5 cm s^{-1} [43]). In fresh waters, the sedimentation rate of dead Cladocera and Copepoda ranged from 80 to 124 m d^{-1}, and from 55 to 112 m d^{-1}, respectively [21]. The average sinking velocity of the species *Arctodiaptomus salinus*, obtained in situ using sediment traps (Lake Shira, Russia), were about 8.5 m d^{-1} [20]. It is interesting that the copepodite stages (C5) of *A. salinus*, which slightly differed from adults in size, had, however, lower sinking velocity (2.0 m d^{-1}), probably due to fat reserves characterizing the diapause state [20].

The mentioned above sinking velocities of dead zooplankton vary in an enormously wide range—from extremely high values obtained during laboratory experiments in vessels with still water to comparatively low values observed in natural bodies of water. Undoubtedly, intense turbulent mixing in the water column and peculiarities of the carcass decomposition (like gas accumulation under carapace) might eventually prevent dead zooplankton from sinking, thus, making it impossible to approximate *sed* from extensive laboratory data obtained in vessels with still water. We believe that the alternative approach applied in this study has provided more accurate estimates of *sed* and demonstrated that carcass sedimentation is not so significant in controlling the FLO dynamics in marine zooplankton.

The estimates of the non-consumptive mortality of copepods obtained in the present study (m = 0.03 to 0.19 d^{-1}), were generally similar to those obtained by other authors for different fresh and marine waters [9,15,20,45]. In the oligotrophic Bay of Calvi (the Mediterranean Sea), they were calculated from experiments with sediment traps and amounted to <0.01–0.05 d^{-1} [45]. In coastal waters of the Mediterranean Sea, non-consumptive mortality of various species of calanoid copepods varied from 0.004 d^{-1} (*Acartia clausi*) to 0.13 d^{-1} (*Paracalanus parvus*) [9]. In Sevastopol Bay, the approximation of m from data on FLO in copepod zooplankton amounted to about 0.05 d^{-1} [15]. In Lake Shira, mortality rate (0.0003 to 0.103 d^{-1}) of the dominant calanoid copepod *A. salinus* was calculated from the numbers of carcasses in sediment traps and water column [20]. Based on these data, the values of m in pelagic copepods do not usually exceed 0.20 d^{-1}.

4.2. Validity and Applicability of Existing Field Methods for Measuring Zooplankton Non-Consumptive Mortality

The main problem that we encountered in the course of the in situ experiments was that sedimentation was not the only process controlling the loss of copepod carcasses in the water column. Moreover, its contribution to the accumulation of carcasses in the trap was minimal even if they were abundant in the water column (SEA05).

At the same time, all currently existing experimental and model methods for studying zooplankton mortality ignore the factors preventing carcasses sinking, such as turbulent mixing, stratification of the water column, and decomposition of dead organisms in the water column. Moreover, the very concept of mortality is often replaced by sedimentation, while its assessment is reduced to a direct account of carcasses in the sediment trap [47,48] and to recalculation of the obtained values into the number of dead organisms that settled per 1 m^2 of the seabed in 24 h [43]. In a number of studies, the non-consumptive mortality of mesozooplankton was presented as "sedimentation losses" (% d^{-1}) and was calculated as the ratio of the sedimentation rate of carcasses into traps (ind. m^{-2} d^{-1}) to the total abundance of zooplankton in the water column [19]. More complex and detailed models (for example, [49]) were also used to calculate mortality rate of zooplankton in many studies, based on the assumption that sedimentation is the main mechanism of carcasses loss, and the other processes such as decomposition and consumption by detritophages can be neglected, since carcasses sink faster than they get consumed or degraded [5,20,46,47].

In later research, more attention was given to factors hindering the sinking of dead zooplankton. In particular, special coefficients were introduced that reflect a combined effect of turbulent mixing, consumption, or microbial degradation [20]. Degradation was even considered as the main factor controlling dynamics of carcasses [9]. Finally, compelling evidence was found that a well-pronounced summer stratification in fresh waters may prevent dead zooplankton from sinking: carcasses turned out not to sink to the bottom for as long as 5 days, being the energy source for pelagic bacteria [21]. The present study is the first attempt to make a more inclusive picture of processes happening inside the sediment trap and the water column above it, while being aware of all the problems related to increased model complexity.

4.3. The Problem of Live Copepods-Swimmers in the Trap

Our experiments showed that the number of swimmers of *O. davisae* found in the trap after its exposure could significantly exceed the number of dead organisms (Table 2). Thus, the swimmers are a potential source of error in further calculations of zooplankton sedimentation rate and, finally, the estimates of vertical matter flow in the water column.

Despite attempts to develop a design of sediment traps preventing swimmers from getting inside [50,51], the problem still remains unresolved. The use of poisons and the joint exposure of poisoned and non-poisoned traps cannot always help, since little is known about the behavioral patterns of plankton swimmers in the trap. Death of the swimmers inside a non-poisoned trap makes them indistinguishable from the carcasses settled in the trap during the exposition, resulting in an overestimation of the sedimentation rate. According to our experience, the use of the poisoned trap units doubled the labor costs for the experiment, but did not provide any additional information about the most important processes—mortality and sedimentation, and did not increase the accuracy of their estimates. However, it allowed us to get information about other variables, such as the flows of the swimmers *mov* and *mov′* (Figure 4).

4.4. Applicability of Other Models to Our Field Data

The dynamics of live and dead copepods, which we had observed during the experiments, seemed to be controlled by a wide range of factors: Weaker sedimentation of carcasses due to water turbulence; decomposition of carcasses in the trap and the above water column; swimming of live copepods into the traps. Consequently, we figured it useful to test the presently known models and methods of zooplankton mortality evaluation against our data in order to estimate and compare the obtained results.

According to Gris et al. [19], zooplankton mortality is evaluated as sedimentation losses (SL, %) over a certain period of time. The adaptation of their formula to our data for the non-poisoned (SL) and poisoned (SL_f) units looks as follows:

$$SL = \frac{(LNT + DNT)100\%}{S\ N_0\ T\ h} \tag{10}$$

$$SL_f = \frac{DNT\ 100\%}{S\ N_0\ T\ h} \tag{11}$$

where LNT and DNT are the numbers of live (swimmers) and dead copepods in the trap (ind.); S is the area of the trap mouth (S = 0.019 m^2 for the two units); N_0 is the total abundance of copepods in the water column at the start of the exposition (ind. m^{-3}); T (day) and h (m) are the time and the depth of the trap deployment, respectively.

According to the results of the autumn experiments, the following estimations of the daily sedimentation losses were obtained: SL = 0.35%, SL_f = 0.87% in the open sea (SEA11) and SL = 3.29%, SL_f = 0.83% in the bay (BAY11). First, a strong discrepancy between SL and SL_f values may be a consequence of the swimming of live copepods into the traps. Second, these estimates are much lower than those obtained in our simulation model (up to 0.3 d^{-1}, Table 3), since the factors preventing carcass sedimentation (decomposition and turbulence) were neglected. In their work, the authors presented the daily sedimentation losses in the epilimnetic cladoceran *Daphnia galeata*, which amounted to 2.3% of the total abundance [19], and were likely underestimated. Moreover, SL cannot be regarded as a measure of zooplankton mortality, as these are different processes.

According to [5,20,46,47,49,52], the non-consumptive mortality of zooplankton (m, d^{-1}) is calculated as follows:

$$m = \frac{\Delta\overline{y} + GN_0(1 - FLO_0)}{T\ N_0\ FLO_0} \tag{12}$$

where N_0 is the initial abundance of copepods in the water column (ind. m^{-3}); FLO_0 is the initial fraction of live organisms in the water column (%); $\Delta\overline{y}$ is a change in the abundance of carcasses during the trap exposition (ind. m^{-3}); T is the duration of the trap exposition (day); G is the specific rate of carcass elimination which is calculated as:

$$G = \frac{v}{h} \tag{13}$$

where h is the depth of the sampling layer (m), v is the sinking velocity of carcasses (m d^{-1}), which we calculated differently for non-poisoned (v_1) and poisoned units (v_2):

$$v_1 = \frac{LNT + DNT}{S\ N_0\ (1 - FLO_0)} \tag{14}$$

$$v_2 = \frac{DNT}{S\ N_0\ (1 - FLO_0)} \tag{15}$$

The change in the abundance of carcasses in the water column during the trap exposition is defined as:

$$\Delta\overline{y} = N_t\ (1 - FLO_t) - N_0(1 - FLO_0) \tag{16}$$

where N_t is the final abundance of copepods in the water column (ind. m^{-3}), FLO_t is the final fraction of live organisms in the water column (%).

Applying this model to our experimental data was impossible because of negative mortality (down to -0.09 d^{-1} in the experiment BAY11) calculated from Equations (12) and (16). Only in SEA11, a positive value (0.005 d^{-1}) was obtained for the species *O. davisae.* The reason for the negative values was a sharp decrease in the carcass abundance in the water column during the experiments. Thus, Gladyshev's model proved to be sensitive to the above mentioned factors, producing greatly underestimated (even negative) estimates of zooplankton non-consumptive mortality.

According to the simplified approach proposed by [9], zooplankton mortality (m) can be calculated based on field data on FLO and the rate of decomposition of carcasses in the water column (measured, for example, under experimental conditions):

$$m = \frac{(1 - FLO_0)}{t_d FLO_0}$$

where t_d is the average time of carcass decomposition under given temperatures. At low water temperature in the autumn experiments (11 to 13 °C), t_d exceeded 10 days and could even reach 20 days (at Q_{10} = 2.3). Accordingly, m calculated from Equation (17) was 0.035 d^{-1} and 0.01–0.07 d^{-1} in the bay (experiment BAY11) and outside it (SEA11), respectively, which is significantly lower than the estimates based on our model (Table 3). The reliability of the results obtained from the Capua's model raises serious doubts because of its extreme simplification: the authors of the method completely excluded from their consideration the most important processes that affect the dynamics of copepod carcasses in the water column.

Thus, ignoring the most important factors controlling the dynamics of dead organisms in the water column (such as turbulent mixing and mobility of water masses, utilization of carcasses by detritophages, and their decomposition by bacteria) makes the methods unsuitable for reliable measurement of zooplankton mortality and carcass sedimentation rates using sedimentation traps. Nevertheless, experiments with short-term (2 to 7 days) exposure of the traps in coastal waters can provide fairly accurate and valuable information on the extent of non-consumptive mortality of zooplankton, if an adequate simulation model is used to interpret the data obtained, taking into account all factors.

5. Conclusions

1. Significant changes in the abundance of copepod carcasses (from 280 to 12,443 ind. m^{-3}) and FLO (53 to 81%) were observed in Sevastopol Bay and adjacent waters over short time periods, which indicated a high variability of zooplankton non-consumptive mortality (m), sedimentation (sed), and decomposition rates of dead organisms (d).
2. Despite the high concentrations of copepod carcasses in the water column, the rates of their enrichment in the traps proved to be extremely low (no more than 20 specimens per day per trap unit), which could be due to intense turbulent mixing of the waters.

The rates of non-consumptive mortality (*m*) and sedimentation (*sed*) of copepods were comparable with each other.

3. The obtained estimates of the sedimentation rate of copepod carcasses (0.012 to 0.39 d^{-1}) were comparable in value with the rate of their microbial decomposition (0.13 and 0.05 d^{-1} in the bay and adjacent waters, respectively), which confirmed the hypothesis on microbial decomposition as one of the key controls of FLO in zooplankton. The influence of sedimentation processes on the dynamics of carcasses in coastal waters seems to be greatly overestimated.
4. The carcass sedimentation rate (*sed*) and the flows of swimmers into the traps (*mov*) were significantly higher in the bay than in the adjacent waters, which may be explained by a difference in hydrological regimes at the stations. Weaker turbulent mixing appeared to increase the contribution of the above processes to the control of FLO in zooplankton.
5. The models used to process and interpret the results of the short-term sedimentation experiments should take into account the zooplankton swimmers and their death in the sedimentation trap. Otherwise, mortality and sedimentation rates may be estimated incorrectly.

Author Contributions: Conceptualization, V.M. and D.L.; validation, D.L. and V.M.; formal analysis, V.M.; investigation, D.L.; data curation, D.L.; writing—original draft preparation, D.L.; writing—review and editing, V.M. and V.E.; supervision, V.M.; project administration, V.E.; funding acquisition, V.M. and V.E. All authors have read and agreed to the published version of the manuscript.

Funding: The research was conducted in the frame of the Russian state assignments No. 121040600178-6 and 121121700354-9 (the program 'Prioritet-2030' of Sevastopol State University, strategic project No. 3), and supported by the RFBR project 21-55-52001.

Institutional Review Board Statement: Not applicable.

Informed Consent Statement: Not applicable.

Data Availability Statement: Not applicable.

Conflicts of Interest: The authors declare no conflict of interest.

References

1. Lomartire, S.; Marques, J.C.; Gonçalves, A.M. The key role of zooplankton in ecosystem services: A perspective of interaction between zooplankton and fish recruitment. *Ecol. Indic.* **2021**, *129*, 107867. [CrossRef]
2. Ohman, M.D.; Hirche, H.-J. Density-dependent mortality in an oceanic copepod population. *Nature* **2001**, *412*, 638–641. [CrossRef] [PubMed]
3. Fasham, M.J.R. (Ed.) *Ocean Biogeochemistry: The Role of the Ocean Carbon Cycle in Global Change, Global Change*; Springer: New York, NY, USA, 2003; Volume XVIII, p. 297.
4. Koval, L.G. *Zoo- and Necrozooplankton of the Black Sea*; Naukova: Kiev, Ukraine, 1984; p. 128. (In Russian)
5. Dubovskaya, O.P.; Tang, K.W.; Gladyshev, M.I.; Kirillin, G.; Buseva, Z.; Kasprzak, P.; Tolomeev, A.P.; Grossart, H.P. Estimating in situ zooplankton non-predation mortality in an oligo-mesotrophic lake from sediment trap data: Caveats and Reality Check. *PLoS ONE* **2015**, *10*, e0131431. [CrossRef] [PubMed]
6. Elliott, D.T.; Tang, K.W. Spatial and temporal distributions of live and dead Copepods in the lower Chesapeake Bay (Virginia, USA). *Estuaries Coasts* **2011**, *34*, 1039–1048. [CrossRef]
7. Farran, G.P. Biscayan plankton collected during a cruise of H.M.S. Research, 1900. Pt. 14: The Copepoda. *Zool. J. Linn. Soc.* **1926**, *36*, 219–310. [CrossRef]
8. Tang, K.W.; Gladyshev, M.I.; Dubovskaya, O.P.; Kirillin, G.; Grossart, H.P. Zooplankton carcasses and non-predatory mortality in freshwater and inland sea environments. *J. Plankton Res.* **2014**, *36*, 597–612. [CrossRef]
9. Di Capua, I.; Mazzocchi, M.G. Non-predatory mortality in Mediterranean coastal copepods. *Mar. Biol.* **2017**, *164*, 198. [CrossRef]
10. Tang, K.W.; Freund, C.S.; Schweitzer, C.L. Occurrence of copepod carcasses in the lower Chesapeake Bay and their decomposition by ambient microbes. *Estuar. Coast. Shelf Sci.* **2006**, *68*, 499–508. [CrossRef]
11. Tolomeev, A.P.; Dubovskaya, O.P.; Kirillin, G.; Buseva, Z.; Kolmakova, O.V.; Grossart, H.-P.; Tang, K.W.; Gladyshev, M.I. Degradation of dead cladoceran zooplankton and their contribution to organic carbon cycling in stratified lakes: Field observation and model prediction. *J. Plankton Res.* **2022**, *44*, 386–400. [CrossRef]
12. Kydd, J.; Rajakaruna, H.; Briski, E.; Bailey, S. Examination of a high resolution laser optical plankton counter and FlowCAM for measuring plankton concentration and size. *J. Sea Res.* **2018**, *133*, 2–10. [CrossRef]

13. Colas, F.; Tardivel, M.; Perchoc, J.; Lunven, M.; Forest, B.; Guyader, G.; Danielou, M.M.; Le Mestre, S.; Bourriau, P.; Antajan, E.; et al. The ZooCAM, a new in-flow imaging system for fast onboard counting, sizing and classification of fish eggs and metazooplankton. *Prog. Oceanogr.* **2018**, *166*, 54–65. [CrossRef]
14. Litvinyuk, D.A.; Altukhov, D.A.; Mukhanov, V.S.; Popova, E.V. Dynamics of live Copepoda in plankton of Sevastopol Bay and open coastal waters (the Black Sea) in 2010–2011. *Mar. Ecol. J.* **2011**, *10*, 56–65. (In Russian)
15. Mukhanov, V.; Litvinyuk, D. Microbial control of live/dead zooplankton ratio in Sevastopol Bay. *Ecol. Montenegrina* **2017**, *11*, 42–48. [CrossRef]
16. Pavlova, E.V.; Melnikova, E.B. Zooplankton in inshore waters of the south-western Crimea (1998–2006). *Mar. Ecol. J.* **2011**, *10*, 33–42. (In Russian)
17. Hirst, A.G.; Kiørboe, T. Mortality of marine planktonic copepods: Global rates and patterns. *Mar. Ecol. Prog. Ser.* **2002**, *230*, 195–209. [CrossRef]
18. Dubovskaya, O.P.; Gladyshev, M.I.; Gubanov, V.G.; Makhutova, O.N. Study of non-consumptive mortality of Crustacean zooplankton in a Siberian reservoir using staining for live/dead sorting and sediment traps. *Hydrobiologia* **2003**, *504*, 223–227. [CrossRef]
19. Gries, T.; Güde, H. Estimates of the nonconsumptive mortality of mesozooplankton by measurement of sedimentation losses. *Limnol. Oceanogr.* **1999**, *44*, 459–465. [CrossRef]
20. Dubovskaya, O.P.; Tolomeev, A.P.; Kirillin, G.; Buseva, Z.; Tang, K.W.; Gladyshev, M.I. Effects of water column processes on the use of sediment traps to measure zooplankton non-predatory mortality: A mathematical and empirical assessment. *J. Plankton Res.* **2018**, *40*, 91–106. [CrossRef]
21. Kirillin, G.; Grossart, H.-P.; Tang, K.W. Modeling sinking rate of zooplankton carcasses: Effects of stratification and mixing. *Limnol. Oceanogr.* **2012**, *57*, 881–894. [CrossRef]
22. Lukashin, V.N.; Klyuvitkin, A.A.; Lisitzin, A.P.; Novigatsky, A.N. The MSL-110 small sediment trap. *Oceanology* **2011**, *51*, 699–703. [CrossRef]
23. Orekhova, N.A.; Varenik, A.V. Carrent hydrochemical regime of the Sevastopol Bay. *Morskoy Gidrofiz. Zhurnal* **2018**, *34*, 134–146. [CrossRef]
24. Gubanov, V.I.; Gubanova, A.D.; Rodionova, N.Y. Diagnosis of water trophicity in the Sevastopol bay and its offshore. In *Current Issues in Aquaculture, Proceedings of the International Scientific Conference, Rostov-on-Don, Russia,28 September–2 October 2015*; FGBNU "AzNIIRKH": Rostov-on-Don, Russia, 2015; pp. 64–67.
25. Tikhonova, E.A.; Burdiyan, N.V.; Soloveva, O.V.; Doroshenko, Y.V. The estimation of the sevastopol bays ecological state on basic chemical and microbiological criteria. *Ecol. Environ. Conserv.* **2018**, *24*, 1574–1584.
26. Aleksandrov, B.; Arashkevich, E.G.; Gubanova, A.D.; Korshenko, A. Black Sea monitoring guidelines: Mesozooplankton. Publ. EMBLAS Project 2014, BSC, 31. Available online: http://emblasproject.org/wp-content/uploads/2017/01/Mesozooplankton_Final-July2015-PA3-f.pdf (accessed on 29 April 2020).
27. Postel, L.; Fock, H.; Hagen, W. Biomass and abundance. In *ICES Zooplankton Methodology Manual*; Harris, R.P., Wiebe, P.H., Lenz, J., Skjoldal, H.R., Huntley, M., Eds.; Academic Press: London, UK, 2000; pp. 83–174.
28. Lytvynyuk, D.A.; Mukhanov, V.S. Advanced method for identifying alive organisms in marine zooplankton stained with neutral red and fluorescein diacetate. *Mar. Ecol. J.* **2012**, *11*, 45–54. (In Russian)
29. Brookes, J.D.; Geary, S.M.; Ganf, G.G.; Burch, M.D. Use of FDA and flow cytometry to assess metabolic activity as an indicator of nutrient status in phytoplankton. *Mar. Freshw. Res.* **2000**, *51*, 817–823. [CrossRef]
30. Litvinyuk, D.; Aganesova, L.; Mukhanov, V. Identifying alive versus dead Copepods in culture of *Calanipeda aquae dulcis* after staining them with neutral red and fluorescein diacetate. *Ekologiyamorya* **2009**, *78*, 65–69. (In Russian)
31. Marie, D.; Partensky, F.; Jacquet, S.; Vaulot, D. Enumeration and cell cycle analysis of natural populations of marine picoplankton by flow cytometry using the nucleic acid stain SYBR Green, I. *Appl. Environ. Microbiol.* **1997**, *63*, 186. [CrossRef] [PubMed]
32. Gasol, J.M.; Del Giorgio, P.A. Using flow cytometry for counting natural planktonic bacteria and understanding the structure of planktonic bacterial communities. *Sci. Mar.* **2000**, *64*, 197–224. [CrossRef]
33. Servais, P.; Casamayor, E.O.; Courties, C.; Catala, P.; Parthuisot, N.; Lebaron, P. Activity and diversity of bacterial cells with high and low nucleic acid content. *Aquat. Microb. Ecol.* **2003**, *33*, 41–51. [CrossRef]
34. Tang, K.W.; Bickel, S.L.; Dziallas, C.; Grossart, H.P. Microbial activities accompanying decomposition of cladoceran and copepod carcasses under different environmental conditions. *Aquat. Microb. Ecol.* **2009**, *57*, 89–100. [CrossRef]
35. Gubanova, A.D.; Garbazey, O.A.; Popova, E.V.; Altukhov, D.A.; Mukhanov, V.S. *Oithona davisae*: Naturalization in the Black Sea, interannual and seasonal dynamics, and effect on the structure of the planktonic copepod community. *Oceanology* **2019**, *59*, 912–919. [CrossRef]
36. Garbazey, O.A.; Popova, E.V.; Gubanova, A.D.; Altukhov, D.A. First report of the occurrence of *Pseudodiaptomus marinus* Sato, 1913 (Copepoda: Calanoida: Pseudodiaptomidae) in the Black Sea (Sevastopol Bay). *Mar. Biol. J.* **2016**, *1*, 78–80. [CrossRef]
37. Richardson, A.J.; Verheye, H.M. The relative importance of food and temperature to copepod egg production and somatic growth in the southern Benguela upwelling system. *J. Plankton Res.* **1998**, *20*, 2379–2399. [CrossRef]
38. Richardson, A.J.; Verheye, H.M. Growth rates of copepods in the southern Benguela upwelling system: The interplay between body size and food. *Limnol. Oceanogr.* **1999**, *44*, 382–392. [CrossRef]

39. Richardson, A.J.; Verheye, H.M.; Herbert, V.; Rogers, C.; Arendse, L.M. Egg production, somatic growth and productivity of copepods in the Benguela Current system and Angola-Benguela Front: BENEFIT Marine Science. *S. Afr. J. Sci.* **2001**, *97*, 251–257. Available online: https://hdl.handle.net/10520/EJC97315 (accessed on 19 June 2022).
40. Persad, G.; Webber, M. The use of Ecopath software to model trophic interactions within the zooplankton community of Discovery Bay, Jamaica. *Open Mar. Biol. J.* **2009**, *3*, 95–104. [CrossRef]
41. Chisholm, L.A.; Roff, J.C. Abundances, growth rates, and production of tropical neritic copepods off Kingston, Jamaica. *Mar. Biol.* **1990**, *106*, 79–89. [CrossRef]
42. Stepanov, V.N.; Svetlichnyyi, L.S. *Research of Hydromechanical Characteristics of Plankton Copepods*; Naukova Dumka: Kiev, Ukraine, 1981; p. 128. (In Russian)
43. Zelezinskaya, L.M. Natural Mortality of Some Forms of Ichthyo and Zooplankton of the Black Sea. Ph.D. Thesis, IBSS UAS, Odessa, Ukraine, 1966; p. 23. (In Russian).
44. Dubovskaya, O.P.; Gladyshev, M.I.; Esimbekova, E.N.; Morozova, I.I.; Gol'd, Z.G.; Makhutova, O.N. Study of possible relation between seasonal dynamics of zooplankton non-consumptive mortality and water toxicity in a pond. *Inland Water Biol.* **2002**, *3*, 39–43. (In Russian)
45. Frangoulis, C.; Skliris, N.; Lepoint, G.; Elkalay, K.; Goffart, A.; Pinnegar, J.K.; Hecq, J.-H. Importance of copepod carcasses versus fecal pellets in the upper water column of an oligotrophic area. *Estuar. Coast. Shelf Sci.* **2011**, *92*, 456–463. [CrossRef]
46. Dubovskaya, O.P. Non-predatory mortality of the crustacean zooplankton, and its possible causes (a review). *Zhurnal Obshch. Biol.* **2009**, *70*, 168–192. (In Russian)
47. Dubovskaya, O.P.; Gladyshev, M.I.; Gubanov, V.G. Seasonal dynamics of number of alive and dead zooplankton in a small pond and some variants of mortality estimation. *J. Gen. Biol.* **1999**, *60*, 543–555. (Translated into English)
48. Ivory, J.A.; Tang, K.W.; Takahashi, K. Use of Neutral Red in short-term sediment traps to distinguish between zooplankton swimmers and carcasses. *Mar. Ecol. Prog. Ser.* **2014**, *505*, 107–117. [CrossRef]
49. Gladyshev, M.I.; Gubanov, V.G. Seasonal dynamics of specific mortality of *Bosmina longirostris* in forest pond determined on the basis of counting of dead individuals. *Dokl. Akad. Nauk.* **1996**, *348*, 127–128.
50. Coale, K.H. Labyrinth of doom: A device to minimize the "swimmer" component in sediment trap collections. *Limnol. Oceanogr.* **1990**, *35*, 1376–1381. [CrossRef]
51. Hansell, D.A.; Newton, J.A. Design and evaluation of a "swimmer"-segregating particle interceptor trap. *Limnol. Oceanogr.* **1994**, *39*, 1487–1495. [CrossRef]
52. Tolomeev, A.P.; Kirillin, G.; Dubovskay, O.P.; Buseva, Z.F.; Gladyshev, M.I. Numerical modeling of vertical distribution of living and dead copepods *Arctodiaptomus salinus* in Salt Lake Shira. *Contemp. Probl. Ecol.* **2018**, *11*, 543–550. [CrossRef]

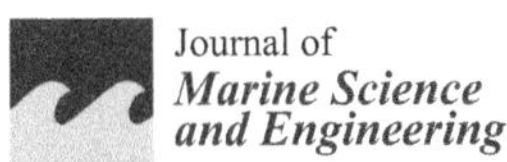
Journal of
*Marine Science
and Engineering*

Article

On the Predation of Doliolids (Tunicata, Thaliacea) on calanoid Copepods

Marion Köster [1] and Gustav-Adolf Paffenhöfer [2,*]

[1] Institut für Mikrobiologie—Mikrobielle Ökologie, Universität Greifswald, 17487 Greifswald, Germany
[2] Skidaway Institute of Oceanography, University of Georgia, Savannah, GA 31411, USA
* Correspondence: gustav.paffenhofer@skio.uga.edu

Abstract: The main goal of this contribution was to determine the effect of predation of the often abundant to dominant doliolid *Dolioletta gegenbauri* (Tunicata, Thaliacea) on the abundance of co-occurring planktonic copepods by feeding on their eggs. Previous oceanographic investigations revealed that doliolids had ingested eggs of small calanoid copepods. The ecological significance of such feeding could not be quantified completely because the environmental abundance of such eggs was not known. In this study, the eggs and nauplii of the neritic calanoid *Paracalanus quasimodo* (Crustacea, Copepoda) were offered to gonozooids and phorozooids of *D. gegenbauri* with a 6–6.5 mm length together with three species of phytoplankton; i.e., simulating diet conditions on the shelf. We hypothesized that copepod eggs of a similar size as food particles would be readily ingested whereas small nauplii, which could escape, would hardly be eaten by the doliolids. Our results revealed that doliolids have the potential to control small calanoids by ingesting their eggs at high rates but not their nauplii or later stages. Late copepodid stages and adults of co-occurring calanoid species could cause less mortality because they prey less on such eggs than doliolids of a similar weight. However, certain abundant omnivorous calanoid species with pronounced perception and/or capture abilities can prey successfully on the nauplii of small calanoids.

Keywords: copepod eggs; nauplii; doliolids; calanoid copepods

Citation: Köster, M.; Paffenhöfer, G.-A. On the Predation of Doliolids (Tunicata, Thaliacea) on calanoid Copepods. *J. Mar. Sci. Eng.* **2022**, *10*, 1293. https://doi.org/10.3390/jmse10091293

Academic Editor: Francesco Tiralongo

Received: 15 August 2022
Accepted: 8 September 2022
Published: 13 September 2022

1. Introduction

Mortality among planktonic copepods in the ocean is particularly pronounced for eggs and nauplii [1–5]). Laboratory studies on *Calanus helgolandicus* [6], on four species of calanoids [7], and on two species of calanoids [8] revealed pronounced predation on eggs and/or nauplii of calanoid copepods.

During a seven-day oceanographic study in January/February of 1990 on the southeastern shelf of the United States of America, about 12% of the fecal pellets of the doliolid *Dolioletta gegenbauri* (Tunicata, Thaliacea) contained eggs of the calanoid genus *Paracalanus* [9]. The researchers' calculations showed that the feeding rates of large gonozooids of nearly 7 mm in length (in situ, about one such large doliolid L^{-1}) resulted in a noticeable suppression of that calanoid genus: the oceanographic data indicated that large parts of the water column where doliolids were abundant showed low abundances of *Paracalanus* copepodid stages and adults. A model of the effects of doliolids on the plankton community structure on the southeastern shelf showed significantly that the presence of doliolids was followed by a larger decrease in copepods than did a decrease in food supply to the copepods [10].

D. gegenbauri has been found during much of the year on the southeastern U.S. shelf at a range of abundances, often surpassing 1000 zooids m^{-3} [9,11]. It is also abundant in other regions of the eastern seaboard of the USA [12] (at 1500 zooids m^{-3}; [13] at thousands of individuals m^{-3}). This doliolid species has been encountered abundantly off the Mississippi Delta [14], in the northern Gulf of Mexico [15], and off southern California [16] and further north [17]. High abundances were found in the Inland Sea of Japan, usually

>2000 zooids m^{-3} with a maximum of 48 zooids L^{-1} [18] (Nakamura 1998), and in the Kuroshio [19,20]).

Our main question after observing oceanographically the effects of doliolids on small copepod abundance [9] was: to what extent can doliolids actually affect the abundances of small calanoids? We designed experiments to offer eggs and nauplii of *Paracalanus quasimodo* to large gono- and phorozooids of *D. gegenbauri* in the presence of environmental concentrations of several phytoplankton species because to comprehensively understand in situ feeding processes, potential food organisms ought to be offered together [7]. We hypothesized that eggs of small planktonic copepods could be readily ingested while they were still in the water column. We also hypothesized that the nauplii would hardly fall prey to doliolids because they would perceive the weak feeding current produced by the doliolids and therefore would escape.

2. Materials and Methods

The doliolid *D. gegenbauri* was collected at different times of the year on the southeastern shelf of the USA, as was the calanoid *Paracalanus quasimodo*, which releases its eggs directly into the water. We utilized a plankton net with a 50 cm mouth diameter and a 200 μm mesh to collect zooplankton gently in oblique tows; i.e., near the surface to near the bottom to near the surface at ship speeds not surpassing 0.5 kn. The 4 L of codend contents were gently immersed in large seawater-filled buckets to avoid damaging the doliolids and copepods. Sorting of doliolids and copepods into freshly collected Niskin bottle water (from 20 m depth) on board the ship occurred in a temperature-controlled room at 20 °C; they were then placed in glass jars with 1.9 and 3.8 L volumes. Back in the laboratory, both the doliolids and copepods were immediately placed in their jars on a plankton wheel rotating at nearly 0.5 rpm. Here, doliolids and copepods were simultaneously offered three species of phytoplankton: the flagellates *Isochrysis galbana* and *Rhodomonas* sp. and the diatom *Thalassiosira weissflogii* at average total concentrations ranging from about 50 to 60 μg C L^{-1} at 20 °C in a light–dark cycle of 12 h:12 h.

The doliolids and copepods were reared in separate jars for the ensuing experiment [21]. Food concentrations were quantified with a Coulter Beckman Multisizer IV (Brea, CA, USA) using an orifice with a 140 μm diameter. Quantifications with the Coulter counter were regularly checked and confirmed with phytoplankton samples that had settled in 10 mL chambers and were counted with a Leitz Dialux (Wetzlar, Germany) inverted microscope. Food concentrations were expressed in units of carbon; i.e., μg C L^{-1}. *I. galbana* had an average cell volume of ~50 μm^3 with 200 μg C mm^{-3} of cell volume; *Rhodomonas* had an ~300 μm^3 cell volume with 160 μg C mm^{-3}; and *T. weissflogii* had an ~1000 μm^3 cell volume with 80 μg C mm^{-3}. Food concentrations in each jar were quantified daily and adjusted in a manner that resulted in an average concentration of ~50 to 60 μg C L^{-1} over 24 h.

Doliolids were kept in the laboratory for weeks to months, growing and reproducing in 3.8 L jars on the rotating wheel. Their water was renewed to about 75% every 4–5 days. *P. quasimodo* was reared through several generations in jars with a 1.9 L volume and also offered those three phytoplankton species, but near a 30 to 40 μg C L^{-1} total. Prior to our experiments, doliolid zooids of a similar size were placed in one jar the day prior to each experiment and we ascertained whether they escaped well.

Each of the six experiments of *D. gegenbauri* feeding on eggs of *P. quasimodo* was started early in the morning using new algal suspensions in GFC-filtered seawater, 50% of which was water in which the doliolids had been previously. Females of *P. quasimodo* produced eggs overnight that were rapidly counted. Each of these experiments lasted 4 to 4.5 h and offered on average 208 eggs L^{-1} (starting concentration) plus phytoplankton to three *D. gegenbauri* gonozooids of a 6–6.5 mm length in a 960 mL jar. The initial and final egg concentrations were counted in 25 mL settling chambers in triplicate. Phytoplankton concentrations were quantified at the start and end of each experiment. Doliolid fecal pellets were collected at the end of each experiment. The production of eggs of *P. quasimodo* started after 17:00 h the previous day when about 10 to 15 females were placed into a

1.9 L jar containing *I. galbana* and *T. weissflogii* at an average total concentration of about 50 μg C L^{-1}. By 08:00 h the next morning, no nauplii had hatched; by 12:00 h, an occasional nauplius was found.

When offering nauplii of *P. quasimodo,* three large phorozooids (two experiments) or gonozooids (four experiments) were placed in a 960 mL jar containing the three phytoplankton species (*I. galbana, Rhodomonas* sp., and *T. weissflogii*) which served as food for both the doliolids and the nauplii. We usually offered 100 nauplii of varying stages or 75 nauplii and 25 copepodid stage I of that copepod species for about 4.5 h to large phorozooids (6–6.5 mm length), and later to large gonozooids (6–6.5 mm length) of that doliolid. Prior to these experiments, those zooids had been in 1.9 L vessels and were offered those three phytoplankton species. The nauplii ranged in age from mainly Nauplius stage III to VI. Since so few nauplii were ingested, we decided to check each doliolid fecal pellet collected at the end of each experiment to determine whether it contained a nauplius.

For each of the feeding experiments on nauplii, 43 to 90 intact zooid fecal pellets of *D. gegenbauri* (i.e., all pellets produced during the experimental period) were individually checked for ingested nauplii, copepodid stages, and exuviae under a light microscope (AxioScope A1, Zeiss, Jena, Germany). Each pellet was transferred with a glass pipette to a slide and covered with a coverslip. The tip of a narrow needle was gently pressed on the coverslip to release the enclosed food items. Microphotographs of intact and smashed pellets, as well as the food items of interest, were captured at 100- to 400-fold magnification using a 5 MP digital CCD camera (AxioCam Mrc 5, Zeiss, Jena, Germany) and the software AxioVision 4.1.

Feeding rates and average food concentrations were determined according to Frost (1972) [22]. Statistical analyses were conducted according to Zar (1974) [23] using the Kruskal–Wallis test, a nonparametric single-factor analysis of variance by ranks for $K \geq 2$ independent samples [24].

3. Results

Doliolids feeding on copepod eggs. We conducted six separate experiments offering on average 208 eggs L^{-1} to three large gonozooids of *D. gegenbauri* (6–6.5 mm length) over an average period of 4.5 h together with three phytoplankton species (Figure 1). The Kruskal–Wallis test showed that the clearance rates for the three phytoplankton species and the copepod eggs did not differ significantly ($p < 0.005$). The eggs were readily ingested (Figure 1). Photographs of fecal pellets from these experiments were taken to document the ingestion of those eggs amidst the phytoplankton cells (Figure 2A,B). Although the concentration of copepod eggs (208 L^{-1}) was high, it did not contain much particulate matter (5.24 μg C L^{-1}) when using data from [25] as compared to the simultaneously offered phytoplankton, which represented the concentrations of a well-developing intrusion onto the southeastern shelf of the US [26].

Doliolids offered nauplii of *P. quasimodo*. As our field results [9] did not show that nauplii of *P. quasimodo* were ingested by large gonozooids of *D. gegenbauri,* we decided to evaluate in specific feeding experiments whether that finding was indeed true. A total of six experiments were conducted in which each of three zooids (6–6.5 mm length) were fed mainly on nauplii and the three phytoplankton species (Figure 3). The clearance rates for phytoplankton were not significantly different from each other (Figure 3, Kruskal–Wallis test, $p < 0.05$) but were significantly higher than those for the nauplii of *P. quasimodo* (Kruskal–Wallis test $p > 0.005$). The *D. gegenbauri* gonozooids in each jar captured a total of between one to four nauplii during the 4–4.5 h experimental periods. The clearance rates for nauplii of *P. quasimodo* were on average 2.0 mL *D. gegenbauri* zooid^{-1} h^{-1} (Figure 3).

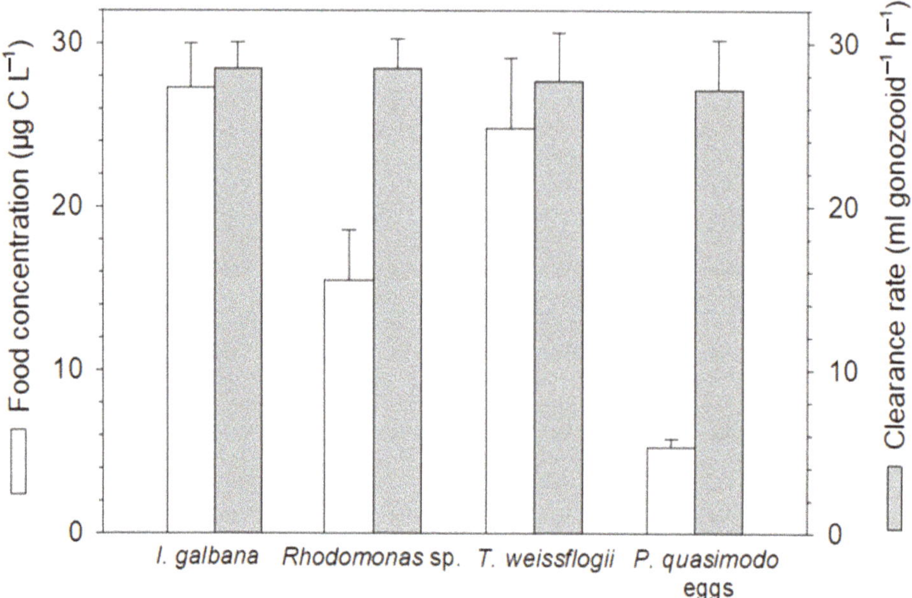

Figure 1. *Dolioletta gegenbauri.* Clearance rates of gonozooids of 6–7 mm length simultaneously offered eggs of *Paracalanus quasimodo* and cells of the three phytoplankton taxa (*Isochrysis galbana*, *Rhodomonas* sp., and *Thalassiosira weissflogii*) at 20 °C.

To recognize nauplii in pellets, they had to be squeezed for microscopical observation (Figure 2C–E). The nauplii we offered ranged from Nauplius stage III to VI. Those clearance rates for nauplii would amount to a clearance rate of 2.0 mL large gonozooid^{-1} h^{-1}, which would be 7.4% of the rates on eggs (Figure 1). Our photographs revealed *P. quasimodo* eggs (Figure 2A,B) and juvenile stages in pellets (Figure 2C–E). Large gonozooids were able to ingest exuviae (e.g., Figure 2F) that were compressed in the feeding net prior to passing the esophagus, which measured 60 to 70 µm in diameter.

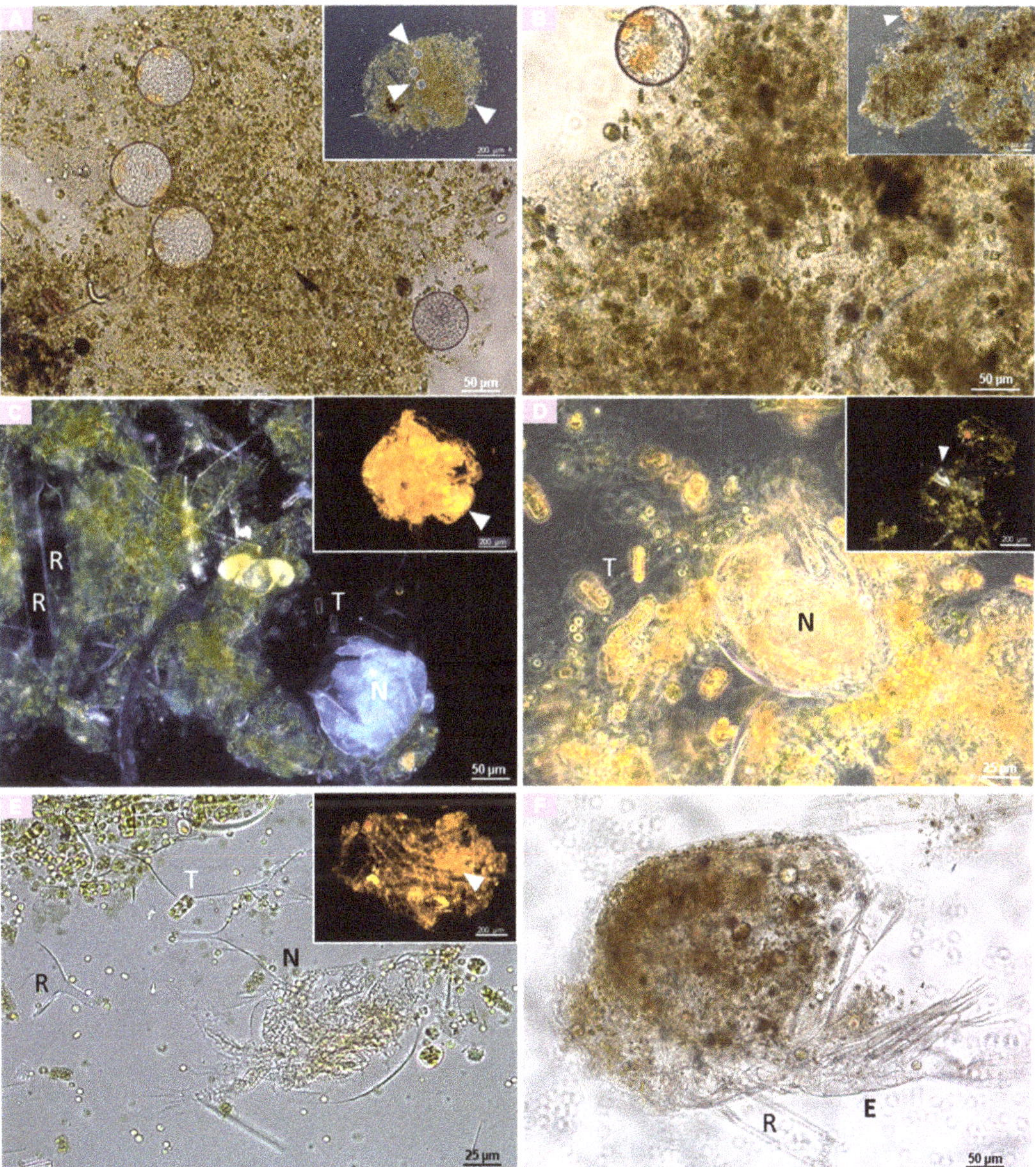

Figure 2. Light microscope photographs of fecal pellets of *Dolioletta gegenbauri* containing intact eggs (**A**,**B**) and juvenile stages (**C**–**E**) of the small calanoid copepod *Paracalanus quasimodo*. The inserts show the intact doliolid pellets. The white arrows indicate the hardly recognizable ingested zooplankton food items (eggs, nauplii, and copepodids of *P. quasimodo*) that were only detectable after the pellets were smashed (see food items at larger magnifications). (**A**,**B**) Phase-contrast micrographs of doliolid fecal pellets containing four and one *P. quasimodo* eggs with diameters ranging from 60 to 65 µm. (**C**–**E**) Dark-field and phase-contrast micrographs of smashed doliolid fecal pellets showing their nauplii (N) "unwrapped". (**F**) Phase-contrast micrograph of a doliolid fecal pellet containing a several hundred micron long exuvia of a copepodid of *P. quasimodo*. T = intact digested cells of *Thalassiosira weissflogii*; *R* = empty cells of *Rhizosolenia alata*. E = exuvia.

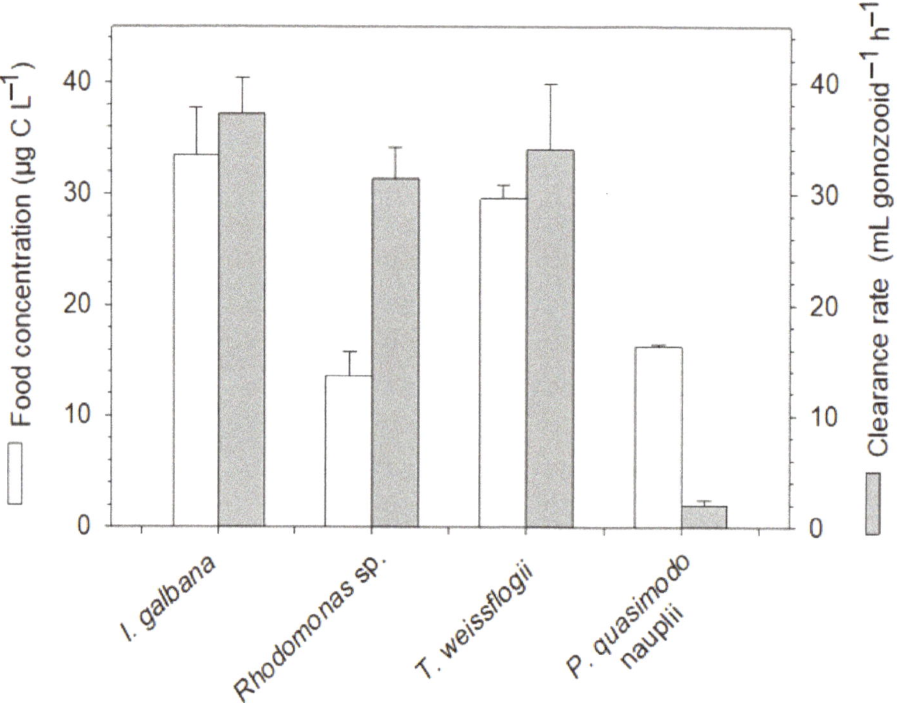

Figure 3. Zooids (phorozooids and gonozooids) of *Dolioletta gegenbauri* (6–6.5 mm body length) feeding simultaneously on three species of phytoplankton and nauplii of *Paracalanus quasimodo*. Arithmetic means and standard errors are given (n = 6).

4. Discussion

We will first discuss the ecological effects of doliolid predation on calanoid eggs and nauplii, then compare those findings with predation results from omnivorous calanoid genera that co-occur with doliolids on subtropical shelves, and end with general conclusions on the effects of doliolid predation on continental shelf food webs.

Feeding of doliolids on eggs and nauplii of small copepods. The assumption of Paffenhöfer et al., 1995 [9] that doliolid feeding could significantly affect the abundance of the calanoid copepod *P quasimodo* came from their field observations on the vertical abundance and distribution of *Paracalanus* spp. copepodids and adults versus that of doliolids. Their abundance was inverse: a higher abundance of *D. gegenbauri* co-occurred with a low abundance of *Paracalanus* spp. Those investigators found no nauplii in the guts or pellets of doliolids. The authors' calculations showed that the doliolid assemblage at times cleared 25% or more of the upper water column of ingestible particles per day. That was considered a conservative estimate.

Our results revealed that large doliolids (≥6 mm length) ingested *P. quasimodo* eggs at rates that were nearly identical to those rates on co-occurring phytoplankton (Figure 1). What did this finding imply? The clearance rate of gonozooids of a 6–6.5 mm length on eggs was 28.6 mL zooid^{-1} h^{-1}, which would amount to 686 mL d^{-1}. As the eggs of *P. quasimodo* would hatch within ~15–18 h of being released, such a doliolid would spend that time at the clearance rate of ~514 mL for that period of 18 h; this implies that about 50% of those eggs would be ingested by one large doliolid L^{-1}. The extent of such predation depends on abundance and size of the zooids. Even a small *D. gegenbauri* zooid of a 4.5 mm length that was able to ingest cells up to 60 µm in diameter and clearing 350 mL d^{-1} at

20 °C [27] would clear 233 mL in 16 h, and thus would ingest near 25% of small copepod eggs. The effect of doliolids of >4 mm length would be a function of their clearance rate as a population, which repeatedly amounted to more than 500 gonozooids/phorozooids m^{-3} in the winter of 1990 [9].

Overall, *D. gegenbauri* should influence to a varying extent the abundance of small calanoids on subtropical shelves. Aside from the genus *Paracalanus*, the calanoids *Temora turbinata, T. stylifera, Clausocalanus furcatus*, and *Centropages furcatus/velificatus* also often occur abundantly on the SE shelf [26] and produce eggs of a similar size, and therefore could fall prey to *D. gegenbauri*.

Having copepod eggs ingested and released within pellets revealed that many of those eggs appeared undamaged (Figure 2A), as we had also observed for diatoms of different sizes [28]. Those pellets remained for several to many hours in the water column [29]. During that time, the nauplii would most likely hatch. We do not know whether the hatching nauplii were affected by the digestion process, as the eggs appeared undamaged externally. Having hatched, the nauplii would attempt to move out of the pellet. We observed one ingested nauplius that attempted this. Our oceanographic study [9] seemed to support the assumption that a high percentage of ingested eggs would not lead to surviving nauplii.

Doliolids produce slow-moving feeding currents by displacing water and particles therein toward their mouth [30]. Even nauplii with continuously moving appendages such as *Paracalanus* most likely will perceive the shear in such feeding currents and escape. Some nauplii may not be able to do so and thus could be ingested (Figure 3). Overall, the clearance rate of *D. gegenbauri* on nauplii of *Paracalanus* that was calculated in our experiments was only 2.0 ml gonozooid/phorozooid^{-1} h^{-1}, implying minimal effects of doliolids on nauplii and later stages.

Comparing the effects of doliolids with those of calanoid copepods on eggs and juveniles of calanoids. The intermittent high abundance of larger doliolids [31] (from about 1000 m^{-3} on) should affect the abundance of egg-releasing smaller calanoids. It appeared that eggs and nauplii were more vulnerable than later juvenile stages of calanoid copepods: Eggs cannot perceive predators and cannot escape, while nauplii can perceive and escape fairly well; however, copepodid stages have superior perception (extended first antennae with 3D-arranged setae) and escape capabilities, and therefore are not as vulnerable to predation by omnivorous copepods as nauplii [32]. Doliolids are effective at preying on calanoid eggs but not on calanoid nauplii. Different species of omnivorous calanoids vary in their capability to prey on eggs and nauplii of other species and their own (Table 1): females of the calanoid *Calanus helgolandicus*, which usually create a feeding current, remove eggs at a far higher rate than its nauplii [6] (Table 1). Those rates are usually lower than those on larger phytoplankton cells by a closely related species [33] (Table 1): while phytoplankton is perceived via chemosensory mechanisms in a calanoid feeding current [34], eggs do not produce a chemical signal like phytoplankton. They may either be perceived by mechanosensory mechanisms, or, if not perceived individually, could arrive at the mouth together with phytoplankton, which provide the signal for ingestion [35].

The almost continuously moving *Temora longicornis* clears eggs at a higher rate than nauplii, whereas two species of the genus *Centropages*, which are considered ambushers, clear nauplii at a higher rate than eggs [7]. Our observations (Paffenhöfer and Knowles, unpubl. observations) revealed that *T. stylifera* females move almost continuously while creating a feeding current. They thus provide a mechanical warning signal to nauplii while ingesting much of the phytoplankton and most likely some calanoid eggs (Table 1). *Centropages velificatus* adults, however, create a feeding current only briefly before starting to sink motionless for longer periods. They are not perceived by slowly swimming nauplii while ingesting only small amounts of phytoplankton due to brief feeding current activity (Table 1): *C. velificatus* adults ingested 7 nauplii day^{-1} at 30 nauplii L^{-1} and 13 nauplii d^{-1} at 60 nauplii L^{-1} [8]. These findings supported the results of Boersma et al., 2014 [7] of feeding on nauplii by congeners, revealing differences between doliolids and co-occurring copepod genera: while omnivorous calanoids can perceive and capture nauplii and eggs, doliolids do not rely on perceiving their prey; they instead create a feeding current that displaces particles with an ~1 to >50 μm diameter toward themselves and ingest them. Most of the copepod rates were obtained with adult females. It remains to be determined to which extent earlier copepodid stages could be carnivorous and to which extent potential predators and their prey operate in similar depth ranges.

Table 1. Average clearance rates on copepod eggs and nauplii and phytoplankton by calanoid copepods and doliolids (n.d.—not determined).

Species and Stage	Temperature (°C)	Clearance Rate (mL^{-1} $Copepod^{-1}$ Day^{-1})			References
		Eggs	Nauplii	Phytoplankton	
Calanus helgolandicus female	13–15	320	>102	n.d.	Bonnet et al., 2004 [6]
Temora longicornis female	10	161	>120	n.d.	Boersma et al., 2014 [7]
Centropages hamatus female	10	94	<195	n.d.	
Centropages typicus female	10	159	<224	n.d.	
Temora stylifera female	20	n.d.	139	<360	Paffenhöfer and Knowles unpubl.results
Centropages furcatus/velificatus female	20	n.d.	230	>60	
Calanus helgolandicus female	15	n.d.	n.d.	530	Paffenhöfer 1971 [33]
Dolioletta gegenbauri	20	648	69	670	This paper
Gonozooids/ Phorozooids					

Data from a previous cruise [36] revealed that during the summer, on average, the genus *Paracalanus* occurred at 1429 copepodid stage II (C II) to adult m^{-3} in warm surface waters and at 3030 CII to adults m^{-3} in cooler bottom layers. At the same time, the potential predators *Centropages furcatus/velificatus* were 287 C II to adult m^{-3} in surface waters and 84 m^{-3} in cooler bottom layers. The doliolid *D. gegenbauri* occurred at a low abundance during those cruises, decreasing from 424 to 8 zooids m^{-3} [37]. This species is usually found during the summer in larger numbers only in cooler bottom layers and in the thermocline.

Data on the actual vertical distribution of nauplii in the ocean are sparse: during the summer of 1979, several cross-shelf transects using a 30-micron mesh for sampling the entire water column with a pump revealed the following [38] when sampling the warm upper mixed layer, the thermocline, and the cold bottom layers consecutively. Nauplii occurred between 7 and 100 L^{-1}. At 11 stations, they were most abundant five times in the upper mixed layer, four times in the thermocline, and twice in the intrusion of cold water. While *Paracalanus* spp. occurred from about 3000 to 11,000 m^{-3} and *Temora turbinata* from near 1000 to 5000 m^{-3} (C I to adults), doliolids were sparse, amounting to 60 to nearly 300 zooids m^{-3} over the following two weeks, and therefore should have had hardly any effect through predation on the copepods' eggs. However, one week later *D. gegenbauri* had increased to 620 to 1230 gono/phorozooids m^{-3} in the thermocline and had intrusion at two stations, which had the potential to affect the populations of *Paracalanus* spp. and *T. turbinata*.

Our time-series findings from a two-month oceanographic coverage (weekly) of the southeastern shelf off northern Florida and southern Georgia showed abundances of 20–250 *D. gegenbauri* zooids m^{-3} on part of the shelf by mid-July and maxima of 500–1000 zooids m^{-3} two to three weeks later, which diminished afterward [26]. In comparison to the winter of 1990 [9], these summer doliolid abundances should have had a limited effect on the populations of small copepods, which were dominated by *T. turbinata* repeatedly reaching maxima between 1000 to 10,000 (mainly juveniles) m^{-3}. There were no data on the sizes of *D. gegenbauri* during all those summer studies. The very high abundances of *T. turbinata* could have diminished doliolid reproduction, as they could have grabbed doliolid larvae and thus killed them (grabbing behavior of Temoridae, pers. observation, G.-A. Paffenhöfer for copepodids and adults).

The clearance rates of *D. gegenbauri* on phytoplankton of about 640 to 670 mL gonozooid (body weight of 35 μg C)$^{-1}$ d^{-1} were slightly above the rates of nearly 600 mL gonozooid (body weight of 35 μg C)$^{-1}$ d^{-1} feeding at 60 μg C L^{-1} of phytoplankton [27] (and also the rates shown in Figure 1) We may attribute such differences to variability when cultivating these doliolids.

Doliolids do not appear to choose particles based on their chemical quality and composition. They could affect calanoid reproduction more than similarly sized calanoids: Our study provided a glimpse into what extent planktonic copepods and doliolids will affect zooplankton communities via predation on copepod eggs. However, the short residence time of small calanoid eggs in the water column prior to hatching (15–18 h) implied only short periods of vulnerability in situ as compared to nauplii, which will exist for about 4–8 days in the water column prior to molting to copepodid stage I (C I). Then they are less vulnerable to predation [32]. In essence, eggs would be mainly vulnerable to doliolid occurrence while nauplii would be sensitive to copepodids and adult calanoids (omnivory) on a subtropical shelf. Other studies revealed that the far less abundant but larger outer-shelf and oceanic salps are considered carnivores [39,40] as observed in their gut contents.

General Conclusions. Earlier results [10,41] and ours indicated that doliolids can have significant influences on food web processes on subtropical continental shelves: they can ingest anything from a nearly one micron width to a >60 micron maximum dimension. Such food particles include detritus such as fecal pellets [41], which are displaced by a gentle current into a doliolid's mouth and settle on the mucous filter. However, in comparison to copepods, the doliolid digestion process is limited [42]. The fecal pellets contain aggregated or not or partly digested small cells [28] (e.g., *Isochrysis galbana*), which as individual cells are not perceived by copepods and can now, as a pellet, be ingested and used by calanoids. At the same time, such doliolid pellets sink slower than similarly sized copepod pellets [29], depriving the seafloor of food particles as those are ingested by suspension feeders in the water column [43] (e.g., heterotrophic dinoflagellates). In comparison to most copepod fecal pellets, the doliolid pellets contain considerable amounts of nitrogen [43], which, compared with copepod pellets, can support growth of those zooplankters ingesting them. Doliolids do not uniformly digest and utilize phytoplankton as many calanoids do, which destroys the cells when they enter the esophagus and then utilizes the cells' contents to a high percentage. Doliolids do not persist permanently in abundance on continental shelves as voracious predators such as hydromedusae with extended tentacles ingest and digest them readily (unpubl. results by L. Frazier and G.-A. Paffenhöfer, shipboard and laboratory observations). As addressed in [1,2], we wanted to inquire to what extent different copepod species (calanoids and cyclopoids) and smaller doliolid zooids actually affected a community's copepod composition via predation on eggs, aside from predation on juveniles (nauplii), of which we already have some knowledge [7,8]. That research will also include obtaining information on the residence times of sinking copepod eggs in the water column and their vertical position when being released.

Author Contributions: Both authors contributed equally. All authors have read and agreed to the published version of the manuscript.

Funding: This research received no funding.

Institutional Review Board Statement: Not applicable.

Informed Consent Statement: Not applicable.

Data Availability Statement: Not applicable.

Acknowledgments: We would like to express our gratitude to Captain Raymond Sweatte and the crew of the R/V *Savannah*, who supported our efforts with competence and an excellent attitude. All experiments were conducted at the Skidaway Institute of Oceanography. Various analyses were conducted at the Department of Microbial Ecology, Institute of Microbiology, University of Greifswald, Germany.

Conflicts of Interest: The authors declare no conflict of interest.

References

1. Kiørboe, T.; Møhlenberg, F.; Tiselius, P. Propagation of planktonic copepods: Production and mortality of eggs. *Hydrobiologia* **1988**, *167*, 219–225. [CrossRef]
2. Peterson, W.T.; Kimmerer, W.J. Processes controlling recruitment of the marine calanoid copepod *Temora longicornis* in Long Island Sound: Egg production, egg mortality, and cohort survival rates. *Limnol. Oceanogr.* **1994**, *39*, 1594–1605. [CrossRef]
3. Ohman, M.D.; Hirche, H.-J. Density-dependent mortality in an oceanic copepod population. *Nature* **2001**, *412*, 638–641. [CrossRef]
4. Frank-Gopolos, T.; Møller, E.F.; Nielsen, T.G. The role of egg cannibalism for the *Calanus* succession in the Disko Bay, Western Greenland. *Limnol. Oceanogr.* **2017**, *62*, 865–883. [CrossRef]
5. Corrales- Ugalde, M.; Sponaugle, S.; Cowen, R.K.; Sutherland, K.R. Seasonal hydromedusan feeding patterns in an Eastern Boundary Current show consistent predation on primary consumers. *J. Plankton Res.* **2021**, *43*, 712–724. [CrossRef]
6. Bonnet, D.; Titelman, J.; Harris, R. *Calanus* the cannibal. *J. Plankton Res.* **2004**, *26*, 937–948. [CrossRef]
7. Boersma, M.; Wesche, A.; Hirche, H.-J. Predation of calanoid copepods on their own and other copepods' offspring. *Mar. Biol.* **2014**, *161*, 733–743. [CrossRef]
8. Paffenhöfer, G.-A.; Knowles, S.C. Omnivorousness in marine planktonic copepods. *J. Plankton Res.* **1980**, *2*, 355–365. [CrossRef]
9. Paffenhöfer, G.-A.; Atkinson, L.P.; Lee, T.N.; Verity, P.G.; Bulluck, L.R., III. Distribution and abundance of thaliaceans and copepods off the southeastern U.S.A. during winter. *Cont. Shelf Res.* **1995**, *15*, 255–280. [CrossRef]
10. Haskell, A.E.G.; Hofmann, E.E.; Paffenhöfer, G.-A.; Verity, P.G. Modeling the effects of doliolids on the plankton community structure of the southeastern U.S. continental shelf. *J. Plankton Res.* **1999**, *21*, 1725–1752. [CrossRef]
11. Deibel, D. Blooms of the pelagic tunicate, *Dolioletta gegenbauri*: Are they associated with Gulf Stream frontal eddies? *J. Mar. Res.* **1985**, *43*, 211–236. [CrossRef]
12. Deevey, G.B. Quantity and composition of the zooplankton of Block Island Sound. *Bull. Bingham Oceanogr. Coll.* **1952**, *13*, 120–164.
13. Ambler, J.W.; Kumar, A.; Moisan, T.A.; Aulenbach, D.L.; Day, M.C.; Dix, S.A.; Winsor, M.A. Seasonal and spatial patterns of *Penilia avirostris* and three tunicate species in the southern Mid-Atlantic Bight. *Cont. Shelf Res.* **2013**, *69*, 141–154. [CrossRef]
14. Di Mauro, R.; Kupchik, M.J.; Benfield, M.C. Abundant plankton-sized microplastic particles in shelf waters of the northern Gulf of Mexico. *Environ. Pollut.* **2017**, *230*, 798–809. [CrossRef]
15. Greer, A.T.; Woodson, C.B.; Smith, C.E.; Guigand, C.M.; Cowen, R.K. Examining zooplankton patch structure and its implications for trophic interactions in the northern Gulf of Mexico. *J. Plankton Res.* **2016**, *38*, 1115–1134. [CrossRef]
16. Crocker, K.M.; Alldredge, A.L.; Steinberg, D.K. Feeding rates of the doliolid, *Dolioletta gegenbauri,* on diatoms and bacteria. *J. Plankton Res.* **1991**, *13*, 77–82. [CrossRef]
17. Mackas, D.L.; Washburn, L.; Smith, S.L. Zooplankton community pattern associated with a California Current cold filament. *J.Geophys. Res.* **1991**, *96*, 14781–14797. [CrossRef]
18. Nakamura, Y. Blooms of tunicates *Oikopleura* spp. and *Dolioletta gegenbauri* in the Seto Inland Sea, Japan, during summer. *Hydrobiologia* **1998**, *385*, 183–192. [CrossRef]
19. Takahashi, K.; Ichikawa, T.; Fukugama, C.; Yamane, Y.; Kahehi, S.; Okazaki, Y.; Kubota, H.; Furuya, K. In situ observations of a doliolid bloom in a warm water filament using a video plankton recorder: Bloom development, fate, and effect on biogeochemical cycles and planktonic food webs. *Limnol. Oceanogr.* **2015**, *60*, 1763–1780. [CrossRef]
20. Ishak, N.H.A.; Tadokoro, A.; Okazaki, Y.; Kahehi, S.; Suyama, S.; Takahashi, K. Distribution, biomass, and species composition of salps and doliolids in the Oyashio-Kuroshio transitional region: Potential impact of massive bloom on the pelagic food web. *J. Oceanogr.* **2020**, *76*, 351–363. [CrossRef]
21. Paffenhöfer, G.-A.; Gibson, D.M. Determination of generation time and asexual fecundity of doliolids (Tunicata, Thaliacea). *J. Plankton Res.* **1999**, *21*, 1183–1189. [CrossRef]
22. Frost, B.W. Effects of size and concentration of food particles on the feeding behavior of the marine planktonic copepod *Calanus pacificus*. *Limnol. Oceanogr.* **1972**, *17*, 805–815. [CrossRef]
23. Zar, J.H. *Biostatistical Analysis;* Prentice-Hall, Inc.: Englewood Cliffs, NJ, USA, 1974; p. 620.
24. Conover, W.J. *Practical Non-Parametric Statistics;* Wiley: New York, NY, USA, 1980; p. 493.
25. Checkley, D.M., Jr. The egg production of a marine planktonic copepod in relation to its food supply: Laboratory studies. *Limnol. Oceanogr.* **1980**, *25*, 430–446. [CrossRef]
26. Paffenhöfer, G.-A.; Lee, T.N. Summer upwelling on the southeastern continental shelf of the U.S.A. during 1981. Distribution and abundance of particulate matter. *Prog. Oceanogr.* **1987**, *19*, 373–401. [CrossRef]
27. Gibson, D.M.; Paffenhöfer, G.-A. Feeding and growth rates of the doliolid *Dolioletta gegenbauri* Uljanin (Tunicata, Thaliacea). *J. Plankton Res.* **2000**, *22*, 1485–1500. [CrossRef]

28. Köster, M.; Sietmann, R.; Meuche, A.; Paffenhöfer, G.-A. The ultrastructure od a doliolid and a copepod fecal pellet. *J. Plankton Res.* **2011**, *33*, 1538–1549. [CrossRef]
29. Patonai, K.; El-Shaffey, H.; Paffenhöfer, G.-A. Sinking velocities of fecal pellets of doliolids and calanoid copepods. *J. Plankton Res.* **2011**, *33*, 1146–1150. [CrossRef]
30. Deibel, D.; Paffenhöfer, G.-A. Cinematographic analysis of the feeding mechanism of the pelagic tunicate *Doliolum nationalis*. *Bull. Mar. Sci.* **1988**, *43*, 404–412.
31. Paffenhöfer, G.-A.; Sherman, B.K.; Lee, T.N. Summer upwelling on the southeastern continental shelf of the U.S.A. during 1981. Zooplankton abundance and distribution. *Prog. Oceanog.* **1987**, *19*, 403–436. [CrossRef]
32. Landry, M.R. Predatory feeding behavior of a marine copepod, *Labidocera trispinosa*. *Limnol. Oceanogr.* **1978**, *23*, 1103–1113. [CrossRef]
33. Paffenhöfer, G.-A. Grazing and ingestion rates of nauplii, copepodids and adults of the marine planktonic copepod *Calanus helgolandicus*. *Mar. Biol.* **1971**, *11*, 286–298. [CrossRef]
34. Paffenhöfer, G.-A.; Lewis, K.D. Perceptive performance and feeding behavior of calanoid copepods. *J. Plankton Res.* **1990**, *12*, 933–946. [CrossRef]
35. Paffenhöfer, G.-A.; Van Sant, K.B. The feeding response of a marine planktonic copepod to quantity and quality of particles. *Mar. Ecol. Prog. Ser.* **1985**, *27*, 55–65. [CrossRef]
36. Paffenhöfer, G.-A. Zooplankton distribution as related to summer hydrographic conditions in Onslow Bay, North Carolina. *Bull. Mar. Sci.* **1980**, *30*, 819–832.
37. Deibel, D.; Paffenhöfer, G.-A. Predictability of patches of neritic salps and doliolids (Tunicata, Thaliacea). *J. Plankton Res.* **2009**, *31*, 1571–1579. [CrossRef]
38. Paffenhöfer, G.-A.; Wester, B.T.; Nicholas, W.D. Zooplankton abundance in relation to state and type of intrusions onto the southeastern United States shelf during summer. *J. Mar. Res.* **1984**, *42*, 995–1017. [CrossRef]
39. Hopkins, T.L.; Torres, J.J. Midwater food web in the vicinity of a marginal ice zone in the western Weddell Sea. *Deep-Sea Res.* **1989**, *36*, 543–560. [CrossRef]
40. Ishak, N.H.A.; Clementson, L.A.; Eriksen, R.S.; van den Enden, R.L.; Williams, G.D.; Swadling, K.M. Gut contents and isotopic profiles of *Salpa fusiformis* and *Thalia democratica*. *Mar. Biol.* **2017**, *164*, 144. [CrossRef]
41. Köster, M.; Paffenhöfer, G.-A. How efficiently can doliolids (Tunicata, Thaliacea) utilize phytoplankton and their own fecal pellets? *J. Plankton Res.* **2017**, *39*, 305–315. [CrossRef]
42. Paffenhöfer, G.-A.; Köster, M. Digestion of diatoms by planktonic copepods and doliolids. *Mar. Ecol. Prog. Ser.* **2005**, *297*, 303–310. [CrossRef]
43. Poulsen, L.K.; Moldrup, M.; Berge, T.; Hansen, P.J. Feeding on copepod fecal pellets: A new trophic role of dinoflagellates as detritivores. *Mar. Ecol. Prog. Ser.* **2011**, *441*, 65–78. [CrossRef]

Journal of
Marine Science and Engineering

MDPI

Article

Copepod Feeding Responses to Changes in Coccolithophore Size and Carbon Content

Jordan Toullec [1,*], Alice Delegrange [1,2], Adélaïde Perruchon [1], Gwendoline Duong [3], Vincent Cornille [1], Laurent Brutier [1] and Michaël Hermoso [1]

1 Laboratoire d'Océanologie et de Géosciences—UMR 8187 LOG, CNRS, Université Littoral Côte d'Opale, F-62930 Wimereux, France
2 Institut national supérieur du professorat et de l'éducation, Académie de Lille—Hauts de France, F-59658 Villeneuve d'Ascq, France
3 Laboratoire d'Océanologie et de Géosciences—UMR 8187 LOG, CNRS, Université de Lille, F-59000 Lille, France
* Correspondence: toullec.jordan@gmail.com

Abstract: Phytoplankton stoichiometry and cell size could result from both phenology and environmental change. Zooplankton graze on primary producers, and this drives both the balance of the ecosystem and the biogeochemical cycles. In this study, we performed incubations with copepods and coccolithophores including different prey sizes and particulate carbon contents by considering phytoplankton biovolume concentration instead of chlorophyll *a* level (Chl *a*) as is usually performed in such studies. The egestion of fecal pellet and ingestion rates were estimated based on a gut fluorescence method. The latter was calibrated through the relationship between prey Chl *a* level and the biovolume of the cell. Chl *a*/biovolume ratio in phytopkanton has to be considered in the copepod gut fluorescent content method. Both coccolithophore biovolume and particulate inorganic/organic carbon ratios affect the food foraging by copepods. Finally, we observed a non-linear relationship between ingestion rates and fecal pellet egestion, due to the presence of calcite inside the copepod's gut. These results illustrate that both prey size and stoichiometry need to be considered in copepod feeding dynamics, specifically regarding the process leading to the formation of fecal pellets.

Keywords: coccolithophore; elemental stoichiometry; copepods; gut content; ingestion rate; fecal pellet egestion; functional response

Citation: Toullec, J.; Delegrange, A.; Perruchon, A.; Duong, G.; Cornille, V.; Brutier, L.; Hermoso, M. Copepod Feeding Responses to Changes in Coccolithophore Size and Carbon Content. *J. Mar. Sci. Eng.* **2022**, *10*, 1807. https://doi.org/10.3390/jmse10121807

Academic Editors: Marco Uttieri, Ylenia Carotenuto, Iole Di Capua and Vittoria Roncalli

Received: 28 October 2022
Accepted: 17 November 2022
Published: 23 November 2022

1. Introduction

By absorbing about 50% of carbon dioxide (CO_2) from the atmosphere, the Ocean plays a major role in the global carbon cycle [1]. The biological carbon pump is sustained by photosynthetic CO_2 fixation by phytoplankton, and by the transfer of both organic and inorganic carbon to the deep sea [2,3]. Zooplankton, as primary consumers, control the carbon transfer through excretion/respiration [4,5] producing fecal pellets that foster the export of particulate carbon flux, as observed through the analysis of sediment traps [4,6–8].

Mesozooplankton (>200 µm) prey assemblages are constituted of heterotrophic microzooplankton (flagellates, cilliates) and autotrophs such as diatoms and coccolithophores [9–11]. Coccolithophores are a key food-source group widely dispersed throughout the world's oceans [12,13]. They produce calcified structures—coccoliths, which have formed a substantial proportion of pelagic sediments since the Late Triassic period (about 200 million years ago). Fossil records show that coccolithophores were a major component of primary producers over this period, and a significant food source for zooplanktonic grazers during this period [14,15].

Both phytoplankton and zooplankton are the first to experience natural environmental shifts such as phenological changes, or anthropogenic changes induced by global warming or ocean acidification [16]. Recently, morphological changes (cell size and shape) and the relative abundance distribution in diatom assemblage were linked to the annual phenology in the North Sea [17]. These findings could have consequences on copepod grazing [18–20]. Indeed, morphological defence of phytoplankton against grazing can be the formation of chain, cell size/shape, and biomineralization [21]. Even though copepods are well designed to break down biomineral structures and sometimes can graze on larger prey than themselves [22,23], diatom frustules limit copepod grazing [24–26] as well as microzooplankton grazing [27]. Moreover, it has been established that grazers could induce diatom silicification [28,29], proving the defensive role of the these biomineral structures. Similarly, coccolithophore build biomineral shells made of calcium carbonate (coccoliths) whose formation is influenced by environmental conditions (see reference above). As for diatom frustules, these coccoliths arranged around the cell forming the coccosphere provide mechanical protection [30], and could play a defensive role against microzooplankton grazing [31,32]. Although suggested, but never demonstrated, this calcified coccosphere could also be considered as an anti-grazing protection against copepods [33–35].

Copepods are characterised by distinct functional feeding traits (they are feeding-current feeders or ambush feeders), and as such, are interesting organisms for studying the trophodynamics towards phytoplankton [36]. Classically, ingestion rates increase with food availability and follow Ivlev's model curves [37]. This relationship is formalised by an optimal foraging theory [38,39]. The modification of copepods feeding behaviour potentially has consequences for the functioning of ecosystems, such as "trophic cascades" with consequences on biogeochemical cycles [33,40–42]. In the context of global warming and ocean acidification, a species-specific difference in coccolithophore response is expected [43] on both cell size and calcification. In this study, the modification of calcite content and cell size on copepod ingestion was explored. As a result of experimental incubations, the prediction of an optimal foraging model (Ivlev's model) was tested through direct observations of copepods' functional responses with different coccolithophore species, characterized by different calcite contents and sizes. Moreover food type and availability affect fecal pellet production rates, pellet volumes, and sinking rate, regarding compactness and mineral ballasting [44,45] (Table 1). We hypothesise that both calcite content and prey volume affect copepod functional responses and by this way, the fecal pellet egestion.

Table 1. Parameters indicating the initial conditions during the experimental incubations (mean ± SD, N = 3).

Experiment	Prey Type per Incubation Batch	Cell Diameter	Chlorophyll *a* per Cell	Organic Carbon per Cell	Inorganic Carbon per Cell	Organic Nitrogen per Cell	TPC/N	POC/N	PIC/POC
		[μm]	[pg Chl *a* $cell^{-1}$]	[pg POC $cell^{-1}$]	[pg PIC $cell^{-1}$]	[pg PON $cell^{-1}$]	[mol:mol]	[mol:mol]	[mol:mol]
1	*G. oceanica*	6.7 ± 0.9	0.16 ± 0.05	13	50.7	4.4 [<dl]	17.0 [ns]	3.5 [ns]	3.9
	G. oceanica + *Tisochrysis* sp.	6.3 ± 0.9 [a]	0.40 ± 0.05 [a]	25.8	17	2.6 [<dl]	19.3 [ns]	11.6 [ns]	0.7
	Tisochrysis sp.	6.1 ± 0.6	0.6 ± 0.1	21.4	0	1.3 [<dl]	19 [ns]	19.0 [ns]	0
2	*G. oceanica*	6.3 ± 0.9	0.17 ± 0.05	21.5 ± 1.3	6.7 ± 1.4	2.5 ± 0.1	13.3 ± 0.8	10.1 ± 0.3	0.3 ± 0.1
3	*E. huxleyi*	4.5 ± 0.5	0.10 ± 0.01	7.7 ± 0.3	5.4 ± 3.2	1.3 ± 0.2 [<dl]	12.1 ± 4.3 [ns]	6.9 ± 0.8 [ns]	0.7 ± 0.4 [ns]
4	*C. braarudii*	17 ± 2	3.5 ± 0.5	110 ± 7	183 ± 3	13.4 ± 0.8	25.6 ± 1.2	9.7 ± 0.7	1.7 ± 0.1
5	*E. huxleyi*	5.13 ± 0.03	0.12 ± 0.01	20 ± 1	2.6 ± 1.2	1.7 ± 0.1	15.8 ± 1.3	13.9 ± 0.3	0.13 ± 0.07

[a] Corresponding to the mix of both cell species. [<dl] Under the detection limit. [ns] Statistically non-significant.

2. Methods and Materials

2.1. Phytoplankton Cultures

For the laboratory experiment setup, three species of calcifying Haptophyceae were used: *Emiliania huxleyi* (strain RCC 1256); *Coccolithus braarudii* (strain RCC 1200); and *Gephyrocapsa oceanica* (strain RCC 1314). They were grown in polycarbonate flasks in 100–400 mL of K/2 + Si media at 15 °C and under a 12:12 h day:night photoperiod (100–150 μE m^{-2} s^{-1}). The culture media were prepared with 0.2 μm filtered seawater (FSW) from the English Channel (33–34 PSU) [46]. The culture media pH was adjusted to 8.2 (total scale) by the addition of NaOH. The cells were maintained in an exponential growth phase by renewing the media every week. In parallel, non-calcifying Haptophyceae species were also cultured, *Tisochrysis* sp. (strain RCC 1350), grown inside a 2 L Erlenmeyer flask with a K/2 + Si medium at 15 °C and under a 12:12 h day:night photoperiod (100–150 μE m^{-2} s^{-1}). These cultures were directly used after dilution with 1 μm of FSW buffered at pH 8.2 for the copepod incubation experiments (Table 1).

2.1.1. Cell Count and Size Measures

Cell numeration and sizing were done using a Beckman Coulter Counter Multisizer 4E apparatus fitted with a 70 μm aperture tube. Sampled cell suspensions were diluted with an isotonic (ISOTON II) solution before being analysed. Cell sizes (cell diameter in μm) were determined by the Gaussian distribution of dominant particles present inside the culture samples (containing phytoplankton) (Table 1).

2.1.2. Cell Chlorophyll *a* (Chl *a*) Content

Amounts of 100 mL of pre-diluted phytoplankton culture were filtered onto pre-combusted (4 h at 450 °C) glass fibre filters (Whatman GF/F) and conserved at −20 °C prior to pigment extractions. The filters were then ground overnight in 6 mL of acetone (90%) for chlorophyllian pigment extraction (Chl *a* and phaeopigments) in the dark at 4 °C. Fluorescence of the extract was measured before and after acidification with 10% HCl using a fluorometer (Turner design Trilogy). Results are expressed in pg Chl a_{eq} $cell^{-1}$ (Table 1).

2.1.3. Particulate Inorganic Carbone (PIC), Particulate Organic Carbon (POC), and Particulate Organic Nitrogen (PON)

Before each incubation, 100 mL phytoplankton culture suspensions (with known cell concentration) were filtered onto pre-combusted (4 h at 450 °C) glass fibre filters (Whatman GF/F). All the filters were then rinsed with 10 mL of FSW. Due to the large number of samples, the filters were not triplicated. The filters were placed inside aluminium foil, dried at 55 °C for 24 h, and analysed for elemental C and N using a Thermo Fisher Flash 2000 elemental analyser [47]. Two batches of glass filters were filtered for each sample, one batch with an acid treatment (providing the POC content) and the other without an acid treatment (providing the PIC + POC content), namely the total particulate carbon content, (TPC). PIC was obtained by subtracting POC from the TPC. The results are expressed in mass per cell (pg $cell^{-1}$), for inorganic carbon, organic carbon, and organic nitrogen (Table 1).

2.2. Copepod Sampling

For the laboratory experiments, two calanoid copepod species (*Temora longicornis* and *Acartia* clausi) were selected due to their abundance in the Eastern English Channel (EEC). Their presence generally matches phytoplankton spring blooms in the coastal areas of the EEC [48]. Each species also exhibits different functional traits [49] regarding their feeding strategies: *A. clausi* (1.1 mm total length) is an omnivorous feeding-current feeder with a clear tendency to herbivory; and *T. longicornis* (1.2 mm total length) is described as both a feeding-current feeder and cruise feeder [49,50].

The copepods were collected from February to May 2021 close to the French coast of the EEC (50°44′27.5 N: 1°34′32.4 E) during cruises on-board the N/O Sepia II (INSU-CNRS) with a WP2 plankton net (200 μm mesh size) fitted with a 2 L filtering cod-end during horizontal net tows (speed < 1 m s^{-1} for less than 10 min) at 1–3 m depth. After each plankton haul, zooplankton samples were immediately diluted in 20 L of surface seawater, then stored in the dark in a cool box and brought back within a few hours to the laboratory. To initiate the rearing phase, a ratio of 1 male per 5 females for calanoid copepods is required [51,52], and this was ensured by selecting about 250 adults of each species under a dissecting microscope. The copepods were placed in polycarbonate beakers of varying volume (from 3 to 7 L according to the number of individuals) containing 1 μm FSW. The copepods were kept at 15 °C, at a salinity of 33–34 PSU and under a 12:12 h day:night photoperiod. They were fed daily under replete food condition. The food supplied consisted of a mixture of microalgae *Rhodomonas salina* (RCC 1507), *Thalassiosira weissflogii* (RCC 1714), *Tisochrysis* sp. (RCC 1350), *Tetraselmis suesica* (RCC 1975), and *Emiliania hyxleyi* (RCC 1256), grown inside a 2 L Erlenmeyer flask with K/2 + Si medium at 15 °C and under a 12:12 h day:night photoperiod (100–150 μE m^{-2} s^{-1}). The media were prepared with autoclaved 1 μm FSW from the EEC. The algal concentrations inside the beakers were from 10^3 to 10^4 cell mL^{-1} [51–53] in order to avoid predation of calanoid copepods on younger stages [54,55]. Seawater was renewed every two days and air was supplied via small bubbles in each rearing beaker.

2.3. Experimental Setup

A total of eleven separate incubations of copepods (each conditions triplicated) were conducted, spread over five assays that allowed the integration of variable predator/prey size ratios and concentration ratios. Phytoplankton cell diameter ranged from 4.5 to 17 μm (Table 1) and concentrations from $1.6 \pm 0.2 \times 10^3$ cell mL^{-1} to $58 \pm 2 \times 10^3$ cell mL^{-1} (Table 2). The corresponding initial food concentrations ranged from 0.49 ± 0.06 to 10.1 ± 2.2 μg Chl *a* L^{-1} and the total cell volume ranged from 0.39 ± 0.03 to 5.55 ± 0.25 mm^3 L^{-1} considering the cell concentrations and their respective cell biovolume, assuming spherical cells (Table 2).

2.3.1. Copepod Selection

For each incubation, adults and copepodite 5 stage were selected corresponding to a mean length of 1097 ± 108 μm (N = 296) and 1216 ± 135 (N = 369) for *Acartia clausi* and *Temora longicornis*, respectively. In order to obtain a significant grazing signal index, copepod abundance inside bottles was high relative to calanoid copepod abundances commonly measured during phytoplankton blooms in the North Atlantic Ocean (typically 4 ind L^{-1} for calanoid copepods such as *T. longicornis, A. clausi* [56]). However, the chosen experimental copepod abundance was comparable to abundances observed in the EEC (up to 11 ind L^{-1}, see Table 2) [57]. These high abundances remained also comparable to values used in most experimental studies ranging from 8 to >15 ind L^{-1} [52,58–60].

2.3.2. Incubation

Twenty-four hours prior to the start of the experiments, 100 reared copepods were isolated in 3 L beakers containing 1 μm FSW without food. This starving phase allowed gut evacuation and maximized the feeding during the incubations. For all experiments, dead and injured individuals were first removed and only healthy-looking and living ones were individually pipetted into a 2350 mL polycarbonate bottle containing prey assemblages. Then, to avoid air bubble introduction the bottles were filled without headspace with FSW adjusted to pH 8.2, and then placed on a rolling table at 3 rpm to allow prey homogenization. Incubation was carried out at 15 °C under a photoperiod regime (12:12 h) for 24 h.

Table 2. Initial grazing experiment incubation setup (mean ± SD, N = 3).

Experiment	Food	Copepod Species	Copepod per Incubation	Replicat	Cell Concentration	Chlorophyll *a* Concentration	Particulate Organic Carbon Concentration	Particulate Inorganic Carbon Concentration	Particulate Organic Nitrogen Concentration
			[ind L^{-1}]	[N]	[10^3 cell mL^{-1}]	[μg Chl *a* L^{-1}]	[μg POC L^{-1}]	[μg PIC L^{-1}]	[μg PON L^{-1}]
1	*G. oceanica*	*T. longicornis*	13–14	3	2.9 ± 0.2	0.5 ± 0.1	37	146	13 [<dl]
	G. oceanica + *Tisochrysis* sp.		13–15	3	3.5 ± 0.3	1.4 ± 0	91	60	9 [<dl]
	Tisochrysis sp.		12–14	3	3.3 ± 0.3	2.0 ± 0.2	72	0	4 [<dl]
2	*G. oceanica*	*A. clausi* *T. longicornis*	11 11	3 3	19.2 ± 0.1	3.2 ± 0.6	441 ± 25	128 ± 27	47 ± 1
3	*E. huxleyi* (low concentration)	*A. clausi* *T. longicornis*	11–14 11–13	3 3	13.7 ± 0.8	1.4 ± 0	104 ± 4	73 ± 43	18 ± 3 [<dl]
4	*C. braarudii*	*A. clausi* *T. longicornis*	14–18 11–17	3 3	2.2 ± 0.9	7.6 ± 0.2	244 ± 16	405 ± 7	30 ± 2
5	*E. huxleyi* (high concentration)	*A. clausi* *T. longicornis*	16–19 11	3 3	57.9 ± 0.2	7.3 ± 0.4	1157 ± 57	151 ± 69	97 ± 7

[<dl] Under the detection limit.

2.4. *Ingestion/Egestion Estimation*

After each incubation, the copepods were carefully retrieved from each bottle by sieving the seawater through an immersed 200 µm mesh. The copepods were placed in 2 mL cryotubes (one per bottle) and then flash frozen in liquid nitrogen and kept at −20 °C until further analysis. Copepod size measurements were performed (as much as possible not withstanding obscurity) under a dissecting microscope (ZEISS Axio Zoom V16), before pigment extraction for gut content quantification (see below). Fecal pellets were recovered after each incubation by filtering the remaining seawater of each bottle onto a 40 µm mesh sieve. Fecal pellets retained on the mesh sieve were resuspended in FSW in a plankton counting chamber (Dolfuss cuvette, 6 mL volume).

2.4.1. Copepod Gut Pigment Content

For gut content analyses, copepods were individually sorted from freshly thawed samples under a cool light stereomicroscope. Individuals were rinsed with 0.2 µm FSW to eliminate phytoplankton cells with aggregates stuck to feeding appendages and were then transferred into 4 mL acetone (90%). Individuals (N = 19 to 42 copepods per extraction) were ground and chlorophyllian pigments (Chl *a* and phaeopigments) were extracted in the dark at 4 °C overnight. Fluorescence of the extract was measured before and after acidification with 10% HCl using a fluorometer (Turner design Trilogy). Copepod gut content was obtained by both Chl *a* and phaeopigment concentrations and values were not corrected for pigment degradation on the recommendation of Durbin and Campbell [61]. Ingestion rates (I, ng Chl a $_{eq}$ ind^{-1} d^{-1}) were derived from gut total pigment content (G_{cop}, ng Chl a $_{eq}$ ind^{-1}) using Equation (1):

$$I = 60 \times G_{cop} \times k \quad (1)$$

where k is the gut evacuation rate (h^{-1}), calculated following the model of Dam and Peterson [62], which accounts for the temperature of incubation, and the specie-dependant allometric constant. In the present study, we carried out our calculations with $k = 0.028$, which corresponds to the allometric constant of evacuation of calanoids at 15 °C ($k = 0.0117 + 0.001794 \times T$).

2.4.2. Copepod Gut Volume Conversion

Phytoplankton species used during the grazing experimental setup did not have the same biovolume and Chl *a* content (see Table 1). In order to compare every gut content for each experiment, we converted the equivalent pigment gut content (ng Chl a $_{eq}$ ind^{-1}) into volume equivalent gut content (µm^3 $_{eq}$ ind^{-1}). A calibration of Chl *a* level (pg Chl a cell^{-1}) over cell biovolume (µm^3) for each phytoplankton species was used. Gut ingestion was then expressed as its prey biovolume equivalent (10^6 µm^3 $_{eq}$ ind^{-1} d^{-1}).

2.4.3. Ivlev's Model

The copepod ingestion functional response toward food availability was calculated by following Ivlev's model [38,39]—Equation (2). This model considers the optimal food foraged by copepods (and more widely by all planktonic active filter feeders), recently described as a *Type II functional response* [63,64].

$$I_{Ivlev} = I_{max} \times \left(1 - e^{(-\alpha \times C_{food})}\right) \quad (2)$$

where I_{max} is the maximum ingestion rate index obtained; α the rate at which saturation is achieved with increasing food levels (slope of the linear regression); I_{Ivlev} is the modelized ingestion rate; and C_{food} is the corresponding food concentration (µg POC L^{-1}, µg Chl a L^{-1} or mm^3 L^{-1}).

2.4.4. Fecal Pellet Production and Size

Fecal pellet production (FP ind^{-1} d^{-1}) was estimated after each experiment by counting the fecal pellets recovered after incubation. For each incubation, between 10 and 186 pellets were measured (length and width in µm) with 5 µm accuracy. Fecal pellets are considered as cylindrical with two half spheres, and volumes were calculated according to Equation (3) [65]:

$$V_{PF} = \pi \times d^2 \times \left(\frac{L}{4} + \frac{d}{6}\right) \tag{3}$$

where d is the pellet diameter (µm), and L is the length of cylindric part of the pellet. Volumes were then converted into equivalent spherical diameter (ESD, mm), according to Equation (4):

$$ESD = \sqrt[3]{\frac{6 \times V}{\pi}} \tag{4}$$

2.5. *Statistical Analyses*

Results are expressed in mean ± standard deviation (SD). When data distribution matched the parametric assumption of normality (tested with a Shapiro–Wilk test, $p < 0.05$), correlation between two variables was analysed using a Pearson correlation test. Otherwise, a Spearman rank correlation test was performed. The statistical effect of the different experimental conditions was tested with a one-way ANOVA, followed by a pairwise Tukey's *post hoc* comparison test. In case of non-normal distribution, multicomparaisons were performed using the Kruslal–Wallis test following Nemeyi *post hoc* test. All the statistical analysis was performed using R software (V 4.1.1).

3. Results

3.1. *Coccolithophore Stochiometry*

Cellular particulate organic carbon (pg POC cell^{-1}), nitrogen (pg PON cell^{-1}), and inorganic carbon (pg PIC cell^{-1}) increase with coccolithophore diameter (Table 1). Cellular content and stoichiometric ratios for each experiment and coccolithophore species are also presented in Table 1. The cells are considered as spheres, whose biovolumes varied from 38 to 83 µm^3 for *E. huxleyi* (RCC 1256), from 133 to 143 µm^3 for *G. oceanica* (RCC 1314), and from 2296 to 2487 µm^3 for *C. braarudii* (RCC 1200).

3.2. *Copepod Ingestion*

For experiment 1, a mixture of coccolithophores (*G. oceanica*, RCC 1314) and non-coccolithophores (*Tisochrysis* sp. RCC 1350) was incubated with the copepods (*T. longicornis*). These two haptophyte species have similar sizes (6.7 µm and 6.1 µm of diameter for *G. oceanica* and *Tisochrysis* sp., respectively). They were mixed to obtain three batches: 100% *G. oceanica*, 50% *G. oceanica* + 50% *Tisochrsis* sp., and 100% *Tisochrysis* sp. with approximately 3000 cell mL^{-1} in total (see Table 2). The cell density (cell mL^{-1}) and total cell volume (mm^3 L^{-1}) were non-significantly different between the three different conditions (Figure 1A) with an average of 3255 ± 292 cell mL^{-1} and 0.44 ± 0.04 mm^3 L^{-1}, respectively. Concerning the Chl *a* concentration, *G. oceanica* incubation contained 0.49 ± 0.06 µg Chl *a* L^{-1}, mix of *G. oceanica* and *Tisochrysis* sp. contained 1.40 ± 0.01 µg Chl *a* L^{-1}, and *Tisochrysis* sp. contained 2.04 ± 0.17 µg Chl *a* L^{-1} (Figure 1C). Particulate matter composition (µg POC, PIC, and PON L^{-1}) was achieved within the three different conditions (Figure 1D) and is presented in Table 2.

The resulting ingestion rates varied from 0 (under detection limit) to 13.1 ± 1.4 ng Chl a_{eq} ind^{-1} d^{-1} with the higher values encountered in the 100% *Tisochryisis* sp. condition. Volume equivalent ingestion rates (Table 1, Figure 2B) varied from 0 (under the detection limit) to higher values for incubation with *Tisochrysis* sp (2.5 ± 0.3 × 10^6 µm$^3_{\text{eq}}$ ind^{-1} d^{-1}). The egestion rates were not significantly different between conditions, with averaged values of 26 ± 7 fecal pellets in d^{-1}, and mean pellet volumes ranging significantly from 0.3 ± 0 with *G. oceanica*, to 1.6 ± 0.6 mm^3 with *Tisochrysis* sp. (Figure 2D).

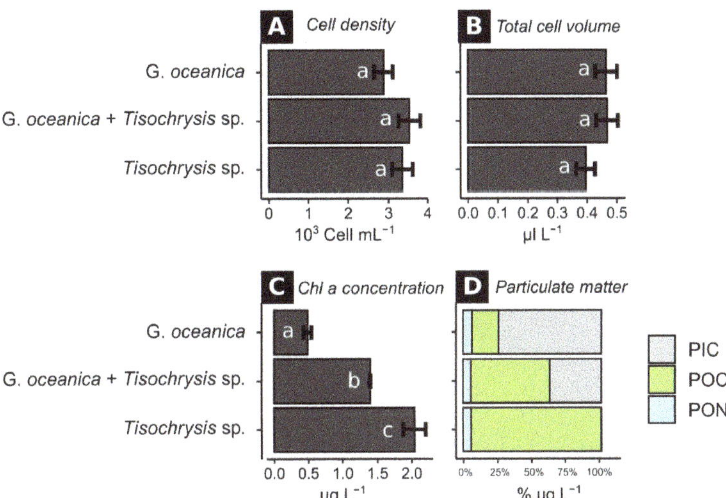

Figure 1. Experiment 1 incubations with *Temora longicornis*. (**A**) Initial cell density for each condition. (**B**) Initial cell volume for each condition for each condition. (**C**) Initial Chl *a* concentration for each condition. (**D**) Initial particulate matter quality for each condition. Groups a, b, and c correspond to statistical groups, according to one-way ANOVA with significant threshold $\alpha = 5\%$ (p-value < 0.05).

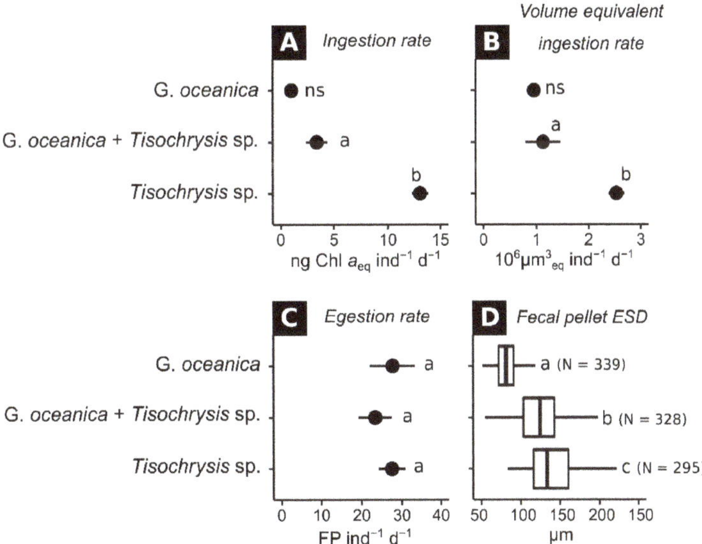

Figure 2. Details of Experiment 1 incubations with *Temora longicornis*. (**A**) Ingestion rate for each condition. (**B**) Volume equivalent ingestion rate for each condition. (**C**) Egestion rate for each condition. (**D**) Fecal pellet equivalent spherical diameter (ESD). Group a, b, and c correspond to statistical group, according to one-way ANOVA with significant threshold $\alpha = 5\%$ (p-value < 0.05). ns = non-significant.

After the incubations, the recovered fecal pellets had both significantly different sizes (Figures 2D and 3) and different opacity: when copepods were fed with 100% *G. oceanica*, fecal pellets were opaque and thick whereas they were light green with *Tisochrysis* sp. Fecal pellets had an intermediate aspect where the copepods were fed with a mix of both species (Figure 3).

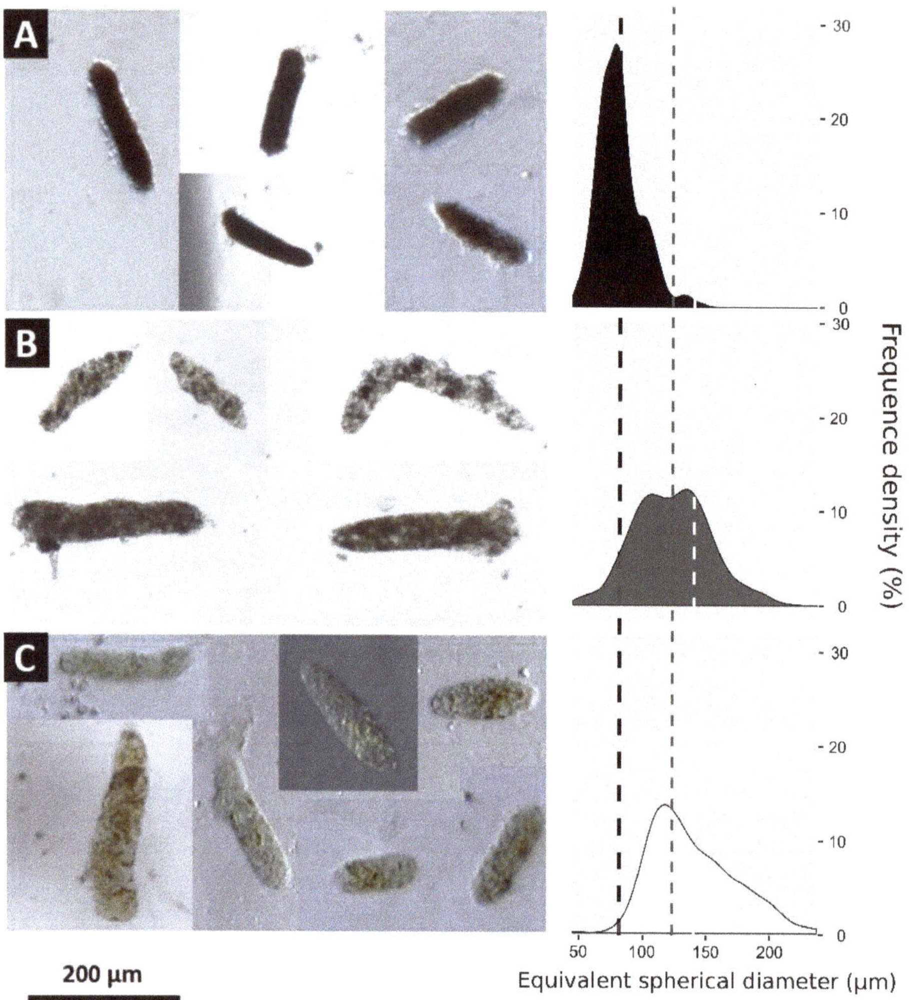

Figure 3. Picture of the recovered *Temora longicornis* fecal pellets of Experiment 1, after different conditions: (**A**) grazing experiment with 100% *G. oceanica;* (**B**) Grazing experiment with a mixture of 50% *G. oceanica* + 50% *Tisochrysis* sp.; and (**C**) Grazing experiment with *Tisochrysis* sp. The scale bar is congruent with figure (**A–C**). The vertical black dashed line corresponds to the mean fecal pellet diameters (µm) recovered after grazing experiment with 100% *G. oceanica* (**A**); the vertical grey dashed line corresponds to the mean fecal pellet diameters (µm) recovered after grazing experiment with a mixture of 50% *G. oceanica* + 50% *Tisochrysis* sp (**B**); the vertical white dashed line corresponds to the mean fecal pellet diameters (µm) recovered after grazing experiment with *Tisochrysis* sp (**C**).

The following figures (Figures 4–8) and results consider all the experiments.

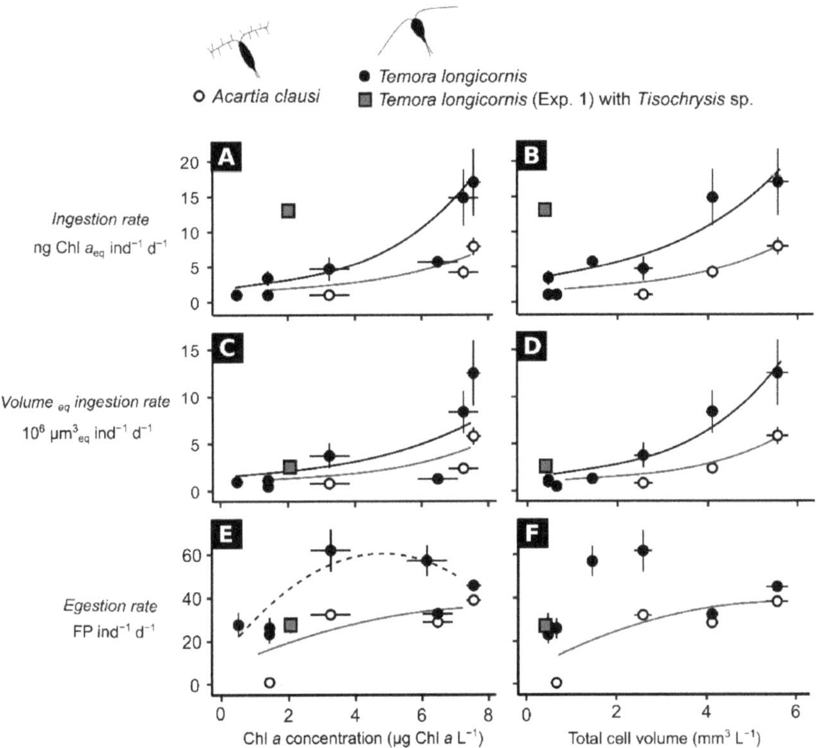

Figure 4. Copepod functional responses over initial Chl *a* concentration (μg Chl *a* L^{-1}) are shown on the left, and over initial total cell volume (mm^3 L^{-1}) are shown on the right. (**A**,**B**) depict the pigment ingestion rate (ng Chl *a* $_{eq}$ ind^{-1} d^{-1}); (**C**,**D**) depict the volume equivalent ingestion rate (10^6 $μm^3$ $_{eq}$ ind^{-1} d^{-1}); and E and F depict the fecal pellet egestion rate (FP ind^{-1} d^{-1}). Black circular dots correspond to incubation with *Temora longicornis*. White circular dots correspond to incubation with *Acartia clausi*. Grey square dots correspond to incubation with *Temora longicornis* and *Tisochrysis* sp. cell (monospecific and mixed with *G. oceanica*). Solid lines represent exponential fit of pigment/volume equivalent ingestion rate over Chl *a* concentration and total cell volume (**A**–**D**). In (**E**,**F**), solid lines correspond to quadratic fit of fecal pellet egestion over Chl *a* concentration and total cell volume. All equations and statistics are displayed in Table 3. In all graphs, *p*-values < 0.001 are displayed by solid lines, however, *p*-values < 0.05 are displayed by dashed lines (see statistical test in the Methods Section).

The functional responses to prey concentration varied significantly between those copepod species with an average lower ingestion and fecal pellet egestion by *A. clausi* compared to those with *T. longicornis* (Figure 4). Including all experiments, the Chl *a* concentration ranged from 0.49 ± 0.06 to 7.6 ± 0.2 μg Chl *a* L^{-1} (Figure 4A,C,E). The total cell volume ranged from 0.39 ± 0.03 to 5.5 ± 0.3 mm^3 L^{-1} (Figure 4B,D,F). In parallel, the ingestion rate values increased from 0 to 9.9 ng Chl *a* $_{eq}$ ind^{-1} d^{-1} for *Acartia clausi* and from 0 to 23.1 ng Chl *a* $_{eq}$ ind^{-1} d^{-1} for *Temora longicornis* (Figure 4A,B). The volume equivalent ingestion rate values ranged from 0.46 to 7.3 × 10^6 $μm^3$ $_{eq}$ ind^{-1} d^{-1} for *Acartia clausi* and from 0.9 to 17 × 10^6 $μm^3$ $_{eq}$ ind^{-1} d^{-1} for *Temora longicornis* (Figure 4C,D). The fecal pellet egestion rate ranged from 4 to 41 FP ind^{-1} d^{-1} for *Acartia clausi* and from 19 to 76 FP ind^{-1} d^{-1} for *Temora longicornis* (Figure 4E,F). All fits and statistical parameters are displayed in Table 3.

Table 3. Regression and statistical parameters within functional relation between food level index and ingestion rate index.

Food Level Index	Grazing Rate Index	Copepod	Statistics	
		T. longicornis	Equation	R^2
Ingestion rate index			*Exponential fit*	
Chl *a* concentration (μg Chl *a* L^{-1})	Pigment ingestion rate (ng Chl *a* $_{eq}$ ind^{-1} d^{-1})		$y = 2.2 \times 10^{-5}\ e^{(1.46x)} + 4.49$	0.64 ***
	Volume equivalent ingestion rate (10^6 $\mu m^3{}_{eq}$ ind^{-1} d^{-1})		$y = 7.4 \times 10^{-7}\ e^{(2.19x)} + 1.6$	0.67 ***
Total cell volume (mm^3 L^{-1})	Pigment ingestion rate (ng Chl *a* $_{eq}$ ind^{-1} d^{-1})		$y = 3.1\ e^{(0.31x)} + 0.82$	0.65 ***
	Volume equivalent ingestion rate (10^6 $\mu m^3{}_{eq}$ ind^{-1} d^{-1})		$y = 3.02\ e^{(0.29x)} - 2.4$	0.85 ***
Egestion rate index			*Quadratic fit*	
Chl a concentration (μg Chl *a* L^{-1})	Egestion rate (FP ind^{-1} d^{-1})		$y = -2.53x^2 + 24.34x$	0.47 *
Total cell volume (mm^3 L^{-1})			$y = -6.05x^2 + 39.3x$	0.38 ns
		A. Clausi		
Ingestion rate index			*Exponential fit*	
Chl *a* concentration (μg Chl *a* L^{-1})	Pigment ingestion rate (ng Chl *a* $_{eq}$ ind^{-1} d^{-1})		$y = 3.9 \times 10^{-8}\ e^{(2.52x)} + 0.99$	0.81 ***
	Volume equivalent ingestion rate (10^6 $\mu m^3{}_{eq}$ ind^{-1} d^{-1})		$y = 1.5 \times 10^{-9}\ e^{(2.9x)} + 0.55$	0.78 **
Total cell volume (mm^3 L^{-1})	Pigment ingestion rate (ng Chl *a* $_{eq}$ ind^{-1} d^{-1})		$y = 0.43\ e^{(0.53x)} + 0.06$	0.85 ***
	Volume equivalent ingestion rate (10^6 $\mu m^3{}_{eq}$ ind^{-1} d^{-1})		$y = 0.126\ e^{(0.683x)} + 0.22$	0.85 ***
Egestion rate index			*Quadratic fit*	
Chl a concentration (μg Chl *a* L^{-1})	Egestion rate (FP ind^{-1} d^{-1})		$y = -1.3x^2 + 13.7x$	0.76 **
Total cell volume (mm^3 L^{-1})			$y = -0.84x^2 + 10.9x$	0.87 ***

ns = non-significant, * *p*-value < 0.05, ** *p*-value < 0.01 and *** *p*-value < 0.001, according to Pearson correlation test.

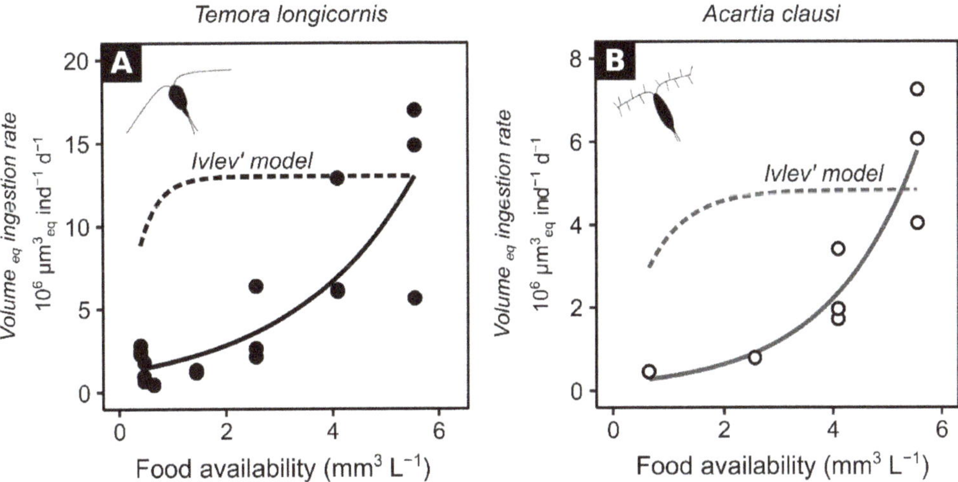

Figure 5. Copepod functional responses over total cell volume (mm^3 L^{-1}), for *Acartia clausi* (**A**) and *Temora longicornis* (**B**). Solid lines represent exponential fit of volume equivalent ingestion rate (10^6 µm^3 $_{eq}$ ind^{-1} d^{-1}) over total cell volume (mm^3 L^{-1}). Dashed lines correspond to Ivlev's model considering the max ingestion rate and the increasing ingestion rate over the food level slope (Equation (2)).

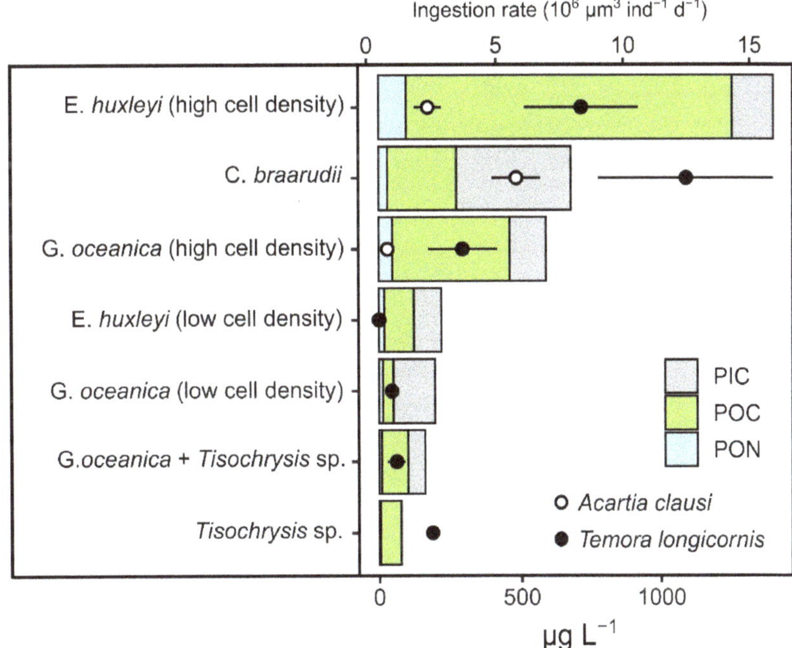

Figure 6. Cumulative barplot of particulate organic nitrogen (PON), particulate organic carbon (POC), and particulate inorganic carbon (PIC) in µg L^{-1} for each experimental incubation (bottom axis). Depicted on the top *x*-axis: the scatterplot of the ingestion rate (10^6 µm^3 $_{eq}$ in^{-1} d^{-1}) with *Acartia clausi* (white dots) and *Temora longicornis* (black dots).

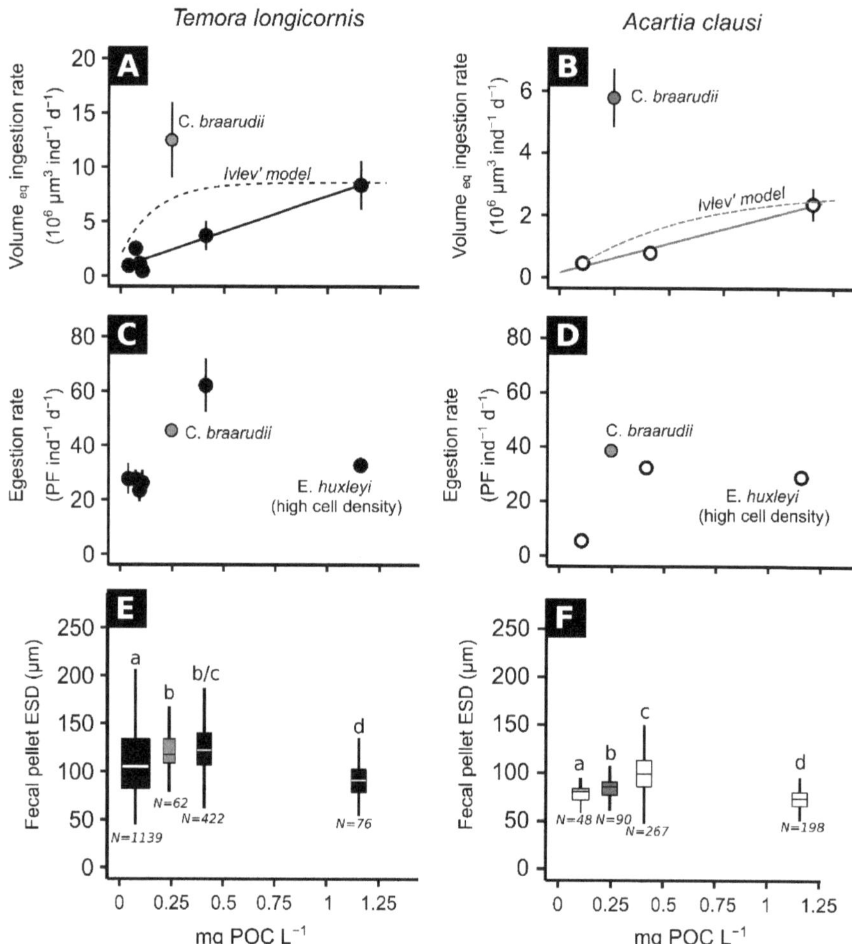

Figure 7. (**A**,**B**) Ingestion rate (10^6 $\mu m^3{}_{eq}$ ind^{-1} d^{-1}); (**C**,**D**) egestion rate (FP ind^{-1} d^{-1}) and (**E**,**F**) fecal pellet ESD (µm) over POC concentration (mg POC L^{-1}) for *Temora longicornis* (**A**,**C**,**E**) and *Acartia clausi* (**B**,**D**,**F**). The grey-boxed dots correspond to Experiment 4 with *C. braarudii*. The solid lines in (**A**,**B**) represent the linear regression between POC concentration and the ingestion rate, excluding the grey circular dot (Experiment 4 with *C. braarudii*). For *T. longicornis*, Pearson $R^2 = 0.86$, N = 12, *p*-value = 0.001, for *A.* clausi, Pearson $R^2 = 0.88$, N = 12, *p*-value = 0.002 Dashed lines correspond to Ivlev's model considering the max ingestion rate and the increasing ingestion rate over the food level slope (Equation (2)). Letters a, b, c, and d (in (**E**,**F**)) correspond to the different statistical groups displayed by the Kruskall–Wallis test.

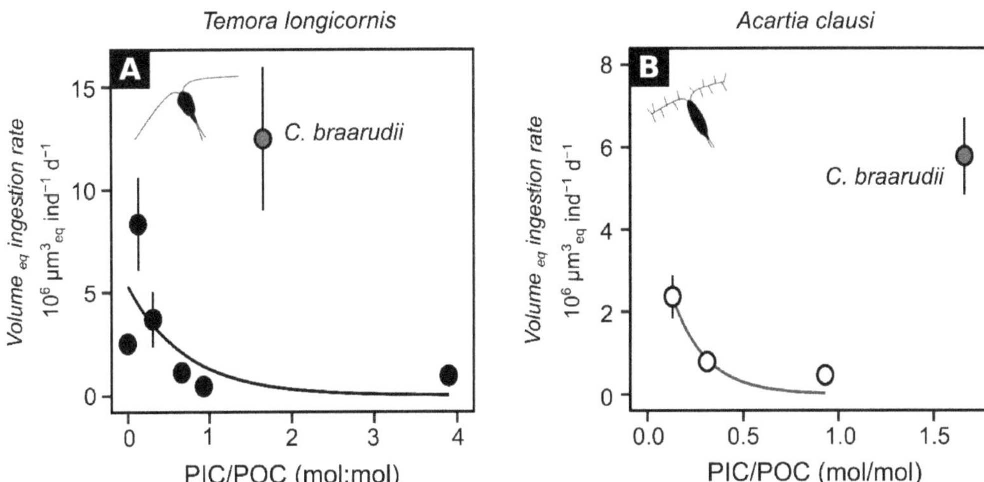

Figure 8. (**A**) Volume equivalent ingestion rate (10^6 μm^3 $_{eq}$ ind^{-1} d^{-1}) over PIC/POC ratios (mol:mol) for incubation with *T. longicornis (Black dots)*; (**B**) volume equivalent ingestion rate (10^6 μm^3 $_{eq}$ ind^{-1} d^{-1}) over PIC/POC ratios (mol:mol) for incubation with *A. clausi* (white dots). Grey dots correspond to incubation with *C. braarudii*. For incubations with *T. longicornis*, Kendall $\tau = -0.59$, N = 18, *p*-value = 0.002, excluding the grey dot (Exp. 4 with *C. braarudii*). For the incubation with *A. clausi*, Kendall $\tau = -0.95$, N = 9, *p*-value < 0.001, excluding the grey dot (Exp. 4 with *C. braarudii*).

The food availability varied between 0.39 ± 0.03 and 5.5 ± 0.3 mm^3 L^{-1}, and the ingestion rate varied from 0.46 to 7.3 × 10^6 μm^3 ind^{-1} d^{-1} for *A. clausi* and from 0.9 to 17.0 × 10^6 μm^3 ind^{-1} d^{-1} for *T. longicornis* (Figures 4 and 5). Exponential fits (solid lines) for both copepods (see Table 3) correspond to *Type III functional response* [63]. The logarithmic fits (dashed lines) were expected (Figure 5), following optimal food foraging (Ivlev's model or *Type II functional response*).

Particulate inorganic carbon (PIC) ranged from 0 (*Tisochrysis* sp. under detection limit) to 433 µg L^{-1} (Figure 6). Particulate organic carbon (POC) ranged from 37 to 1157 µg L^{-1} (Figure 6). Particulate organic nitrogen (PON) ranged from 4 to 97 µg L^{-1} (Figure 6). During Experiment 3 (C. *braarudii*) the PIC concentration represented 55% of the total particulate pool (with 433 µg PIC L^{-1} over 300 µg POC L^{-1} and 55 µg PON L^{-1}). Except for Experiment 5, the copepod ingestion rate increased with increasing ambient particulate content (Figure 6).

When considering all the experiments—except the one with C. *braarudii* (grey dots in Figures 7 and 8)—POC concentration and equivalent volume ingestion rates are positively correlated for both copepod species (*T. longicornis*: y = 0.006x + 0.81; R^2 = 0.71; *p*-value < 0.001 Pearson correlation test, and with *A. clausi*: y = 0.002x + 0.17; R^2 = 0.77; *p*-value = 0.002 Pearson correlation test). Logarithmic fits were expected following theoretical optimal food foraging (Figures 4, 5 and 7), dashed lines: Ivlev's model or *Type II functional response*. Nevertheless we observed linear regression between POC concentration and equivalent volume ingestion rates which are described in the literature as *Type I functional response* [63]. Fecal pellet egestion ranged from 4 to 41 FP ind^{-1} d^{-1} for *Acartia clausi* and from 19 to 76 FP ind^{-1} d^{-1} for *Temora longicornis* (Figure 7 C,D). Despite a positive correlation between POC concentration and equivalent volume ingestion rates for both copepods, fecal pellet egestion was not correlated to POC concentration (Figure 7C,D). Moreover, within incubations with *T. Longicornis* and *E. Huxleyi* with high cell density (Exp. 5), despite high volume equivalent ingestion rate (8.3 ± 3.9 × 10^6 μm^3 $_{eq}$ ind^{-1} d^{-1}), fecal pellet egestion remained low (32 ± 3 FP ind^{-1} d^{-1}). After incubation with *T. longicornis*, the mean fecal pellet ESD ranged from 90 ± 17 to 124 ± 25 mm, whereas, after incubation with *A. clausi*, the mean fecal pellet ESD ranged from 71 ± 11 to 97 ± 20 mm.

Variations of cell calcite content were expressed as PIC/POC ratios (mol:mol) for each experiment (see Tables 1 and 2). Volume equivalent ingestion rate (10^6 $\mu m^3{}_{eq}$ ind^{-1} d^{-1}) over PIC/POC ratios (mol:mol) decreased non-linearly for both *T. longicronis* and *A. clausi* (Figure 8), when excluding experiments performed with *C. braarudii*.

4. Discussion

Copepod and more widely zooplankton food foraging is the main prey/predator quantifiable interaction. Within marine planktonic ecosystems, these trophic relationships have direct consequences on population dynamics for both preys and predators [66,67]. Indeed, copepod behavioural adaptations affect both ecological dynamics and biogeochemical cycles, such as primary production and sinking particles fluxes [4,68,69]. Numerical models have shown that copepods could also affect the phytoplankton prey population diversity via a top-down control [70], and also the seasonal succession of plankton communities [19].

4.1. Equivalent Volume Ingestion Estimation

Pigment ingestion rates based on copepod gut fluorescent content [62,71] represent a fast and easy workable way to estimate grazing. Regarding incubation times, which were equivalent in all incubations, pigment destruction inside the gut increased with gut ingestion [72], suggesting that loss of fluorescence is equivalent among the different samples from the different incubations. However, ingested preys could present significant variation in fluorescence, especially in situ; due to ingestion of non-chlorophyllian prostists (ciliates, heterotropic flagellates, nauplii) and algae (diatoms, haptophyceae) in varying proportions. In order to compare the ingestion rates derived from all the experiments (regardless of phytoplankton cell Chl *a* content and their biovolume), we used a conversion of gut content considering Chl *a* and biovolume (see Methods Section). This allowed us to get a better correlation between ingestion and total cell volume (r^2= 0.85 *** for both A. *clausi* and T. *longicornis*, Table 3) than the pigment ingestion rate in accordance to Chl *a* concentration (r = 0.64 *** and 0.81 *** for T. *longicornis* and A. *clausi*, respectively; Table 3, Figure 5). Regarding these findings, we assume that the probability of prey/predator contact is more dependent on total cell volume than the number of particles (cell L^{-1}), or biomass (g Chl *a* L^{-1} or g POC L^{-1}). Thus, it can be assumed that the total cell volume per litre (mm^3 L^{-1}) represents a better index of the prey-encounter rate. This suggests that, at equivalent total cell concentrations, the same ingestion rate pattern, expressed in volume equivalent Chl *a* ($\mu m^3{}_{eq}$ ind^{-1} d^{-1}), could be expected with large cells at low concentration as well as with small cells at high concentration. However, gut analysis neither takes into account pigment degradation inside the copepod's gut [61,72] before ingestion nor sloppy feeding (cell fragmentation without ingestion, see pictures in Jansen, 2008). Considering the very short gut passage time (less than an hour) and the relative evidence of viable cell preservation inside fecal pellets [73,74], the pigment degradation could be neglected (the same condition of sample preservation and treatment).

4.2. Adaptive Functional Response

Classically, the ingestion rate index based on gut content over food availability, which provides Ivlev's model curve [37], represents the optimal foraging behaviour, even regarding incubation time and pigment destruction inside the gut [72]. In this study, both prey/predator size ratios and prey stoichiometry modulate the ingestion rate index. A *Type III functional response* was obtained with both *A. clausi* and *T. longicornis* when considering food availability by total cell volumes and Chl *a* concentration (Figures 5 and 6). Indeed, ingestion rates increased exponentially according to food availability [63,64]. This functional relationship reflects a switching adaptation considering the food quality. The maximum food level reached 5 mm^3 L^{-1}, and is comparable to the maximum food availability in the literature (4 mm^3 L^{-1}) in Kiørboe et al. [64], when the copepod's ingestion saturation occurs (e.g., *Acartia tonsa, Temora longicornis, Centropages hamatus*, and *Oithona davisae*). At high food concentration (7.6 ± 0.2 µg Chl *a* L^{-1}), we assume a saturation of the feeding activity. Indeed, with more than 7 µg Chl *a* L^{-1}, the bottles showed green coloration. In our study, we exceeded 5 mm^3 L^{-1} at 15 °C. Thus, performing additional experiments at higher food concentrations would have had no benefit. POC, PIC, and PON quota per cell compare well to those presented in the literature, with a magnitude from 1 to 10^2 pgC $cell^{-1}$ and 10^{-1} to 10 pgN $cell^{-1}$ [75,76]. For both *T. longicornis* and *A. clausi*, the ingestion rates increased linearly with POC concentrations (Figure 7A,B), when excluding experiments with *C. braarudii*, 17 µm diameter. This suggests a *Type I functional response* [63] corresponding to a linear increase of ingestion according to food availability. This is the most common behaviour attributed to planktonic active filter feeders (such as copepods). However, calanoid copepods (such as *Acartia* spp., *Temora* spp., *Centropages* spp., and *Calanus* spp.) commonly present *Type I* and *II* functional responses, mainly corresponding to the Ivlev model [77–80]. Ivlev's model is shaped like *Type II functional response*, which corresponds to the optimal feeding behaviour towards high-quality food availability. In this study, any relationship (either considering Chl *a* concentration, total volume, or POC concentration) fits with Ivlev's model (or *Type II functional response*) suggesting an anti-grazing propriety of the coccolithophores as a food source alone. Within the six incubations with *C. braarudii*, regarding food availability as equivalent carbon, Chl *a* or total volume, we obtained higher ingestion rates than for smaller coccolithophore, which can be explained by an intense gut accumulation of algae material because of the large cell size. Calcite cell content through PIC/POC ratio (mol:mol) for coccolithophore cells of similar sizes (Figure 8) could partially explain a non-optimal ingestion pattern observed in our experiments. These results suggest that the coccosphere (i.e., calcified exosqueletton around the cell) could be a structure protecting the coccolithophore from grazing by copepods, such as diatom frustules, as previously proposed [33,34].

4.3. Calcite Obstruction and Potential Dissolution Inside Copepod Guts

While the copepods ingested large coccolithophore (*C. braarudii*), we measured high ingestion rates despite low carbon concentration and low fecal pellet egestion. This observation indicates a decoupling between ingestion rate and gut passage time [80], probably due to high calcite ingestion and a decrease in gut pH resulting from calcite dissolution. This phenomenon may explain an importance paradox in ocean zooplankton mediated calcite dynamics. Indeed, considering a global oceanic alkalinity budget, there is a loss of calcite between the production by calcifier organisms in the euphotic zone and the estimated calcite flux below the lysocline [81]. This calcite loss could be attributed to biological activities and more specifically the dissolution mediated by zooplankton grazing or transport. Several studies have even shown a loss of calcite after zooplankton gut passage, a striking feature of the sedimentary record that relies on the observation of well-preserved coccoliths within zooplankton fecal pellets [79,82–85]. However, numerical models using a timeframe and pH inside copepod guts suggest a moderate calcite dissolution inside the gut [86]. Langer et al. [87] showed that calcite dissolution during copepod gut passage was below 8% of the weight of the coccoliths of *Calcidiscus leptoporus* inside fecal pellets, but these coccoliths were intact and showed no evidence of any dissolution [87]. In addition, Antia et al. [88] successfully observed that coccolith dissolution/fragmentation occurs inside zooplankton guts and microzooplankton vacuoles. During the first experiment, we observed a decoupling between ingestion rates (both pigment ingestion rates and equivalent volume ingestion rates) and fecal pellet egestion (Figure 3). Taking all the experiments collectively, this fact was also noticed in Experiment 5, with a high cell concentration of *E. huxleyi*. In addition, despite high measured ingestion rates, few fecal pellets were produced (Figures 5, 7 and 8). This decoupling between ingestion and egestion could be the result of a modulation of the residence time in the gut. Hence, fecal pellet size seems to depend on the ingestion rate index and prey quality (Figures 1, 6 and 8). Indeed, the fecal pellet size variation could depend on gut passage time as well [89]. By considering all these points, both coccolithophore biovolume and relative calcite content may modulate coccolith dissolution due to gut passage variations.

4.4. Consequences for Vertical Particle Flux in the Pelagic Realm

In this study, we observed a loss of fecal pellet production with high prey concentrations, despite high ingestion rates. The number of egested particles (fecal pellets) seems to be dependant, not only on the food quantity, but also on the quality of the ingested food. The size of egested particles could be increased by both the number of ingested particles and their quality (inclusion of calcite, silica frustules, etc.). Prey/predator size ratio and relative carbon content [90] suggest that these environmental food conditions may provide predictable constraints to copepod biogeography size distribution in the ocean [91]. This therefore suggest that size and primary producer stoichiometry could influence oceanic carbon flux patterns through fecal pellet egestion by copepods. This may result in a decrease of carbon passive flux due to fecal pellets sinking in the water column. In addition, if we consider that fecal pellets follow Stoke's law of sedimentation (as suggested by Komar et al. [92]), the ballast effect of calcified coccoliths inside fecal pellets should foster the sedimentation rate much more than changes in the size of the pellets [93]. Hence, modification of fecal pellet egestion patterns in addition to ballast effect of calcite could be an important process driving the particle flux in the water column.

5. Conclusions

In this study, we demonstrated that copepod ingestion rates based on the volume equivalent of cells is better scaled to total prey volume concentration (mm^3 L^{-1}). The Chl *a*/biovolume calibration developed in this study highlights the importance of considering the Chl *a* level inside the gut fluorescent content, regarding food types ingested by wild copepods, such as non-chlorophyllian preys (e.g., microzooplankton, heterotrophic flagelles, nauplii, etc.). Our results highlight an exponential increase of ingestion rates according to food availability (*Type III functional response*), which is in contrast to the optimal Ivlev model (*Type II functional response*) corresponding to optimal food foraging. This parametric pattern supports the role of food quality in the feeding behaviour of copepods, such as coccolithophore defence structures (calcified coccospheres). We demonstrated this aspect by showing the relationship between calcite content (PIC/POC ratio) and the ingestion rate index. Finally, we observed a decoupling between ingestion rates and fecal pellet egestion, which may be the consequence of an "obstruction" effect of calcite inside the copepod's gut. This "obstruction" may be the result of varying gut passage times—modulating the intensity of calcite dissolution. These results suggest that both prey allometry and stoichiometry need to be considered with copepod feeding dynamics, specifically regarding fecal pellet production, and the sedimentary flux, which is an important component of the biological carbon pump.

Author Contributions: Contributed to conception and design: J.T. and A.D. Contributed to acquisition of data: J.T., A.D., A.P. and G.D. Contributed to the copepod rearing maintenance: J.T., A.D. and A.P. Contributed to analysis and interpretation of data: J.T., A.D. and M.H. Drafted and/or revised the article: J.T., A.D. and M.H. Approved the submitted version for publication: J.T., A.D., A.P., G.D., V.C., L.B. and M.H. All authors have read and agreed to the published version of the manuscript.

Funding: This work was supported by the ANR CARCLIM (https://anr.fr/Projet-ANR-17-CE01-0004). Jordan Toullec's postdoctorate was equally funded by ANR CARCLIM and by Université Littoral Côte d'Opale. This work was also supported by the CPER MARCO and the SFR Campus de la Mer.

Institutional Review Board Statement: Not applicable.

Informed Consent Statement: Not applicable.

Data Availability Statement: All data are displayed within the manuscript or may be obtained directly from the authors.

Acknowledgments: We would like to thank the Sepia II crew for their help at sea, during zooplankton collection, and more specifically for the multiple WPII plankton hauls during *Phaeocystis globosa* sticky bloom event. We are also grateful to Michel Laréal for his help and modifications made to the rolling table.

Conflicts of Interest: The authors declare no conflict of interest.

References

1. Sanders, R.; Henson, S.A.; Koski, M.; De La Rocha, C.L.; Painter, S.C.; Poulton, A.J.; Riley, J.; Salihoglu, B.; Visser, A.; Yool, A.; et al. The Biological Carbon Pump in the North Atlantic. *Prog. Oceanogr.* **2014**, *129*, 200–218. [CrossRef]
2. Henson, S.; Le Moigne, F.; Giering, S. Drivers of Carbon Export Efficiency in the Global Ocean. *Glob. Biogeochem. Cycles* **2019**, *33*, 891–903. [CrossRef] [PubMed]
3. Le Moigne, F.A. Pathways of organic carbon downward transport by the oceanic biological carbon pump. *Front. Mar. Sci.* **2019**, *6*, 634. [CrossRef]
4. Turner, J.T. Zooplankton fecal pellets, marine snow, phytodetritus and the ocean's biological pump. *Prog. Oceanogr.* **2015**, *130*, 205–248. [CrossRef]
5. Steinberg, D.K.; Landry, M.R. Zooplankton and the Ocean Carbon Cycle. *Annu. Rev. Mar. Sci.* **2017**, *9*, 413–444. [CrossRef]
6. Laurenceau-Cornec, E.; Trull, T.W.; Davies, D.M.; Bray, S.G.; Doran, J.; Planchon, F.; Carlotti, F.; Jouandet, M.-P.; Cavagna, A.-J.; Waite, A.M.; et al. The relative importance of phytoplankton aggregates and zooplankton fecal pellets to carbon export: Insights from free-drifting sediment trap deployments in naturally iron-fertilised waters near the Kerguelen Plateau. *Biogeosciences* **2015**, *12*, 1007–1027. [CrossRef]
7. Belcher, A.; Iversen, M.; Manno, C.; Henson, S.A.; Tarling, G.A.; Sanders, R. The role of particle associated microbes in remineralization of fecal pellets in the upper mesopelagic of the Scotia Sea, Antarctica. *Limnol. Oceanogr.* **2016**, *61*, 1049–1064. [CrossRef]
8. Estapa, M.; Valdes, J.; Tradd, K.; Sugar, J.; Omand, M.; Buesseler, K. The neutrally buoyant sediment trap: Two decades of progress. *J. Atmos. Ocean. Technol.* **2020**, *37*, 957–973. [CrossRef]
9. Calbet, A. Mesozooplankton grazing effect on primary production: A global comparative analysis in marine ecosystems. *Limnol. Oceanogr.* **2001**, *46*, 1824–1830. [CrossRef]
10. Irigoien, X.; Harris, R.P.; Verheye, H.M.; Joly, P.; Runge, J.; Starr, M.; Pond, D.; Campbell, R.; Shreeve, R.; Ward, P.; et al. Copepod hatching success in marine ecosystems with high diatom concentrations. *Nature* **2002**, *419*, 387–389. [CrossRef]
11. Irigoien, X.; Flynn, K.J.; Harris, R.P. Phytoplankton blooms: A 'loophole' in microzooplankton grazing impact? *J. Plankton Res.* **2005**, *27*, 313–321. [CrossRef]
12. Brown, C.W.; Yoder, J.A. Coccolithophorid blooms in the global ocean. *J. Geophys. Res. Oceans* **1994**, *99*, 7467–7482. [CrossRef]
13. Eikrem, W.; Medlin, L.K.; Henderiks, J.; Rokitta, S.; Rost, B.; Probert, I.; Throndsen, J.; Edvardsen, B. Haptophyta. In *Handbook of the Protists*; Archibald, J.M., Simpson, A.G.B., Slamovits, C.H., Margulis, L., Melkonian, M., Chapman, D.J., Corliss, J.O., Eds.; Springer International Publishing: Cham, Switzerland, 2016; pp. 1–61. ISBN 978-3-319-32669-6.
14. Rost, B.; Riebesell, U. Coccolithophores and the biological pump: Responses to environmental changes. In *Coccolithophores*; Springer: Berlin/Heidelberg, Germany, 2004; pp. 99–125.
15. Henderiks, J.; Bartol, M.; Pige, N.; Karatsolis, B.-T.; Lougheed, B.C. Shifts in phytoplankton composition and stepwise climate change during the middle Miocene. *Paleoceanogr. Paleoclimatol.* **2020**, *35*, e2020PA003915. [CrossRef]
16. Hays, G.C.; Richardson, A.J.; Robinson, C. Climate change and marine plankton. *Trends Ecol. Evol.* **2005**, *20*, 337–344. [CrossRef]
17. Kléparski, L.; Beaugrand, G.; Edwards, M.; Schmitt, F.G.; Kirby, R.R.; Breton, E.; Gevaert, F.; Maniez, E. Morphological traits, niche-environment interaction and temporal changes in diatoms. *Prog. Oceanogr.* **2022**, *201*, 102747. [CrossRef]
18. Kiørboe, T.; Visser, A.W. Predator and prey perception in copepods due to hydromechanical signals. *Mar. Ecol. Prog. Ser.* **1999**, *179*, 81–95. [CrossRef]
19. Visser, A.W.; Fiksen, Ø. Optimal foraging in marine ecosystem models: Selectivity, profitability and switching. *Mar. Ecol. Prog. Ser.* **2013**, *473*, 91–101. [CrossRef]
20. Djeghri, N.; Atkinson, A.; Fileman, E.S.; Harmer, R.A.; Widdicombe, C.E.; McEvoy, A.J.; Cornwell, L.; Mayor, D.J. High prey-predator size ratios and unselective feeding in copepods: A seasonal comparison of five species with contrasting feeding modes. *Prog. Oceanogr.* **2018**, *165*, 63–74. [CrossRef]
21. Pančić, M.; Kiørboe, T. Phytoplankton defence mechanisms: Traits and trade-offs. *Biol. Rev.* **2018**, *93*, 1269–1303. [CrossRef]
22. Jansen, S. Copepods grazing on *Coscinodiscus wailesii*: A question of size? *Helgol. Mar. Res.* **2008**, *62*, 251–255. [CrossRef]
23. Friedrichs, L.; Hörnig, M.; Schulze, L.; Bertram, A.; Jansen, S.; Hamm, C. Size and biomechanic properties of diatom frustules influence food uptake by copepods. *Mar. Ecol. Prog. Ser.* **2013**, *481*, 41–51. [CrossRef]
24. Liu, H.; Chen, M.; Zhu, F.; Harrison, P.J. Effect of Diatom Silica Content on Copepod Grazing, Growth and Reproduction. *Front. Mar. Sci.* **2016**, *3*, 89. [CrossRef]
25. Pančić, M.; Torres, R.R.; Almeda, R.; Kiørboe, T. Silicified cell walls as a defensive trait in diatoms. *Proc. R. Soc. B Biol. Sci.* **2019**, *286*, 20190184. [CrossRef] [PubMed]

26. Xu, H.; Shi, Z.; Zhang, X.; Pang, M.; Pan, K.; Liu, H. Diatom frustules with different silica contents affect copepod grazing due to differences in the nanoscale mechanical properties. *Limnol. Oceanogr.* **2021**, *66*, 3408–3420. [CrossRef]
27. Zhang, S.; Liu, H.; Ke, Y.; Li, B. Effect of the Silica Content of Diatoms on Protozoan Grazing. *Front. Mar. Sci.* **2017**, *4*, 202. [CrossRef]
28. Pondaven, P.; Gallinari, M.; Chollet, S.; Bucciarelli, E.; Sarthou, G.; Schultes, S.; Jean, F. Grazing-induced Changes in Cell Wall Silicification in a Marine Diatom. *Protist* **2007**, *158*, 21–28. [CrossRef]
29. Petrucciani, A.; Chaerle, P.; Norici, A. Diatoms Versus Copepods: Could Frustule Traits Have a Role in Avoiding Predation? *Front. Mar. Sci.* **2022**, *8*, 804960. [CrossRef]
30. Jaya, B.N.; Hoffmann, R.; Kirchlechner, C.; Dehm, G.; Scheu, C.; Langer, G. Coccospheres confer mechanical protection: New evidence for an old hypothesis. *Acta Biomater.* **2016**, *42*, 258–264. [CrossRef]
31. Harvey, E.L.; Bidle, K.D.; Johnson, M.D. Consequences of strain variability and calcification in *Emiliania huxleyi* on microzooplankton grazing. *J. Plankton Res.* **2015**, *37*, 1137–1148. [CrossRef]
32. Mayers, K.M.J.; Poulton, A.J.; Bidle, K.; Thamatrakoln, K.; Schieler, B.; Giering, S.L.C.; Wells, S.R.; Tarran, G.A.; Mayor, D.; Johnson, M.; et al. The Possession of Coccoliths Fails to Deter Microzooplankton Grazers. *Front. Mar. Sci.* **2020**, *7*, 569896. [CrossRef]
33. Nejstgaard, J.C.; Gismervik, I.; Solberg, P.T. Feeding and reproduction by *Calanus finmarchicus*, and microzooplankton grazing during mesocosm blooms of diatoms and the coccolithophore *Emiliania huxleyi*. *Mar. Ecol. Prog. Ser.* **1997**, *147*, 197–217. [CrossRef]
34. Zondervan, I.; Zeebe, R.E.; Rost, B.; Riebesell, U. Decreasing marine biogenic calcification: A negative feedback on rising atmospheric pCO_2. *Glob. Biogeochem. Cycles* **2001**, *15*, 507–516. [CrossRef]
35. Monteiro, F.M.; Bach, L.T.; Brownlee, C.; Bown, P.; Rickaby, R.E.M.; Poulton, A.J.; Tyrrell, T.; Beaufort, L.; Dutkiewicz, S.; Gibbs, S.; et al. Why marine phytoplankton calcify. *Sci. Adv.* **2016**, *2*, e1501822. [CrossRef]
36. Kiørboe, T. How zooplankton feed: Mechanisms, traits and trade-offs. *Biol. Rev.* **2011**, *86*, 311–339. [CrossRef]
37. Kiørboe, T.; Møhlenberg, F.; Nicolajsen, H. Ingestion rate and gut clearance in the planktonic copepod *Centropages hamatus* (Lilljeborg) in relation to food concentration and temperature. *Ophelia* **1982**, *21*, 181–194. [CrossRef]
38. Ivlev, V.S. *Experimental Ecology of Nutrition of Fishes. Moscow, Pishchepromizdat. Translated by D. Scott*; Yale University Press: New Haven, CT, USA, 1955.
39. Kooij, R.E.; Zegeling, A. A Predator–Prey Model with Ivlev's Functional Response. *J. Math. Anal. Appl.* **1996**, *198*, 473–489. [CrossRef]
40. Stibor, H.; Vadstein, O.; Diehl, S.; Gelzleichter, A.; Hansen, T.; Hantzsche, F.; Katechakis, A.; Lippert, B.; Løseth, K.; Peters, C.; et al. Copepods act as a switch between alternative trophic cascades in marine pelagic food webs. *Ecol. Lett.* **2004**, *7*, 321–328. [CrossRef]
41. Moriceau, B.; Iversen, M.H.; Gallinari, M.; Evertsen, A.-J.O.; Le Goff, M.; Beker, B.; Boutorh, J.; Corvaisier, R.; Coffineau, N.; Donval, A.; et al. Copepods Boost the Production but Reduce the Carbon Export Efficiency by Diatoms. *Front. Mar. Sci.* **2018**, *5*, 82. [CrossRef]
42. Pan, Y.; Zhang, Y.; Sun, S. Phytoplankton–zooplankton dynamics vary with nutrients: A microcosm study with the cyanobacterium *Coleofasciculus chthonoplastes* and cladoceran *Moina micrura*. *J. Plankton Res.* **2014**, *36*, 1323–1332. [CrossRef]
43. Liu, Y.-W.; Rokitta, S.D.; Rost, B.; Eagle, R.A. Constraints on coccolithophores under ocean acidification obtained from boron and carbon geochemical approaches. *Geochim. Cosmochim. Acta* **2021**, *315*, 317–332. [CrossRef]
44. Besiktepe, S.; Dam, H. Coupling of ingestion and defecation as a function of diet in the calanoid copepod *Acartia tonsa*. *Mar. Ecol. Prog. Ser.* **2002**, *229*, 151–164. [CrossRef]
45. Butler, M.; Dam, H. Production rates and characteristics of fecal pellets of the copepod *Acartia tonsa* under simulated phytoplankton bloom conditions: Implications for vertical fluxes. *Mar. Ecol. Prog. Ser.* **1994**, *114*, 81–91. [CrossRef]
46. Keller, M.D.; Selvin, R.C.; Claus, W.; Guillard, R.R.L. Media for the Culture of Oceanic Ultraphytoplankton 1, 2. *J. Phycol.* **1987**, *23*, 633–638. [CrossRef]
47. Aminot, A.; Kérouel, R. *Hydrologie des Écosystèmes Marins: Paramètres et Analyses*; Editions Quae: Versailles, France, 2004; ISBN 978-2-84433-133-5.
48. Seuront, L.; Vincent, D. Increased seawater viscosity, *Phaeocystis globosa* spring bloom and *Temora longicornis* feeding and swimming behaviours. *Mar. Ecol. Prog. Ser.* **2008**, *363*, 131–145. [CrossRef]
49. Benedetti, F.; Gasparini, S.; Ayata, S.-D. Identifying copepod functional groups from species functional traits. *J. Plankton Res.* **2016**, *38*, 159–166. [CrossRef]
50. Lombard, F.; Koski, M.; Kiørboe, T. Copepods use chemical trails to find sinking marine snow aggregates. *Limnol. Oceanogr.* **2013**, *58*, 185–192. [CrossRef]
51. Vincent, D.; Slawyk, G.; L'Helguen, S.; Sarthou, G.; Gallinari, M.; Seuront, L.; Sautour, B.; Ragueneau, O. Net and gross incorporation of nitrogen by marine copepods fed on ^{15}N-labelled diatoms: Methodology and trophic studies. *J. Exp. Mar. Biol. Ecol.* **2007**, *352*, 295–305. [CrossRef]
52. Toullec, J.; Vincent, D.; Frohn, L.; Miner, P.; Le Goff, M.; Devesa, J.; Moriceau, B. Copepod Grazing Influences Diatom Aggregation and Particle Dynamics. *Front. Mar. Sci.* **2019**, *6*, 751. [CrossRef]
53. Berggreen, U.; Hansen, B.; Kiørboe, T. Food size spectra, ingestion and growth of the copepod *Acartia tonsa* during development: Implications for determination of copepod production. *Mar. Biol.* **1988**, *99*, 341–352. [CrossRef]

54. Bonnet, D.; Titelman, J.; Harris, R. *Calanus* the cannibal. *J. Plankton Res.* **2004**, *26*, 937–948. [CrossRef]
55. Boersma, M.; Wesche, A.; Hirche, H.-J. Predation of calanoid copepods on their own and other copepods' offspring. *Mar. Biol.* **2014**, *161*, 733–743. [CrossRef]
56. Schultes, S.; Sourisseau, M.; Le Masson, E.; Lunven, M.; Marié, L. Influence of physical forcing on mesozooplankton communities at the Ushant tidal front. *J. Mar. Syst.* **2013**, *109–110*, S191–S202. [CrossRef]
57. Grattepanche, J.-D.; Breton, E.; Brylinski, J.-M.; Lecuyer, E.; Christaki, U. Succession of primary producers and micrograzers in a coastal ecosystem dominated by *Phaeocystis globosa* blooms. *J. Plankton Res.* **2011**, *33*, 37–50. [CrossRef]
58. Sautour, B.; Castel, J. Feeding behaviour of the coastal copepod *Euterpina acutifrons* on small particles. *Cah Biol Mar* **1993**, *34*, 239–251.
59. Vincent, D.; Hartmann, H.J. Contribution of ciliated microprotozoans and dinoflagellates to the diet of three copepod species in the Bay of Biscay. *Hydrobiologia* **2001**, *443*, 193–204. [CrossRef]
60. Sarthou, G.; Vincent, D.; Christaki, U.; Obernosterer, I.; Timmermans, K.R.; Brussaard, C.P. The fate of biogenic iron during a phytoplankton bloom induced by natural fertilisation: Impact of copepod grazing. *Deep Sea Res. Part II Top. Stud. Oceanogr.* **2008**, *55*, 734–751. [CrossRef]
61. Durbin, E.G.; Campbell, R.G. Reassessment of the gut pigment method for estimating in situ zooplankton ingestion. *Mar. Ecol. Prog. Ser.* **2007**, *331*, 305–307. [CrossRef]
62. Dam, H.G.; Peterson, W.T. The effect of temperature on the gut clearance rate constant of planktonic copepods. *J. Exp. Mar. Biol. Ecol.* **1988**, *123*, 1–14. [CrossRef]
63. Jeschke, J.M.; Kopp, M.; Tollrian, R. Consumer-food systems: Why type I functional responses are exclusive to filter feeders. *Biol. Rev.* **2004**, *79*, 337–349. [CrossRef]
64. Kiørboe, T.; Saiz, E.; Tiselius, P.; Andersen, K.H. Adaptive feeding behavior and functional responses in zooplankton. *Limnol. Oceanogr.* **2018**, *63*, 308–321. [CrossRef]
65. Hillebrand, H.; Dürselen, C.-D.; Kirschtel, D.; Pollingher, U.; Zohary, T. Biovolume calculation for pelagic and benthic microalgae. *J. Phycol.* **1999**, *35*, 403–424. [CrossRef]
66. Villiot, N.; Poulton, A.J.; Butcher, E.T.; Daniels, L.R.; Coggins, A. Allometry of carbon and nitrogen content and growth rate in a diverse range of coccolithophores. *J. Plankton Res.* **2021**, *43*, 511–526. [CrossRef]
67. Menden-Deuer, S.; Lessard, E.J. Carbon to volume relationships for dinoflagellates, diatoms, and other protist plankton. *Limnol. Oceanogr.* **2000**, *45*, 569–579. [CrossRef]
68. Holling, C.S. The Functional Response of Predators to Prey Density and its Role in Mimicry and Population Regulation. *Memoirs Èntomol. Soc. Can.* **1965**, *97*, 5–60. [CrossRef]
69. Murdoch, W.W. Stabilizing effects of spatial heterogeneity in predator-prey systems. *Theor. Popul. Biol.* **1977**, *11*, 252–273. [CrossRef]
70. Cavan, E.L.; Henson, S.A.; Belcher, A.; Sanders, R. Role of zooplankton in determining the efficiency of the biological carbon pump. *Biogeosciences* **2017**, *14*, 177–186. [CrossRef]
71. Belcher, A.; Manno, C.; Ward, P.; Henson, S.A.; Sanders, R.; Tarling, G.A. Copepod faecal pellet transfer through the meso- and bathypelagic layers in the Southern Ocean in spring. *Biogeosciences* **2017**, *14*, 1511–1525. [CrossRef]
72. Prowe, A.E.F.; Pahlow, M.; Dutkiewicz, S.; Follows, M.; Oschlies, A. Top-down control of marine phytoplankton diversity in a global ecosystem model. *Prog. Oceanogr.* **2012**, *101*, 1–13. [CrossRef]
73. Wang, R.; Conover, R.J. Dynamics of gut pigment in the copepod Temora longicornis and the determination of in situ grazing rates. *Limnol. Oceanogr.* **1986**, *31*, 867–877. [CrossRef]
74. Kiørboe, T.; Tiselius, P.T. Gut clearance and pigment destruction in a herbivorous copepod, Acartia tonsa, and the determi-nation of in situ grazing rates. *J. Plankton Res.* **1987**, *9*, 525–534. [CrossRef]
75. Jansen, S.; Bathmann, U. Algae viability within copepod faecal pellets: Evidence from microscopic examinations. *Mar. Ecol. Prog. Ser.* **2007**, *337*, 145–153. [CrossRef]
76. Agustí, S.; González-Gordillo, J.I.; Vaqué, D.; Estrada, M.; Cerezo, M.I.; Salazar, G.; Gasol, J.M.; Duarte, C.M. Ubiquitous healthy diatoms in the deep sea confirm deep carbon injection by the biological pump. *Nat. Commun.* **2015**, *6*, 7608. [CrossRef] [PubMed]
77. Harris, R.P.; Paffenhöfer, G.-A. Feeding, growth and reproduction of the marine planktonic copepod *Temora longicornis* Müller. *J. Mar. Biol. Assoc. UK* **1976**, *56*, 675–690. [CrossRef]
78. Price, H.J.; Paffenhöfer, G.-A. Effects of concentration on the feeding of a marine copepod in algal monocultures and mixtures. *J. Plankton Res.* **1986**, *8*, 119–128. [CrossRef]
79. Harris, R.P. Zooplankton grazing on the coccolithophore *Emiliania huxleyi* and its role in inorganic carbon flux. *Mar. Biol.* **1994**, *119*, 431–439. [CrossRef]
80. Tirelli, V.; Mayzaud, P. Relationship between functional response and gut transit time in the calanoid copepod *Acartia clausi*: Role of food quantity and quality. *J. Plankton Res.* **2005**, *27*, 557–568. [CrossRef]
81. Milliman, J.D.; Troy, P.J.; Balch, W.M.; Adams, A.K.; Li, Y.-H.; Mackenzie, F.T. Biologically mediated dissolution of calcium carbonate above the chemical lysocline? *Deep Sea Res. Part Oceanogr. Res. Pap.* **1999**, *46*, 1653–1669. [CrossRef]
82. Roth, P.H.; Mullin, M.M.; Berger, W.H. Coccolith Sedimentation by Fecal Pellets: Laboratory Experiments and Field Observations. *GSA Bull.* **1975**, *86*, 1079–1084. [CrossRef]
83. Honjo, S. Coccoliths: Production, transportation and sedimentation. *Mar. Micropaleontol.* **1976**, *1*, 65–79. [CrossRef]

84. Honjo, S.; Roman, M.R. Marine copepod fecal pellets: Production, preservation and sedimentation. *J Mar Res* **1978**, *36*, 45–57.
85. Samtleben, C.; Bickert, T. Coccoliths in sediment traps from the Norwegian Sea. *Mar. Micropaleontol.* **1990**, *16*, 39–64. [CrossRef]
86. Jansen, H.; Wolf-Gladrow, D.A. Carbonate dissolution in copepod guts: A numerical model. *Mar. Ecol. Prog. Ser.* **2001**, *221*, 199–207. [CrossRef]
87. Langer, G.; Nehrke, G.; Jansen, S. Dissolution of *Calcidiscus leptoporus* coccoliths in copepod guts? A morphological study. *Mar. Ecol. Prog. Ser.* **2007**, *331*, 139–146. [CrossRef]
88. Antia, A.N.; Suffrian, K.; Holste, L.; Müller, M.N.; Nejstgaard, J.C.; Simonelli, P.; Carotenuto, Y.; Putzeys, S. Dissolution of coccolithophorid calcite by microzooplankton and copepod grazing. *Biogeosci. Discuss.* **2008**, *5*, 1–23.
89. Wilson, S.E.; Steinberg, D.K.; Buesseler, K.O. Changes in fecal pellet characteristics with depth as indicators of zooplankton repackaging of particles in the mesopelagic zone of the subtropical and subarctic North Pacific Ocean. *Deep Sea Res. Part II Top. Stud. Oceanogr.* **2008**, *55*, 1636–1647. [CrossRef]
90. Almeda, R.; Someren Gréve, H.; Kiørboe, T. Prey perception mechanism determines maximum clearance rates of planktonic copepods. *Limnol. Oceanogr.* **2018**, *63*, 2695–2707. [CrossRef]
91. Beaugrand, G.; Reid, P.C.; Ibañez, F.; Lindley, J.A.; Edwards, M. Reorganization of North Atlantic Marine Copepod Biodiversity and Climate. *Science* **2002**, *296*, 1692–1694. [CrossRef]
92. Komar, P.D.; Morse, A.P.; Small, L.F.; Fowler, S.W. An analysis of sinking rates of natural copepod and euphausiid fecal pellets 1. *Limnol. Oceanogr.* **1981**, *26*, 172–180. [CrossRef]
93. White, M.M.; Waller, J.D.; Lubelczyk, L.C.; Drapeau, D.T.; Bowler, B.C.; Balch, W.M.; Fields, D.M. Coccolith dissolution within copepod guts affects fecal pellet density and sinking rate. *Sci. Rep.* **2018**, *8*, 9758. [CrossRef]

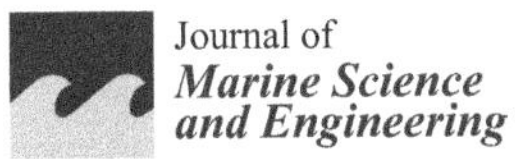
Journal of *Marine Science and Engineering*

Article

The Infection of Paracalanid Copepods by the Alveolate Parasite *Ellobiopsis chattoni* Caullery, 1910 in a Subtropical Coastal Area

José E. Martinelli Filho [1], Felipe Gusmão [2], Flavio A. Alves-Júnior [3,4] and Rubens M. Lopes [5,*]

1 Biological Oceanography Laboratory, Geosciences Institute, Centre for Advanced Studies of Biodiversity, Federal University of Pará (UFPA), Campus Universitário do Guamá, Belém 66075-110, Brazil
2 Instituto do Mar, Federal University of São Paulo (UNIFESP), Santos 11060-001, Brazil
3 Centro Universitário Brasileiro-UNIBRA, Rua Padre Inglês, 257, Boa Vista, Recife 50050-230, Brazil
4 Carcinology Laboratory (Labcrust) Prof. Dra. Kátia Cristina de Araújo Silva, Federal Rural University of Amazonia (UFRA), Avenida Tancredo Neves, 2501, Terra Firme, Belém 66077-830, Brazil
5 Department of Biological Oceanography, Oceanographic Institute, University of São Paulo (USP), São Paulo 05508-120, Brazil
* Correspondence: rubens@usp.br

Abstract: Paracalanid copepods, common in tropical zooplankton communities, are known hosts for a variety of parasites. Nevertheless, relatively little is known about the prevalence and consequences of parasitism in these copepods. In this study, we analyzed the relationship between two paracalanid copepods, *Parvocalanus crassirostris* and *Paracalanus* spp., with a common parasite, the alveolate protist *Ellobiopsis chattoni*, in a subtropical environment on the south-east Brazilian coast. We assessed the frequency and abundance of parasites in juveniles and adult male and female copepods. We observed that 22 out of 4014 *Paracalanus* spp. (0.55%) and 98 out of 3920 *P. crassirostris* were infected (2.5%). *E. chattoni* were rarely found in other taxa (about 0.05% for *Oithona* spp. and *Acartia lilljeborgii*). The parasites were most frequently attached to cephalosome appendages (73.6%), with up to four cells per copepod. The parasites were more prevalent in adults than juveniles, and adult females were more frequently infected than males. *E. chattoni* had a likely negative impact on copepod growth because the infected females were smaller than the non-infected females ($p < 0.001$). Females are usually bigger and live longer than males, which could account for their high frequency of infection.

Keywords: parasitism; marine zooplankton; *Parvocalanus crassirostris*; *Paracalanus*; Ellobiopsidae; South Atlantic

Citation: Martinelli Filho, J.E.; Gusmão, F.; Alves-Júnior, F.A.; Lopes, R.M. The Infection of Paracalanid Copepods by the Alveolate Parasite *Ellobiopsis chattoni* Caullery, 1910 in a Subtropical Coastal Area. *J. Mar. Sci. Eng.* **2022**, *10*, 1816. https://doi.org/10.3390/jmse10121816

Academic Editors: Marco Uttieri, Ylenia Carotenuto, Iole Di Capua and Vittoria Roncalli

Received: 11 October 2022
Accepted: 22 November 2022
Published: 25 November 2022

1. Introduction

Copepods are the dominant metazoans in the marine pelagic environment, playing fundamental roles in biogeochemical cycles and energy transfer to higher trophic levels. Ecological investigations in recent decades have shifted the focus on copepods from simply being major phytoplankton consumers [1] to important components of microbial food webs [2] and as hosts for gut-specific bacteria [3] and eukaryotic parasites [4]. A wide range of symbiotic relationships between marine pelagic copepods and other organisms such as *Vibrio cholerae* [5], parasitic dinoflagellates [6], and epicarid isopods [7] has been documented, several of which may cause severe injuries to copepod hosts, as reported for certain parasitic protists [4,8]. However, very little is known about the ecological aspects and consequences of parasitic protists such as ellobiopsids on copepods [9–11].

It has been suggested that juvenile copepods encounter *Ellobiopsis* spores during feeding activity, when the infection process takes place [10,12–14]. Copepod life cycles can be influenced by parasitic protists, which may induce castration, the mutation of sexual characteristics, or even sex changes during the later copepodid stages [15,16]. The parasites may also influence the metabolic and reproductive rates of the copepod hosts [17], as observed in *Undinula vulgaris* [10] and *Calanus helgolandicus* [11]. However, the effect of *Ellobiopsis* on copepod host biology and reproduction is poorly investigated [10,11], despite the many reports on the occurrence of *Ellobiopsis* spp. attached to copepods (e.g., [18]).

Ellobiopsis has a multinucleate, alveolate parasite body divided into two sections, a trophomere and a gonomere. The trophomere is the cell vegetative phase, bearing a root-like structure that penetrates the tissue of the host, and the gonomere is distally located and responsible for the sporulation process; both grow outside the host body [19]. A single *Ellobiopsis* may reach a biovolume as large as 1/15 of its host [10]. *Ellobiopsis* probably has a low infection specificity as more than 25 pelagic copepod species, and a few decapod zoea such as *Portunus*, have been documented as hosts [12,17,19,20].

Ellobiopsis hosts are virtually unknown in the South Atlantic, except for a single publication [9]; most records come from the North Atlantic, the North Sea, and the Mediterranean [11,21–25]. Here, we aimed to describe the copepod hosts for *E. chattoni* Caullery 1910, the abundance and frequency of infestation of this parasite in a subtropical coastal region in the South-West Atlantic, and the potential effects of the parasite on the host species.

2. Materials and Methods

Zooplankton was sampled off Ubatuba, São Paulo state, on the south-east Brazilian coast, at two coastal stations located at 23°30′59″ S; 45°06′10″ W (station A) and 23°31′27″ S and 45°04′54″ W (station B), using a small motorboat. Sampling took place mostly in the morning on alternate days during the austral summer and winter of 2009 and 2010 as well as the summer of 2011 for a total of 23 sampling periods (Table S1). Station B (10 m depth) was sampled on two occasions (26 January 2009 and 18 July 2009) when high waves, winds, and associated risks prevented sampling at the regular station, A (~30 m depth). The stations were ~2 km apart and under the influence of the same water mass. Plankton was sampled for 2 to 3 min by subsurface horizontal tows with a 100 μm mesh-size plankton net equipped with a calibrated flowmeter to estimate the filtered volume. All samples were preserved in a 4% formaldehyde–seawater buffered solution.

The copepods were identified following the taxonomic literature for the South and Central Atlantic [26–30] and counted to estimate the abundance (ind m^{-3}) and frequency (%) of infected specimens by developmental stages and sex. These were determined according to the usual morphological characteristics [26,27,31,32] and split into juvenile copepodids (CI-V), adult females (F), and adult males (M). A minimum of 30 individuals of each dominant copepod taxon were counted in sample aliquots to estimate the copepod abundances [28]. The infected copepods were enumerated in larger subsamples (from 1/10 to whole samples) because of their low prevalence in the samples (Table S1). Naupliar abundances were not estimated as the 100 μm plankton net failed to quantitatively capture the larval stages of most copepod species in the region [33].

The prosome length and width of the copepods and the parasite size (major length and width, excluding the stalk) were measured under a stereomicroscope equipped with a digital camera with the aid of Image J software after staining the samples with Bengal rose. The *Paracalanus* specimens were grouped as *Paracalanus* spp. because the identification accuracy was limited by the image resolution and the small body size of these copepods. A microscope analysis of selected individuals suggested that *P. quasimodo* was the dominant *Paracalanus* species in the samples, followed by *P. indicus* and occasional occurrences of *P. aculeatus* and an unidentified *Paracalanus* spp. [33].

Ellobiopsis chattoni parasites were identified based on their tube-like structure in the distal part of the forming gonomere and the occurrence of a single gonomere, an exclusive characteristic of the species (Figure 1) [21,22,24,25]. The gonomeres were assigned to three categories according to [14,21]: absent; immature (visible cell constriction, but at different developmental stages); and mature (fully developed). The initial developmental stages of parasites may have been underestimated in our samples due to the image resolution and parasite size (<15 μm length without the stalk). The frequency of occurrence of the parasites by attachment location on the copepods (antenna, mouth appendages, prosome, etc.) was analyzed for paracalanid copepods.

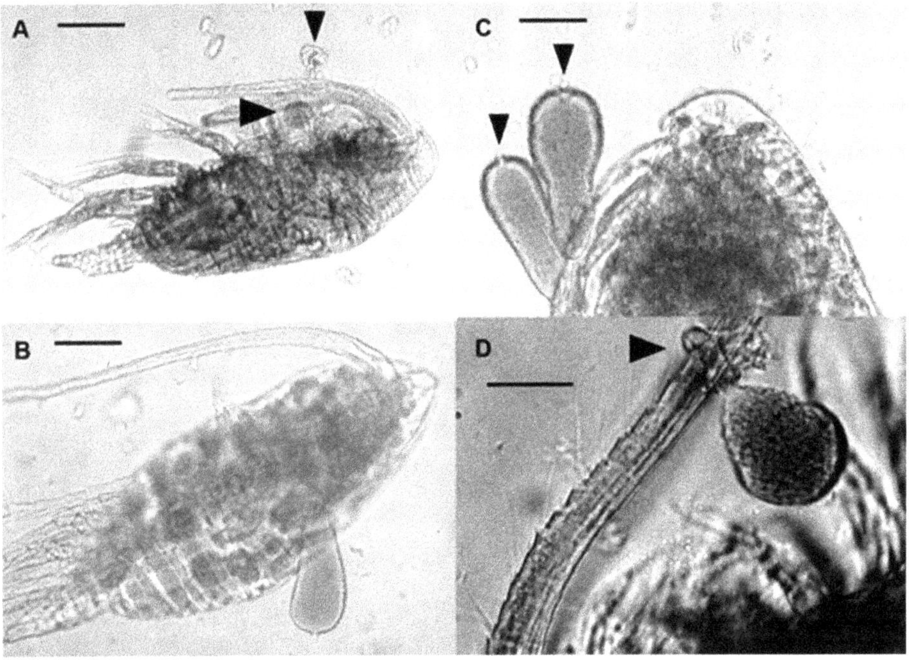

Figure 1. Paracalanid copepods infected by *Ellobiopsis chattoni*: (**A**) *Paracalanus indicus* with two adhered parasites; (**B**) *P. indicus* showing *E. chattoni* adhering to a mouth appendage; (**C**,**D**) *Parvocalanus crassirostris* with parasites attached to the antenna. (**A**,**B**) Scale = 200 µm; (**C**) scale = 80 µm; (**D**) scale = 60 µm.

Comparisons between the sexes and developmental stages were performed on the paracalanid copepods to verify a possible differential occurrence of *E. chattoni* with these categories. The data were tested for normality and an equal variance by the Shapiro–Wilk and Lilliefors tests, respectively. The abundance data were not fit for parametric tests, so non-parametric statistical tests were used instead. A comparison of the multiple groups was made using a Kruskal–Wallis test, followed by a Student–Newman–Keuls test when the differences were significant; Mann–Whitney tests were used for pairwise comparisons. All statistical analyses were considered with a 5% significance level [34].

The effect of the hosts on the copepod size was tested by comparing the prosome length of parasitized adult copepods with the same number of random non-parasitized copepods from the same sex and sample. The host and parasite biovolumes (μm^3) were estimated by $V = 4/3\ a\ b^2$, where a was the copepod or protist length and b was their width, assuming that both the copepods and *E. chattoni* had a spherical ellipsoid format. The Spearman test was used to assess the correlation between the copepod and parasite body volumes.

3. Results

A total of 7934 paracalanids (4014 *Paracalanus* spp. and 3920 *Parvocalanus crassirostris*), 19,219 *Oithona* spp., and 8302 *Acartia lilljeborgii* were inspected. An infection was observed in 98 *Parvocalanus crassirostris* (2.5% of the total), 22 *Paracalanus* spp. (0.5% of the total), 8 oithonids (3 *Dioithona oculata*, 3 *Oithona hebes*, and 2 *O. plumifera*), and 4 specimens of *Acartia lilljeborgii* (~0.05%). The most frequently infected developmental stages were adult females in *Parvocalanus crassirostris* and copepodids in *Paracalanus* spp. (Table 1). *Acartia lilljeborgii, Dioithona oculata, Oithona hebes*, and *O. plumifera* are herein reported for the first time as hosts for *E. chattoni*.

Table 1. The total amount of examined and infected copepods for each sex and developmental stage for *Parvocalanus crassirostris* and *Paracalanus* spp. and the respective frequency of infection (F.I.). M: adult male; F: adult female; C: juvenile copepodid, stages I to V.

N	*Parvocalanus crassirostris*				*Paracalanus* spp.			
	Total	**M**	**F**	**C**	**Total**	**M**	**F**	**C**
Observed	3920	597	1211	2014	4014	260	302	3430
Infected	98	14	61	23	22	7	1	14
F.I. (%)	2.5	2.4	5	1.1	0.6	2.7	0.3	0.4

A total of 61 copepod taxa were identified (Table S2). *Paracalanus* spp. was the most abundant copepod taxon, varying from 59 ind m^{-3} on 21 July 2010 to 3982 ind m^{-3} on 26 January 2009. The estimated abundance of infected *Paracalanus* spp. was 2 ± 3 ind m^{-3} for juveniles, 11 ± 11 ind m^{-3} for males, and 2 ± 2 ind m^{-3} for females (Table 2). *Parvocalanus crassirostris* was the third most abundant copepod species during this study, with a large abundance range between 41 ind m^{-3} on 22 January 2010 and 3282 ind m^{-3} on 26 January 2009. The estimated abundance of infected *P. crassirostris* was 6 ± 11 ind m^{-3} for juveniles, 4 ± 6 ind m^{-3} for males, and 16 ± 28 ind m^{-3} for females (Table 2).

Table 2. Abundance estimates of infected *Parvocalanus crassirostris* and *Paracalanus* spp. (ind. m^{-3}) for each sampling occasion. C: juvenile copepodid, stages I to V; F: adult females; M: adult males; F + M: all adults.

Season	Period	Species	C	F	M	M + F
Summer 2009	01/26	*P. crassirostris*	9	84	19	103
	01/26	*Paracalanus* spp.	8	4	25	29
Winter 2009	07/18	*P. crassirostris*	2	13	6	19
	07/18	*Paracalanus* spp.	0	0	1	1
	07/22	*P. crassirostris*	32	7	4	11
	07/22	*Paracalanus* spp.	2	3	20	23
	07/26	*P. crassirostris*	1	7	1	8
	07/26	*Paracalanus* spp.	0	1	3	4
Winter 2010	07/21	*P. crassirostris*	0	2	0	2
	07/21	*Paracalanus* spp.	0	0	1	1
	07/26	*P. crassirostris*	0	1	1	2
	07/26	*Paracalanus* spp.	0	0	2	2
	07/30	*P. crassirostris*	2	9	0	9
	07/30	*Paracalanus* spp.	4	4	26	30
Summer 2011	01/24	*P. crassirostris*	0	2	0	2
	01/24	*Paracalanus* spp.	1	1	6	7

Parasitized copepods were observed during all winter and summer seasons sampled (Table S1). The seasonal and interannual differences could not be reliably estimated because of the low number of observations during the summer of both years. The females of *Parvocalanus crassirostris* were more frequently infected than males ($p = 0.04$), representing 61 of the 98 infected specimens, and the adults were more frequently infected than the immature copepodid stages ($p = 0.04$). The differences between the sex and developmental stages were not significant for *Paracalanus* spp. and were not tested for *Acartia lilljeborgii* and *Oithona* spp. due to the low number of infected copepods. The prosome length between the uninfected and infected females of *P. crassirostris* was significantly different (Table 3). The *Parvocalanus crassirostris* and *E. chattoni* biovolume showed a significant but weakly positive correlation (r^2 = 0.21), indicating that larger parasites were often associated with larger copepods. However, this was not observed for *Paracalanus* spp. (Figure 2).

Table 3. Results of the Mann–Whitney test comparing the prosome length (μm) between uninfected and infected copepods for adult males and females of *Parvocalanus crassirostris* and *Paracalanus* spp. F: adult females; M: adult males.

Species	Category	Mean ± S.D. Uninfected	Mean ± S.D. Infected	Sample Number (N)	p-Value
Parvocalanus crassirostris	M	322.2 ± 17.6	307.7 ± 29.2	14	0.383
	F	398.8 ± 24.5	381.7 ± 22.9	60	<0.001
Paracalanus spp.	M	434.4 ± 66.7	376.9 ± 26.9	7	0.128
	F	-	-	1	-

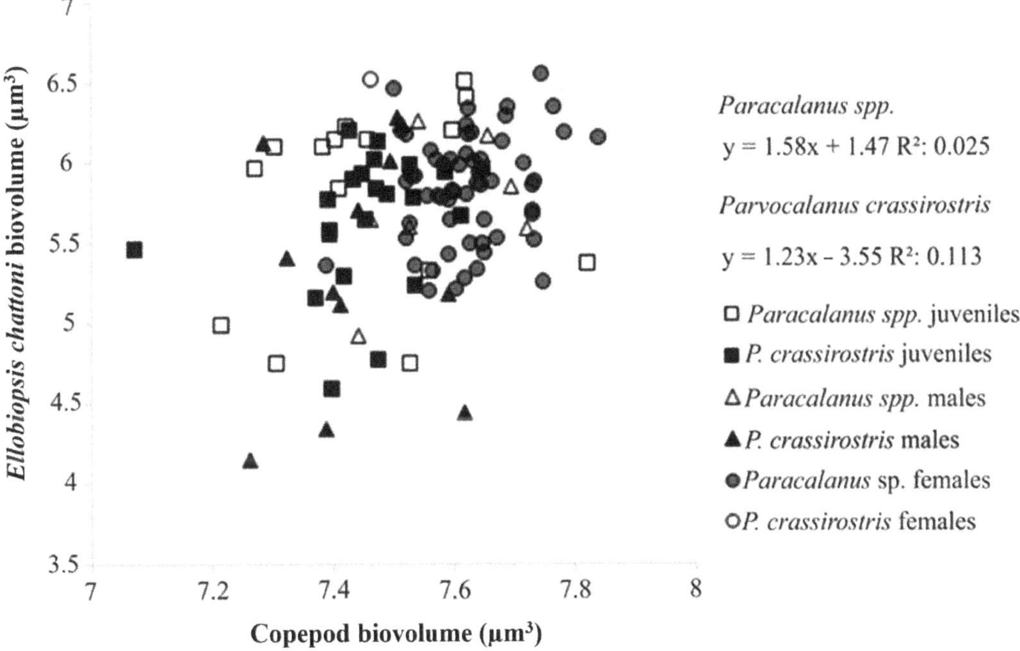

Figure 2. Relationship between the biovolume (μm^3, log scale) of infected paracalanid copepods (*Paracalanus* spp. and *Parvocalanus crassirostris*) and the respective attached *Ellobiopsis chattoni* parasites.

In *P. crassirostris*, 77 specimens carried a single parasite cell and 18 carried 2 parasite cells. The occurrence of three or four *E. chattoni* cells on a single copepod was rare, occurring on three individuals only. For *Paracalanus* spp., 20 specimens had a single *E. chattoni* attached; more than one parasite cell occurred only on two copepod hosts. No difference was found in the size of *P. crassirostris*, whether the copepods were carrying one or more parasites.

Nearly 80% of the parasites were attached to cephalosome appendages, including maxillipeds, maxillae, and antennae (Figure 1). Other less frequent adhesion sites were the swimming legs and the urosome (Table 4). The preferential adhesion site did not differ between the sexes or developmental stages.

Table 4. *Ellobiopsis chattoni* adhesion sites on the host copepod species (*Parvocalanus crassirostris* and *Paracalanus* spp.). C: juvenile copepodid, stages I to V; F: adult females; M: adult males; %: frequency of adhesion at each site.

	Parvocalanus crassirostris				*Paracalanus* spp.			
	C	F	M	%	C	F	M	%
Mouth appendages	4	32	8	42.7	6	1	5	46.2
Antennae	4	30	3	35.9	5	1	1	26.9
Prosome	2	12	6	19.4	3	-	3	23.1
Others	-	1	1	1.9	-	-	1	3.8

The *Ellobiopsis chattoni* size varied widely, with a minimum and maximum length of 17 and 153 μm, respectively. The average length and width were 75.7 ± 30.2 μm and 44.1 ± 14 μm, respectively, and the average biovolume was 804,117 ± 714,671 μm^3. From a total of 143 detected parasites, the gonomere was absent in 38.5%, immature in 39.2%, and mature in 22.3% of the specimens.

4. Discussion

The description of *Ellobiopsis* dates from more than a century ago [21], but knowledge of the biology and taxonomy of this genus has slowly advanced in the last decades [14,35]. The literature on *Ellobiopsis* is mostly restricted to its occurrence and host prevalence. The parasite has been recorded in association with several species of calanoid copepods [18–20,36–38], occasionally on Harpacticoida [37], and on decapod larvae [19]. The incidence of *Ellobiopsis* on freshwater hosts is uncertain and probably related to yet-unidentified parasites [4,39].

Parvocalanus crassirostris, *Paracalanus* spp., and *Oithona* sp. have previously been reported as hosts [35,36]. This study is the first account for *Acartia lilljeborgii*, *Dioithona oculata*, *Oithona hebes*, and *O. plumifera* as hosts for *E. chattoni*; it is the second *Ellobiopsis* record for the South Atlantic to date [9].

Despite the growing list of hosts, geographical variations seem to exist regarding the association of *Ellobiopsis* with copepods. For instance, *Calanoides carinatus* was infected on the Namibian shelf [40], but an infection was absent for such species at the Bay of Biscay, Spain, where *Calanus helgolandicus* was the main host, followed by sporadic infections on *Ctenocalanus vanus*, *Temora longicornis*, and *Pseudocalanus elongatus* [11]. Here, the parasite was observed on six different taxa, but numerically important only for paracalanids. In [18] only the parasitism of *Ellobiopsis* for *Bestiolina similis* was described. The geographical distribution of *E. chattoni* is widespread, and has been observed in the Arabian Sea [18,41], on the coast of Tanzania [10], the North Sea [24], the Norwegian Sea [42], the Mediterranean [21], Indian coastal waters [43,44], and Alaska [45], indicating the possibility of a species complex [14].

The *E. chattoni* found in Ubatuba were generally smaller (~20–160 μm) than the 250–750 μm range reported in other studies [9,11–22,39]. Although the stalks were not considered in our length measurements, other investigations did the same [10,14,43]; hence, that alone should not explain the variability in the parasite size among the regions. Such size differences might result from our study reporting infections in host species smaller than those reported in the literature, particularly from the northern hemisphere, or parasite genotypic variability.

It cannot be ruled out that *E. chattoni* represents a species complex because of the high variation in host specificity between localities, the wide geographical distribution, and the substantial variation in the parasite size. Recent studies (e.g., [18]) lack a molecular identification of the parasite and have, therefore, relied on morphological characteristics. A molecular approach, in addition to diverse sampling locations and hosts, should be considered in future studies to determine whether *E. chattoni* is an independent species or a species complex [18].

Despite the apparently higher number of infected copepods during winter compared with summer, the seasonality could not be properly assessed in this study due to an unbalanced sampling frequency between the seasons. *Ellobiopsis* infections have been reported to temporally vary, but without a consistent seasonal pattern when different locations are considered. For instance, the infection rate was higher during the summer in the Clyde Sea [24], from late autumn through winter in the Mediterranean Sea [12], and during winter in the estuarine and coastal waters of Iraq [37].

The infection rates observed here for paracalanid copepods were within a lower range of those recorded elsewhere for copepods carrying *Ellobiopsis* [13,20,38,42]. *P. crassirostris* and *Paracalanus* spp. were the prevalent hosts off Ubatuba whilst other abundant copepods such as *A. lilljeborgii* or oithonids remained virtually devoid of parasites, with a few exceptions.

The positive relationship between the host and parasite size could be explained by the higher proportion of infected females in comparison with the males, which are smaller in paracalanids. Nevertheless, infected *P. crassirostris* females had smaller body sizes compared with the non-infected specimens. As copepods do not molt after reaching maturity, size differences are most likely related to a development impairment in the pre-adult stages. An ellobiopsid infection might negatively impact individual energy input and expenditure by affecting the copepod motility and feeding efficiency; thus, an early infected individual would reach maturity with a small body size. As observed in other studies [11,24,43], most parasites were found adhered to cephalosome appendages such as the mouth parts and antennae, which generate feeding currents for prey capture [46], suggesting that the adhesion and infection mechanisms are connected to copepod feeding behavior.

The maximum number of parasites registered here was four (in a few individual hosts), but up to 15 *Ellobiopsis* have been observed on a single copepod; a few of these parasites were smaller than 15 μm [24]. The small size of immature parasites at early developmental stages could represent a potential explanation for our findings as we relied on a stereomicroscope to inspect nearly 8000 paracalanid copepods, preventing a more detailed microscopic analysis.

E. chattoni has been assigned as the causative agent of tumor-like anomalies (TLAs) in copepods [47]. However, from the 132 infected copepods analyzed in our samples, none displayed TLAs, which was in line with the results from a laboratory study showing a TLA absence in copepods during and after *Ellobiopsis* gonomere sporulation [43]. In addition, TLA protrusions have not been observed in marine copepods infected by *Ellobiopsis* [9,11,14,39]. An *Ellobiopsis* infection has been suggested to raise copepod mortality [47] but, instead of TLAs, a reduced feeding efficiency and an increased visibility to predators [48,49] are more plausible explanations.

5. Conclusions

This is the second report of *E. chattoni* as a copepod parasite in the South Atlantic, nearly one century after its first account in the region [9]. Four previously unreported host copepod species were found in this study, including acartiids and oithonids, but the parasites were mostly prevalent in paracalanids, particularly in *Parvocalanus crassirostris*. The size of females was impacted by *E. chattoni* infections, which suggested that copepod fitness in general is affected by the presence of parasites [48]. Further experimental studies are needed to verify the negative effects of the parasite on growth, egg production, and the fitness of paracalanid hosts as well as the selection and specificity of the hosts.

Supplementary Materials: The following supporting information can be downloaded at: https://www.mdpi.com/article/10.3390/jmse10121816/s1, Table S1: Zooplankton sampling dates from 2009 to 2011. Aliquot sizes analyzed to estimate parasitized copepod abundances, date and time of sampling, tidal period, and copepod taxa infected by *Ellobiopsis chattoni* are provided. *: Sampling performed at the protected station (B); -: absence of infected copepods. Table S2: pelagic copepod species found off Ubatuba, south-east Brazil from 2009–2011. *: Host species for *Ellobiopsis chattoni*.

Author Contributions: J.E.M.F. and R.M.L. were responsible for the conceptualization, methodology, investigation, funding acquisition, and resources; J.E.M.F. performed the fieldwork and was responsible for the data curation; all authors contributed equally to the formal analysis, writing, and visualization, and approved the publication. All authors have read and agreed to the published version of the manuscript.

Funding: J.E.M.F. was funded by Pró-Reitoria de Pesquisa e Pós-graduação, Universidade Federal do Pará (grant no. 04 and 09/2014). R.M.L. is a CNPq fellow (315033/2021-5).

Institutional Review Board Statement: Not applicable.

Informed Consent Statement: Not applicable.

Data Availability Statement: Not applicable.

Acknowledgments: We acknowledge the staff of the Clarimundo de Jesus coastal station from the Instituto Oceanográfico, Universidade de São Paulo for their support during sampling, as well as Bruno Simi and Luis Paulo P. Lima for laboratory assistance.

Conflicts of Interest: The authors declare no conflict of interest.

References

1. Frost, B.W. Effects of size and concentration of food particles on the feeding behavior of the marine planktonic copepod *Calanus pacificus*. *Limnol. Oceanogr.* **1972**, *17*, 805–815. [CrossRef]
2. Azam, F.; Fenchel, T.; Field, J.G.; Gray, J.S.; Meyer-Reil, L.A.; Thingstad, F. The ecological role of water-column microbes in the sea. *Mar. Ecol. Prog. Ser.* **1983**, *10*, 25–263. [CrossRef]
3. Wäge, J.; Strassert, J.F.H.; Landsberger, A.; Loicj-Wilde, N.; Schmale, O.; Stawiarski, B.; Kreikemeyer, B.; Michel, G.; Labrenz, M. Microcapillary sampling of Baltic Sea copepod gut microbiomes indicates high variability between individuals and the potential for methane production. *FEMS Microbiol. Ecol.* **2019**, *95*, fiz024. [CrossRef] [PubMed]
4. Skovgaard, A. Dirty Tricks in the Plankton: Diversity and Role of Marine Parasitic Protists. *Acta Protozool.* **2014**, *53*, 51–62.
5. Martinelli-Filho, J.E.; Lopes, R.M.; Rivera, I.N.G.; Colwell, R.R. Are natural reservoirs important for cholera surveillance? The case of an outbreak in a Brazilian estuary. *Lett. Appl. Microbiol.* **2016**, *63*, 183–188. [CrossRef]
6. Gómez, F.; Artigas, L.F.; Gast, R.J. Molecular phylogeny of the parasitic dinoflagellate *Syltodinium listii* (Gymnodiniales, Dinophyceae) and generic transfer of *Syltodinium undulans* comb. nov. (= *Gyrodinium undulans*). *Eur. J. Protistol.* **2019**, *71*, 125636. [CrossRef] [PubMed]
7. Willians, J.D.; An, J. The cryptogenic parasitic isopod *Orthione griffenis* Markham, 2004 from the eastern and western Pacific. *Integr. Comp. Biol.* **2009**, *49*, 114–126. [CrossRef]
8. Ho, J.-S.; Perkins, P.S. Symbionts of marine Copepoda: An overview. *Bull. Mar. Sci.* **1985**, *37*, 586–598.
9. Steuer, A. Über *Ellobiopsis elongata* n. sp. aus dem Südatlantik. *Note Dell'Istituto Italo-Ger. Biol. Mar. Rovigno d'Istria* **1932**, *5*, 1–6.
10. Wickstead, J.H. A new record of *Ellobiopsis chattoni* (Flagellata incertae sedis) and its incidence in a population of *Undinula vulgaris* var. *major* (Crustacea, Copepoda). *Parasitology* **1963**, *53*, 293–296. [CrossRef]
11. Albaina, A.; Irigoien, X. Fecundity limitation of *Calanus helgolandicus*, by the parasite *Ellobiopsis* sp. *J. Plankton Res.* **2006**, *28*, 413–418. [CrossRef]
12. Skovgaard, A.; Saiz, E. Seasonal occurrence and role of protistan parasites in coastal marine zooplankton. *Mar. Ecol. Prog. Ser.* **2006**, *327*, 37–49. [CrossRef]
13. Walkusz, W.; Rolbiecki, L. Epibionts (*Paracineta*) and parasites (*Ellobiopsis*) on copepods from Spitsbergn (Kongsfjorden area). *Oceanologia* **2007**, *49*, 369–380.
14. Gómez, F.; López-García, P.; Nowaczyk, A.; Moreira, D. The crustacean parasites *Ellobiopsis* Caullery, 1910 and *Thalassomyces* Niezabitowski, 1913 form a monophyletic divergent clade within the Alveolata. *Syst. Parasitol.* **2009**, *74*, 65–74. [CrossRef] [PubMed]
15. Cattley, J.G. Sex reversal in copepods. *Nature* **1948**, *161*, 937. [CrossRef] [PubMed]
16. Gusmão, L.F.M.; Mckinnon, A.D. Sex ratios, intersexuality and sex change in copepods. *J. Plankton Res.* **2009**, *31*, 1101–1117. [CrossRef]
17. Fields, D.M.; Runge, J.A.; Thompson, C.; Shema, D.; Bjelland, R.M.; Durif, C.M.E.; Skiftesvik, A.B.; Browman, H.I. Infection of the planktonic copepod *Calanus finmarchicus* by the parasitic dinoflagellate, *Blastodinium* spp.: Effects on grazing, respiration, fecundity and fecal pellet production. *J. Plankton Res.* **2015**, *37*, 211–220. [CrossRef]
18. Purushothaman, A.; Thomas, L.C.; Padmakumar, K.B. First record of epibiotic parasitc dinoflagellate *Ellobiopsis chattoni* on copepod *Bestiolina similis* from northeastern Arabian Sea. *Symbiosis* **2020**, *80*, 239–244. [CrossRef]
19. Shields, J.D. The parasitic dinoflagellates of marine crustaceans. *Ann. Rev. Fish Dis.* **1994**, *4*, 241–271. [CrossRef]
20. Artüz, M.L. Parasites (*Ellobiopsis chattoni* Caullery, 1910) on Copepoda with two new host records, form Sea of Marmara, Turkey. *Mar. Biodivers. Rec.* **2016**, *9*, 11. [CrossRef]

21. Caullery, M. *Ellobiopsis chattoni*, n. g., n. sp., parasite de *Calanus helgolandicus* Claus, appartenant probablement aux Péridiniens. *Bull. Sci. Fr. Belg.* **1910**, *44*, 201–214.
22. Chatton, E. Les Péridiniens parasites. Morphologie, reproduction, éthologie. *Arch. Zool. Exp. Gen.* **1920**, *59*, 1–475.
23. Hovasse, R. "*Parallobiopsis coutieri*" Collin. Morphologie, cytologie, évolution, affinités des Ellobiopsidés. *Bull. Sci. Fr. Belg.* **1926**, *60*, 409–446.
24. Jepps, M.W. On the protozoan parasites of *Calanus finmarchicus* in the Clyde Sea area. *Q. J. Microsc. Sci.* **1937**, *79*, 589–658. [CrossRef]
25. Boschma, H. Ellobiopsidae. *Cons. Int. Explor. Mer.* **1956**, *65*, 1–4.
26. Björnberg, T.K.S. Developmental stages of some tropical and subtropical planktonic marine copepods. In *Studies on the Fauna of Curaçao and Other Caribbean Islands*; Martinus Nijhoff: The Hague, The Netherlands, 1972; p. 136.
27. Björnberg, T.K.S.; Lopes, R.M.; Björnberg, M.H.G.C. Chave para a identificação de náuplios de copépodos planctônicos marinhos do Atlântico Sul-Ocidental. *Nauplius* **1994**, *2*, 1–16.
28. Boltovskoy, D. (Ed.) *Atlas del Zooplancton del Atlantico Sudoccidental y Métodos de Trabajo con el Zooplancton Marino*; INIDEP: Mar Del Plata, Argentina, 1981.
29. Boltovskoy, D. (Ed.) *South Atlantic Zooplankton*; Backhuys Publishers: Leiden, The Netherlands, 1999; p. 1706.
30. Boxshall, G.A.; Halsey, S.H. *An Introduction to Copepod Diversity*; Ray Society: London, UK, 2004; p. 966.
31. Ferrari, F.D.; Dhams, H.-U. Post embryonic development of the Copepoda. *Crustaceana Monogr.* **2007**, *8*, 1–226.
32. Huys, R.; Boxshall, G.G. *Copepod Evolution*; The Ray Society: London, UK, 1991; pp. 1–468.
33. Martinelli Filho, J.E. Variação Temporal e Crescimento do Zooplâncton no Litoral Norte do Estado de São Paulo, com Ênfase em Estágios Imaturos de Copépodes. Ph.D. Thesis, Universidade de São Paulo, São Paulo, Brazil, 2013.
34. Zar, J.H. *Biostatistical Analysis*, 2nd ed.; Prentice Hall: Upper Saddle River, NJ, USA, 1999; p. 663.
35. Gómez, F.; Horiguchi, T. Ultrastructural features of the basal dinoflagellate *Ellobiopsis chattoni* (Ellobiopsidae, Alveolata), parasite of copepods. *CICIMAR Oceánides* **2014**, *29*, 1–10. [CrossRef]
36. Fhami, A.M.; Hussain, M. Two groups of parasites on the body of copepods. *J. Arid Environ.* **2003**, *54*, 149–153. [CrossRef]
37. Khalaf, T.A.; Awad, A.H.H.; Morad, M.S.S. New record of Chromistan parasites of copepods and rotifers in Iraqi marine and brackish waters. *Mesopot. J. Mar. Sci.* **2016**, *31*, 1–14.
38. Cleary, A.C.; Callesen, T.A.; Berge, J.; Gabrielsen, T.M. Parasite–copepod interactions in Svalbard: Diversity, host specificity, and seasonal patterns. *Polar Biol.* **2022**, *45*, 1105–1118. [CrossRef]
39. Skovgaard, A. Tumour-like anomalies on copepods may be wounds from parasites. *J. Plankton Res.* **2004**, *26*, 1129–1131. [CrossRef]
40. Schweikert, M.; Elbrächter, M. First ultrastructural investigations on *Ellobiopsis* spec. (incertae sedis) a parasite of copepods. *Endocytobiosis Cell Res.* **2006**, *17*, 73.
41. Sewell, R.B. The epibionts and parasites of the planktonic copepods of the Arabian Sea. *Sci. Rep. Murray Exped.* **1951**, *9*, 255–394.
42. Timofeev, S.F. The effect of the parasitic dinoflagellate *Ellobiopsis chattoni* (Protozoa: Mastigophora) on the winter mortality of the calanoid copepod *Calanus finmarchicus* (Crustacea: Copepoda) in the Norwegian Sea. *Parazitologiya* **2002**, *36*, 158–162.
43. Jagadeesan, L.; Jyothibabu, R. Tumour-like anomaly of copepods—An evaluation of the possible causes in Indian marine waters. *Environ. Monit. Assess.* **2016**, *188*, 244. [CrossRef] [PubMed]
44. Santhakumari, V.; Saraswathy, M. On the Ellobiopsidae, parasitic protozoa from zooplankton. *Mahasagar Bull. Nat. Inst. Oceanogr.* **1979**, *12*, 83–92.
45. Hoffman, E.G.; Yancey, R.M. Ellobiopsidae of Alaskan coastal waters. *Pac. Sci.* **1966**, *20*, 70–78.
46. Strickler, J.R. Calanoid copepods, feeding currents, and the role of gravity. *Science* **1982**, *218*, 158–160. [CrossRef] [PubMed]
47. Bhandare, C.; Ingole, B.S. First evidence of tumor-like anomaly infestation in copepods from the Central Indian Ridge. *Indian J. Mar. Sci.* **2008**, *37*, 227–232.
48. Bass, D.; Rueckert, S.; Stern, R.; Cleary, A.C.; Taylor, J.D.; Ward, G.M.; Huys, R. Parasites, pathogens, and other symbionts of copepods. *Trends Parasitol.* **2021**, *37*, 875–889. [CrossRef] [PubMed]
49. Willey, R.L.; Cantrell, P.A.; Threkeld, R.T. Epibiotic euglenoid flagellates increase the susceptibility of some zooplankton to fish predation. *Limnol. Oceanogr.* **1990**, *35*, 952–959. [CrossRef]

Journal of
Marine Science and Engineering

Article

Response of the Black Sea Zooplankton to the Marine Heat Wave 2010: Case of the Sevastopol Bay

Alexandra Gubanova [1,*], Katerina Goubanova [2], Olga Krivenko [1], Kremena Stefanova [3], Oksana Garbazey [1], Vladimir Belokopytov [4], Tatiana Liashko [1] and Elitsa Stefanova [3]

1 A.O. Kovalevsky Institute of Biology of the Southern Seas, Russian Academy of Sciences, Leninsky Prospekt, 38, 119991 Moscow, Russia
2 Centro de Estudios Avanzadosen Zonas Áridas (CEAZA), La Serena 1700000, Chile
3 Institute of Oceanology-Bulgarian Academy of Sciences, 9000 Varna, Bulgaria
4 Federal Research Center Marine Hydrophysical Institute of Russian Academy of Sciences, 299011 Sevastopol, Russia
* Correspondence: adgubanova@ibss-ras.ru

Abstract: Global warming is increasing the frequency and severity of the marine heat waves, which poses a serious threat to the marine ecosystem. This study analyzes seasonal and interannual dynamics in the abundance and structure of the mesozooplankton community in Sevastopol Bay based on bi-monthly routine observations over 2003–2014. The focus is on the impact of the summer 2010 marine heat wave (MHW2010) on crustaceans belonging to different ecological groups. As a response to the MHW2010, three warm-water species (*O. davisae, A. tonsa* and *P. avirostris*) exhibiting the maximum seasonal density in latter summer showed a sharp increase in the annual abundance and their share in the mesozooplankton community. The increase in the annual abundance in 2010 of the eurythermal species *P. parvus* and *P. polyphemoides* exhibiting seasonal peaks in spring and autumn is not related to the MHW2010 but can be explained by a rise of temperature in the first part of the year. *O. davisae* and *A. tonsa* showed the most pronounced response among the species to the MHW2010, confirming that non-native species exhibited great flexibility as an adaptive response to environmental changes, especially in the case of climate warming. Among crustaceans observed in this study, *O. davisae* can be considered as an indicator of the environmental conditions associated with the warming of the Black Sea and the Mediterranean basin as a whole.

Keywords: marine heat waves; mesozooplankton; copepod; crustacean; Sevastopol Bay; Black Sea

Citation: Gubanova, A.; Goubanova, K.; Krivenko, O.; Stefanova, K.; Garbazey, O.; Belokopytov, V.; Liashko, T.; Stefanova, E. Response of the Black Sea Zooplankton to the Marine Heat Wave 2010: Case of the Sevastopol Bay. *J. Mar. Sci. Eng.* **2022**, *10*, 1933. https://doi.org/10.3390/jmse10121933

Academic Editors: Marco Uttieri, Ylenia Carotenuto, Iole Di Capua and Vittoria Roncalli

Received: 1 November 2022
Accepted: 25 November 2022
Published: 7 December 2022

Publisher's Note: MDPI stays neutral with regard to jurisdictional claims in published maps and institutional affiliations.

1. Introduction

Marine heat waves (MHWs) are extreme warm oceanic events that persist for days to months and can have devastating impacts on marine ecosystem often with ecological and socioeconomic consequences. Over the last decades, the MHWs have been increasing in frequency, intensity and duration worldwide, and these trends are projected to continue in the future as a consequence of anthropogenic climate change [1,2]. The Black Sea is an example of semi-closed sea experiencing a rapid warming, which is considered an amplified precursor of the changes to expect in the greater oceans [3]. The increasing warming rate of the Sea Surface Temperature (SST) in the Black Sea in the two last decades with respect to the previous two decades was associated with an increasing rate in MHWs frequency: the annual mean SST trends was 0.4 °C/decade in 1982–2000 and 0.7 °C/decade in 2001–2020, and the corresponding average frequencies of MHW were estimated as about 0.6 events/year and about 3 events/year, respectively [4]. Among the most intense and prolonged MHWs is the summer 2010 event associated with the extreme atmospheric heat wave that hit western Russia as result of the strong atmospheric blocking [5]. Given that MHW are expected to rise in magnitude, frequency and duration in the future, it is important to evaluate the response of pelagic communities to extreme MHWs, especially in shallow coastal areas which are more sensitive to temperature variations than open seas [6].

Marine mesozooplankton is a suitable candidate for investigation of the ecosystem response to climate variability and extremes, and multi-year mesozooplankton time series provide useful information about climate–ecosystem interactions [7,8]. Indeed, mesozooplankton plays a pivotal role in marine ecosystems, providing a link between primary producers and secondary consumers in food webs. Therefore, all changes in the food chain, from the bottom to the top, are reflected in the mesozooplankton. Mesozooplankton comprises poikilothermic animals, sensitive to temperature changes, which is one of the most important factors, driving its temporal and spatial distribution. Mesozooplankton species have a short lifespan, six to nine generations of copepods per year in the Black Sea, and can provide an early signal of environmental changes [9,10].

Despite its major role in marine ecosystems, only a few studies have investigated the response of coastal populations of zooplankton species to MHWs. Rhian Evans and co-authors demonstrated that the 2015–2016 Tasman MHW caused a shift in the abundance and compositions of the zooplankton community resulting in an increase (decrease) in warm- (cold-) water species of copepods [11]. Similarly, Caitlin A.E. Mckinstry and co-authors showed an elevated abundance of warm waters copepods in response to the 2014–2015 MHW in the low Cook Inlet, Alaska [12].

A specific feature of the Black Sea is its low biodiversity. In general, it is 3.5–4 times poorer than that in the Mediterranean Sea, where the copepod species are generally functionally redundant [13]. This redundancy should compensate for the loss of ecosystem functions of the Mediterranean zooplankton communities caused by climate change [14]. In the Black Sea, there are currently 12 species of marine planktonic copepods, three of which are invasive, namely *Acartia tonsa*, *Oithona davisae*, and *Pseudadiaptomus marinus* [15,16]. In this regard, changes in environment caused by climate impact or anthropogenic pressure lead to noticeable effects in the zooplankton community and the ecosystem of the Black Sea as a whole. Thus, the Black Sea, and in particular Sevastopol Bay, is a suitable model for assessing the environmental impacts of climate change on marine biodiversity.

Zooplankton of the Black Sea include species of various origins and, hence, they are different in ecology and biology [17,18]. Cold-water assemblage copepods are considered boreal-Atlantic relics that inhabited the Black Sea during the past cooling period. They dwell in a deep layer of the open sea in summer and appear in surface waters and coastal areas during the cold season. Thermophilic species colonized the Black Sea as it warmed in the last stages of the Quaternary. They survive the cold season at the dormancy stage, rapidly increase in abundance in the warm season and peak in abundance from late July to October. Individual representatives of this group (namely ctenophores *Mnemiopsis leidyi*, *Beroe ovata* and copepods *Acartia tonsa*, *Oithona davisae*) were established in the Black Sea quite recently, during the last 50 years. Finally, some copepods belong to the eurythermal assemblage and are quite numerous in the plankton all year round [10,15].

Our goal in this paper is to assess the response of the crustacean mesozooplankton populations in the Sevastopol Bay (northern Black Sea) to the 2010 summer MHW, which was one of the most intense and longest MHWs observed in the region. Based on long-term (2003–2014) routine observations of zooplankton, we aim at identifying species sensitive to extreme warm temperature anomalies observed during summer 2010 and documenting corresponding changes in composition, abundance, structure and seasonal variations.

2. Materials and Methods

2.1. Area of Investigation

Sevastopol Bay is located in the northern part of the Black Sea at the southwestern tip of the Crimean Peninsula (Figure 1). It is about 7 km long and 1 km wide at its widest point, and it has an average depth of 12 m. It is a semi-enclosed estuarine-type bay having a restricted water exchange with the open sea because of its large length and the mole built at the entrance.

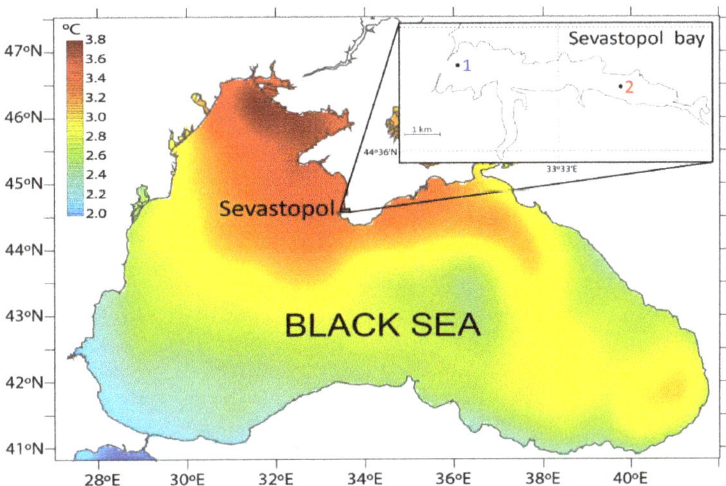

Figure 1. Sampling stations in Sevastopol Bay and spatial pattern of the monthly average sea surface temperature (SST, °C) anomalies in the Black Sea in August 2010 with respect to the 1985–2014 monthly climatology.

The salinity ranges within 14–17.5 ppt and is largely controlled by freshwater input from the Chernaya River that flows into the head of the bay [19]. The bay is suffering from a heavy anthropogenic pressure associated with discharges of industrial and domestic wastewaters as well as stormwater runoff. Based on the eutrophication E-TRIX-index assessments made in 2011–2012, the trophic level in the Sevastopol Bay was characterized as a transitional from medium to high [20]. Both pollution and trophic levels gradually decrease from the head of the bay to its mouth [21]. Being a port area, Sevastopol Bay is also heavily affected by maritime traffic, which contributes to invasions of alien species.

2.2. SST Data

SST in Sevastopol Bay is investigated based on a long-term series of continuous 6 h measurements from 1950 to 2014 provided by the Sevastopol Hydrometeorological Station. Additionally, the SST product from the OSTIA archive (Operational Sea Surface Temperature and Sea Ice Analysis) with a spatial resolution of 0.054° (about 5 km) was used to analyze the spatial distribution of the summer 2010 warm anomaly in the Black sea. The archive includes daily fields averaged using optimal interpolation and is based on satellite data from sensors AVHRR, AMSRE, AATSR, SEVERI and TMI, as well as on data received from drifting and moored buoys [22].

2.3. Sampling and Zooplankton Processing

The analysis of the mesozooplankton community is based on a long-term data set collected between 2003 and 2014 at two stations (Figure 1), one located in the mouth of Sevastopol Bay (station 1) and the other one in the middle of the bay (station 2). Throughout the entire period, sampling and samples processing were made according to the same methods, allowing a reliable assessment of seasonal and interannual changes.

Zooplankton samples were taken twice per month in the morning using a Juday plankton net (with a mouth area of 0.1 m^2 and a mesh size of 150 mm) from the whole water column: 10–0 m at station 1, and 9–0 m at station 2. Samples were fixed with formaldehyde solution (4% final conc.) and processed in the laboratory using the standard methodology for zooplankton. The sample was homogenized before taking aliquot. A calibrated 1 mL Stempel pipette was used for sub-sampling. Quantitative and qualitative processing was carried out in Bogorov's chamber under a stereomicroscope. At least 2 aliquots were calculated for each sample. In the sub-sample(s), all crustaceans were counted until each of the three dominant species reached 100 individuals. The entire specimen was examined for rare species [23,24].

2.4. Data Analysis

2.4.1. SST Data

The monthly seasonal cycle of SST (or monthly mean climatology) was calculated based on observed mean monthly data averaged for each calendar month over the 30-year period from 1985 to 2014. The monthly SST anomalies were then calculated by subtracting the monthly seasonal cycle from the observed monthly means. In order to obtain the monthly mean climatology on a daily basis, the monthly climatology was interpolated from 12 calendar months to 365 calendar days using a spline interpolation.

To describe the summer 2010 MHW characteristics, we used the MHW definition from [25]: the mean daily SST exceeds a seasonally varying threshold (95th percentile) for at least 5 consecutive days; successive heatwaves with gaps of 2 days or less are considered part of the same event. The seasonally varying 95th percentile was calculated as monthly 95th percentile climatology for each calendar month over the period 1985–2014 and then interpolated to 365 calendar days using a spline interpolation.

2.4.2. Zooplankton

Based on the long-term observation dataset described in Section 2.3, seasonal and interannual variability in zooplankton abundance was analyzed at both stations for 12 species of crustacean zooplankton representing different ecological groups (warm-water, cold-water and eurythermal).

The seasonal variability of zooplankton species abundance was analyzed based on normalized average monthly values calculated as:

$$N_{ij}^{\sigma} = \left(\overline{N_{ij}} - \overline{\overline{N_j}}\right) / \sigma\overline{\overline{N_j}} \quad (1)$$

where $\overline{N_{ij}}$—average monthly values of the abundance for each i-th month and j-th species; $\overline{\overline{N_j}}$ and $\sigma\overline{\overline{N_j}}$ the average long-term value of the abundance and its standard deviation for the j-th species, respectively.

Interannual abundance variability was estimated as the deviation of the numbers of each zooplankton species (N_{ij}) from the corresponding average monthly values (for each i-th month and j-th species) according to:

$$\delta N_{ij} = N_{ij} - \overline{N_{ij}} \quad (2)$$

Furthermore, the values of abundance anomalies relative to the annual variation (δN_{ij}) were averaged over the 2003 to 2014 ($\overline{\delta N_{ij}}$), and the standard deviation $\sigma(\overline{\delta N_{ij}})$ of the obtained time series were calculated. The average annual values of abundance anomalies relative to the annual variation normalized to this value ($^{\sigma}_{\delta}\overline{N}_{ij}$) were used to characterize the interannual variability of particular zooplankton species and to assess the significance of differences in their abundance among years.

Hereafter, the normalized average annual values of abundance anomalies relative to the annual variation $^{\sigma}_{\delta}\overline{N}$ will be referred to as the indicator of interannual variability.

3. Results

3.1. SST Variability and the Summer 2010 MHW

The SST of Sevastopol Bay has a marked seasonal cycle with a cold season from January to March and a warm season occurring form June to September (Figure 2a, blue line). The minimum and maximum are observed in February (6.6 °C) and August (24.8 °C), respectively. During the study period, SST shows a clear warming trend from 2003 to 2010, the latter year being the hottest one (Figure 2b). The month-to-month variability was rather low between 2003 and 2010. The period 2011–2014 exhibits stronger variability, which is associated in particular with two cold events (February 2012 and October 2013) when the monthly anomalies fell below −2 °C and the hottest event in May 2013 when the monthly anomaly reached 3.4 °C. The minimum and maximum are observed in February (6.6 °C) and August (24.8 °C), respectively.

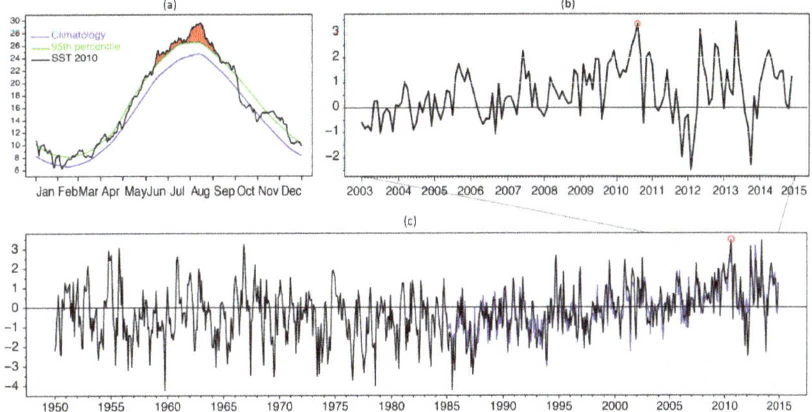

Figure 2. SST series in Sevastopol Bay (°C): (**a**) The climatology (blue), 95th percentile marine heat wave (MHW) threshold (green) and time series (black) for 2010. The red filled area indicates the period of time associated with the MHW; (**b**) Monthly averaged anomalies over the study period 2003–2014 with respect to the 1985–2014 monthly climatology as recorded by in situ observations; (**c**) As in (**b**) but over the period 1950–2014. Black and blue lines in (**c**) show the in situ and satellite observations, respectively.

During the study period (Figure 2b), SST shows a clear warming trend from 2003 to 2010, the latter year being the hottest one. The month-to-month variability was rather low between 2003 and 2010. The period 2011–2014 exhibits stronger variability, which is associated in particular with two cold events (February 2012 and October 2013) when the monthly anomalies fell below −2 °C and the hottest event in May 2013 when the monthly anomaly reached 3.4 °C. Figure 2c further evidences that the study period (2003–2014) corresponds to the warmest duodecad and 2010 is the warmest year since at least 1950 when began regular SST observations in Sevastopol Bay.

We now focus on the summer 2010 MHW event. Figure 2a shows that during almost all of 2010, the daily SST anomalies were above the climatological values. The summer MHW event starts on June 11th and ends on 9th September, lasting in total 3 months. The peak of the event was observed from 1 August to 19 August, when the daily SST anomalies exceeded 4 °C and reached 5 °C on 15 August. Figure 1 shows that in August 2010, abnormally high SSTs were recorded throughout the entire Black Sea area. The mean monthly anomalies ranged from about 2 °C in the Bosphorus region to almost 3.8 °C near the Dnieper-Bug Estuary in the northwestern shelf. Crimea's southwestern coast was marked by extremely high values (more than 3 °C), which closely matched the in situ observations in Sevastopol Bay (Figure 2).

3.2. Species Composition and Ecological Groups of Zooplankton

During the study period, 12 species of crustaceans representing three different ecological groups (thermophilic, eurythermal and cold-water) were found in Sevastopol Bay. Table 1 summarizes the annual average abundance of each species. Copepods *Acartia tonsa*, *Centropages ponticus*, *Oithona davisae* and cladocerans *Penilia avirostris*, *Pseudevadne tergestina*, *Evadne spinifera* are representatives of a thermophilic assemblage. Among them, *A. tonsa* and *O. davisae* are non-indigenous species, which appeared in the Black Sea in 1970s and

2005, respectively. *Pseudevadne tergestina* and *Evadne spinifera* were not numerous and were not considered here.

Table 1. Interannual variations in abundance of the numerous crustaceans (ind. m^{-3} ± standart error) at the mouth of Sevastopol Bay (station 1) and in the middle of the bay (station 2).

Station 1.												
Year	**2003**	**2004**	**2005**	**2006**	**2007**	**2008**	**2009**	**2010**	**2011**	**2012**	**2013**	**2014**
Samples Number	21	21	19	24	21	24	23	23	22	22	21	22
Penilia avirostris	238 ± 117	433 ± 268	2391 ± 1596	792 ± 630	272 ± 166	334 ± 198	590 ± 370	1164 ± 597	825 ± 477	483 ± 284	518 ± 421	378 ± 241
Pleopis polyphemoides	227 ± 151	526 ± 246	741 ± 506	360 ± 114	383 ± 142	799 ± 513	372 ± 136	865 ± 374	795 ± 426	362 ± 219	376 ± 226	202 ± 97
Acartia clausi	582 ± 194	263 ± 56	232 ± 51	203 ± 52	142 ± 44	922 ± 408	198 ± 53	362 ± 162	389 ± 87	125 ± 35	573 ± 126	238 ± 42
Acartia tonsa	129 ± 47	480 ± 246	116 ± 53	99 ± 42	52 ± 26	62 ± 24	8 ± 7	72 ± 35	3 ± 1	13 ± 6	38 ± 18	13 ± 7
Centropages ponticus	62 ± 25	92 ± 54	37 ± 15	171 ± 50	346 ± 174	72 ± 31	321 ± 104	268 ± 109	177 ± 64	310 ± 174	471 ± 159	248 ± 57
Oithona davisae	NA *	NA *	22 ± 14	1892 ± 1056	2344 ± 1717	3256 ± 1349	5770 ± 1763	17,236 ± 5400	5043 ± 1384	4678 ± 2729	4211 ± 1159	7140 ± 3107
Oithona similis	30 ± 10	58 ± 20	23 ± 9	31 ± 7	63 ± 22	24 ± 8	160 ± 46	43 ± 10	122 ± 35	156 ± 57	127 ± 33	87 ± 21
Paracalanus parvus	173 ± 57	178 ± 41	377 ± 176	564 ± 169	524 ± 146	638 ± 207	1786 ± 354	1830 ± 567	1280 ± 249	474 ± 125	1261 ± 386	839 ± 165
Pseudocalanus elongatus	204 ± 74	193 ± 68	121 ± 44	120 ± 37	234 ± 102	55 ± 21	189 ± 61	64 ± 25	325 ± 141	118 ± 43	224 ± 67	180 ± 72
Station 2.												
Year	**2003**	**2004**	**2005**	**2006**	**2007**	**2008**	**2009**	**2010**	**2011**	**2012**	**2013**	**2014**
Samples number	18	20	15	21	21	22	22	22	22	16	21	21
Penilia avirostris	284 ± 117	133 ± 100	1265 ± 1170	75 ± 29	73 ± 45	80 ± 62	664 ± 369	1202 ± 749	258 ± 221	348 ± 168	215 ± 144	2362 ± 2151
Pleopis polyphemoides	1051 ± 612	966 ± 475	959 ± 641	1636 ± 845	1401 ± 678	912 ± 501	688 ± 398	1056 ± 511	2027 ± 1086	1662 ± 820	1060 ± 777	617 ± 295
Acartia clausi	338 ± 162	308 ± 71	159 ± 50	372 ± 157	137 ± 45	472 ± 131	198 ± 43	232 ± 62	270 ± 63	83 ± 17	485 ± 117	375 ± 106
Acartia tonsa	1777 ± 1245	690 ± 361	599 ± 264	366 ± 215	292 ± 162	261 ± 152	30 ± 17	883 ± 609	73 ± 25	197 ± 110	483 ± 237	189 ± 130
Centropages ponticus	61 ± 36	40 ± 20	25 ± 10	36 ± 15	196 ± 98	32 ± 14	184 ± 63	79 ± 28	284 ± 130	183 ± 85	186 ± 53	122 ± 32
Oithona davisae	NA *	NA *	301 ± 168	4849 ± 2224	6867 ± 3128	16,312 ± 5456	22,069 ± 4345	41,754 ± 12,337	25,059 ± 6520	13,174 ± 6807	8946 ± 2017	18,131 ± 7869
Oithona similis	69 ± 50	120 ± 60	20 ± 7	49 ± 19	111 ± 36	22 ± 7	151 ± 60	36 ± 15	115 ± 37	247 ± 146	177 ± 67	193 ± 76
Paracalanus parvus	61 ± 17	85 ± 28	174 ± 60	364 ± 113	295 ± 65	484 ± 211	674 ± 160	731 ± 192	686 ± 137	214 ± 59	474 ± 91	548 ± 102
Pseudocalanus elongatus	324 ± 195	305 ± 118	224 ± 87	303 ± 145	262 ± 84	130 ± 40	598 ± 228	152 ± 64	370 ± 126	254 ± 115	241 ± 128	426 ± 184

* NA—not available.

The seasonal pattern dynamics of warm-water species are shown in Figure 3. All these species survived in the cold season in the Black Sea at a dormant stage. Most of them produced resting eggs in response to low temperatures, and *O. davisae* maintained in plankton at the stage of fertilized females. The populations of warm-water species began to grow rapidly in late May (at a temperature of 16–18 °C) and peaked in August–October (Figure 3A–C).

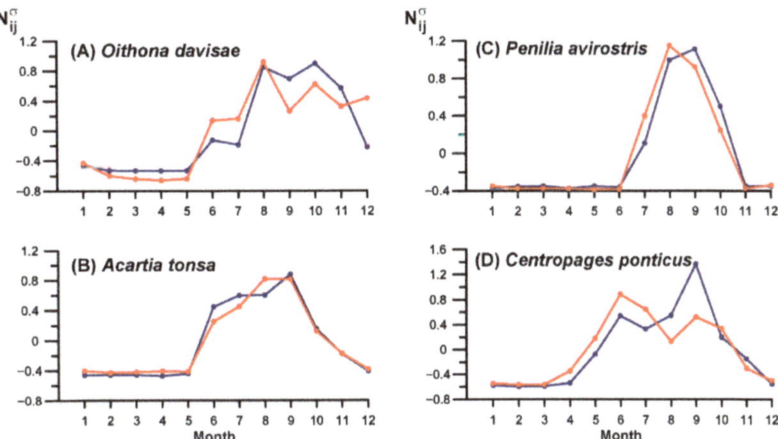

Figure 3. The patterns of seasonal dynamics of warm-water crustaceans in Sevastopol Bay at station 1 (blue) and station 2 (red) presented as normalized average monthly values (see details of the method in Section 2.4): (**A**)—*Oithona davisae;* (**B**)—*Acartia tonsa;* (**C**)—*Centropages ponticus;* (**D**)—*Penilia avirostris.*

Such seasonal pattern was typical for all the above-mentioned species with the exception of *C. ponticus* (Figure 3D). The first peak of *C. ponticus* abundance occurred in June, which was followed by a slight decline in the hottest months of July and August. The more pronounced peak took place in September (Figure 3D). So, despite the fact that centropages is typically a warm-water species, it preferred temperatures not higher than 23 °C.

The eurythermal assemblage of crustaceans was represented by copepods *A. clausi*, *P. parvus* and cladocera *P. polyphemoides*. All these species are numerous in the Sevastopol Bay plankton community all year round. Seasonal pattern of *A. clausi* demonstrated two pronounced picks, in early spring (March) and autumn (September–November), respectively (Figure 4A). *P. parvus* peaked in November–December (Figure 4B). For both species, there was a summertime decline in abundance. (Figure 4A,B). *P. polyphemoides* showed strong picks only in May–June (Figure 4C).

Copepods *P. elongatus, O. similis, C. euxinus* belong to a cold-water assemblage. Two of them, *P. elongatus* and *O. similis,* are important components of the zooplankton of Sevastopol Bay in cold seasons. *C. euxinus*, an inhabitant of the open Black Sea, was found in small amounts in the bay, usually during winter. It was not taken under consideration in the present analysis. *P. elongatus* and *O. similis* were abundant in Sevastopol Bay during January–April and November–December, whereas in June–October, their density was the lowest (Figure 5A,B).

3.3. Interannual Variation in Abundance

For the study period 2003–2014, the maximum abundance of crustaceans was recorded in 2010: 22,000 ind. m^{-3} at the mouth of the bay (station 1) and 46,000 ind. m^{-3} in its middle (station 2), which is nearly three times the long-term average values for the whole period (about 7000 ind. m^{-3} and about 17,000 ind. m^{-3}, respectively). At both stations, the crustaceans were numerically dominated by warm water species in 2010 (Table 1; Figure 6a). They amounted more than 19,000 ind. m^{-3} at station 1 and about 44,000 ± ind. m^{-3} at station 2 (85% and 95% of the total crustacean abundance, respectively). The warm-water assemblage prevailed not only among crustaceans but also in the mesozooplankton community as a whole, both by season (in the summer–autumn months) and by year (on average per year). Warm-water species abundance increased year after year from 2006 to 2010. This was attributable to the introduction and rapid growth of the *O. davisae* population, which is a new copepod species discovered in the Black Sea in 2005. Note also that a general positive trend in warm-water species abundance over 2003–2010 coincides with a strong warming SST trend in this period (Figure 2c).

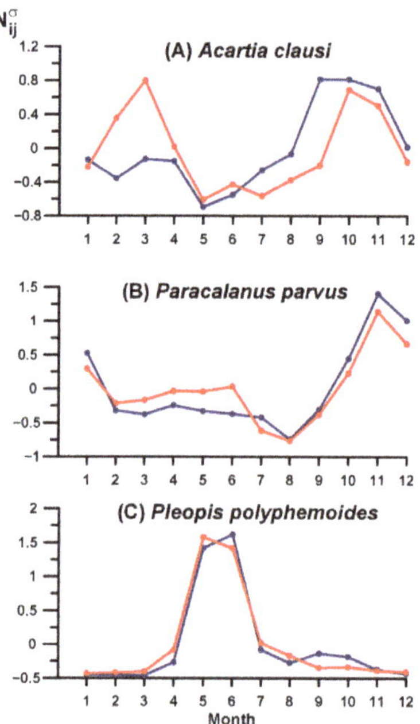

Figure 4. The patterns of seasonal dynamics of eurythermalcrustaceansin Sevastopol Bay at station 1 (blue) and station 2 (red) presented as normalized average monthly values (see details of the method in Section 2.4): (**A**)—*Acartia clausi*; (**B**)—*Paracalanus parvus*; (**C**)—*Pleopis polyphemoides*.

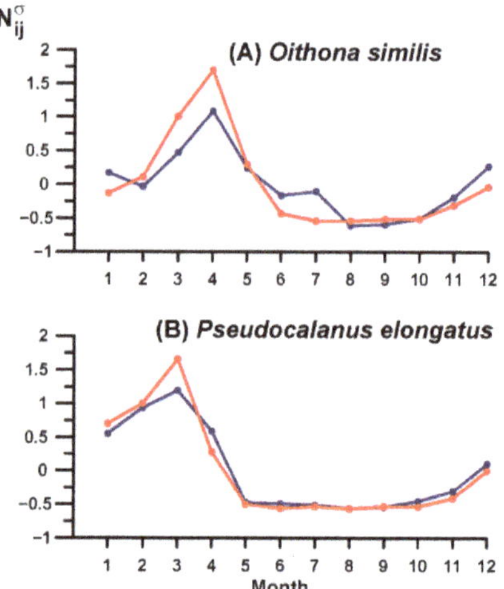

Figure 5. The patterns of seasonal dynamics of cold-water crustaceans in Sevastopol Bay at station 1 (blue) and station 2 (red) presented as normalized average monthly values (see details of the method in Section 2.4): (**A**)—*Oithona similis*; (**B**)—*Pseudocalanus elongatus*.

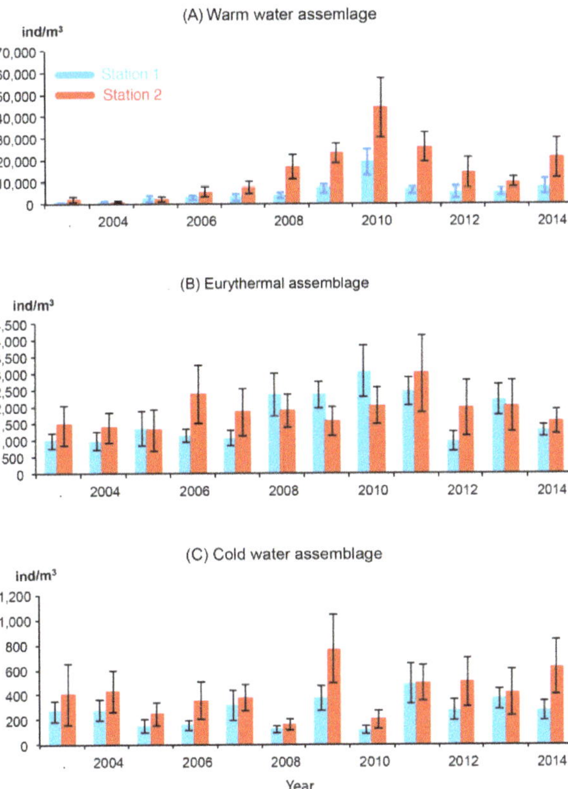

Figure 6. Interannual variability of annual average abundance of the: (**A**)—warm-water crustacean assemblage, (**B**)—euthermal crustacean assemblage; (**C**)—cold-water crustacean assemblage in Sevastopol Bay at station 1 (blue bars) and station 2 (red bars).

The density of eurythermal and cold-water species varied slightly during the study period (Figure 6B,C). The eurythermal crustacean abundance ranged from 960 to 3057 ind. m^{-3} at station 1 and from 1292 to 2983 ind. m^{-3} at station 2, whereas the abundance of cold-water crustacean ranged from 119 to 487 ind. m^{-3} at station 1 and from 162 to 767 ind. m^{-3} at station 2. A minor rise in the abundance of eurythermal species was observed in 2010, notably at the mouth of the bay (Figure 6B). In the middle of the bay (station 2), the abundance of eurythermal species reached the highest values in 2011. On the contrary, the density of cold-water crustaceans was the lowest in 2010: 119 ind. m^{-3} at station 1 and 199 ind. m^{-3} at station 2 (Figure 6C).

3.4. Key Species Variability

Warm-water assemblages of crustaceans in Sevastopol Bay were represented by four key species: *O. davisae*, *A. tonsa*, *C. ponticus* and *P. avirostris*. Non-indigenous copepod *O. davisae* was detected in the Black Sea at the end of 2005 and contributed the most to the total abundance of crustaceans in Sevastopol Bay during the following years. Since 2006, its abundance increased annually by about 1.5 times and reached 5770 ± 1763 ind. m^{-3} at station 1 and 22,069 ± 4345 ind. m^{-3} at station 2 in 2009 (Table 1). The population explosion occurred in 2010 (17,236 ± 5400 ind. m^{-3} at station 1 and 41,754 ± 12,337 ind. m^{-3} at station 2), and since 2011, species density stabilized near the values of 2009 (Table 1). At both stations, the indicator of interannual variability of abundance ($^{\sigma}_{\delta}\overline{N}$) reached 3σ in 2010 (Figure 7A).

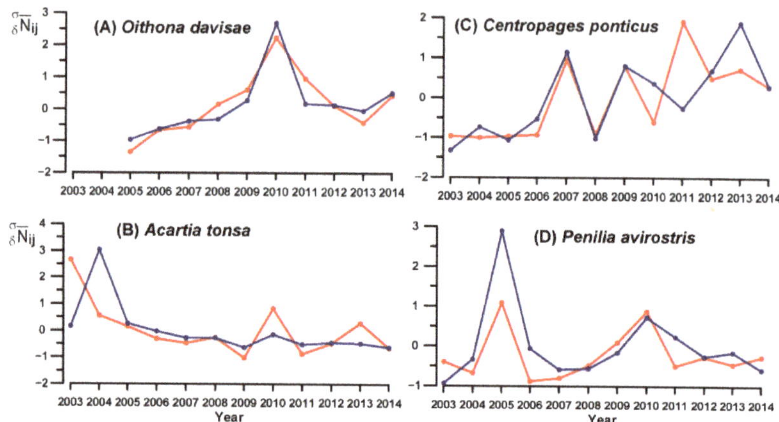

Figure 7. Pattern of interannual variability of abundance of the warm-water crustaceans in Sevastopol Bay at station 1 (blue) and station 2 (red) presented as normalized annual anomalies (see details of the method in Section 2.4): (**A**)—*Oithona davisae;* (**B**)—*Acartia tonsa;* (**C**)—*Centropages ponticus;* (**D**)—*Penilia avirostris.*

Positive anomalies in the abundance of *O. davisae* in Sevastopol Bay were observed throughout the entire breeding season of 2010 (Figure 8A). The largest abundance of *O. davisae* was recorded at both stations (86,000 ind. m^{-3} at station 1 and 145,000 ind. m^{-3} at station 2) in August 2010 during the peak of the summer MHW when the highest daily SST reached 29.6 °C. Moreover, *O. davisae* contributed hugely to the total abundance of crustaceans in August 2010: 78% at the mouth and 90% in the middle of the bay.

Another warm-water species *A. tonsa* was more abundant in 2003–2005, before the introduction of *O. davisae* (Figure 7B), showing the annual average density between 116 and 480 ind. m^{-3} at station 1 and between 599 and 1777 ind. m^{-3} at station 2. Since 2006, the density of *A. tonsa* decreased steadily, dropping to 52–99 (261–366) ind. m^{-3} at station 1 (station 2) in 2006–2008, and further to 3–38 (30–483) ind. m^{-3} at station 1 (station 2) in 2009 and 2011–2014 (Table 1). However, in 2010, the average annual abundance of *A. tonsa* increased sharply with respect to the previous years (up to 72 ± 35 ind. m^{-3} at station 1 and 883 ± 609 ind. m^{-3} at station 2). The maximum abundance was observed in July–August during the peak of the MHW (Figure 8B): 563 ind. m^{-3} and 12,000 ind. m^{-3} at the mouth and in the middle of the bay, respectively.

The warm-water species *P. avirostris* exhibited two pronounced peaks in its density in 2005 and 2010 (Table 1, Figure 7D). The peak of 2005 at the mouth of the bay (station 1) was the most significant and showed the indicator of interannual variability above 3σ. The average abundance in 2010 amounted to 1164 ind. m^{-3} ± 374 at station 1 and to 1202 ind. m^{-3} ± 749 at station 2. *P. avirostris* density reached the maximum in July–August (Figure 8D): the abundance was 7000–8000 ind. m^{-3} in July and 3000–4000 ind. m^{-3} in August (Table 1). The indicator of interannual variability reached 1σ for these peaks (Figure 7D).

Unlike the three warm-water species described above, the pattern of interannual fluctuations in *C. ponticus* abundance was more heterogenous, showing different behaviors at two stations (Figure 7C). The annual average density of *C. ponticus* reached its maximum in 2013 at station 1 (471 ± 159 ind. m^{-3}) and 2011 at station 2 (284 ± 130 ind. m^{-3}) (Table 1). Both stations did not experience a density peak in 2010 showing the values of 268 ± 109 ind. m^{-3} and 79 ± 28 ind. m^{-3} at station 1 and station 2, respectively. In August 2010, negative anomalies of *C. ponticus* abundance were observed at both stations (Figure 8D).

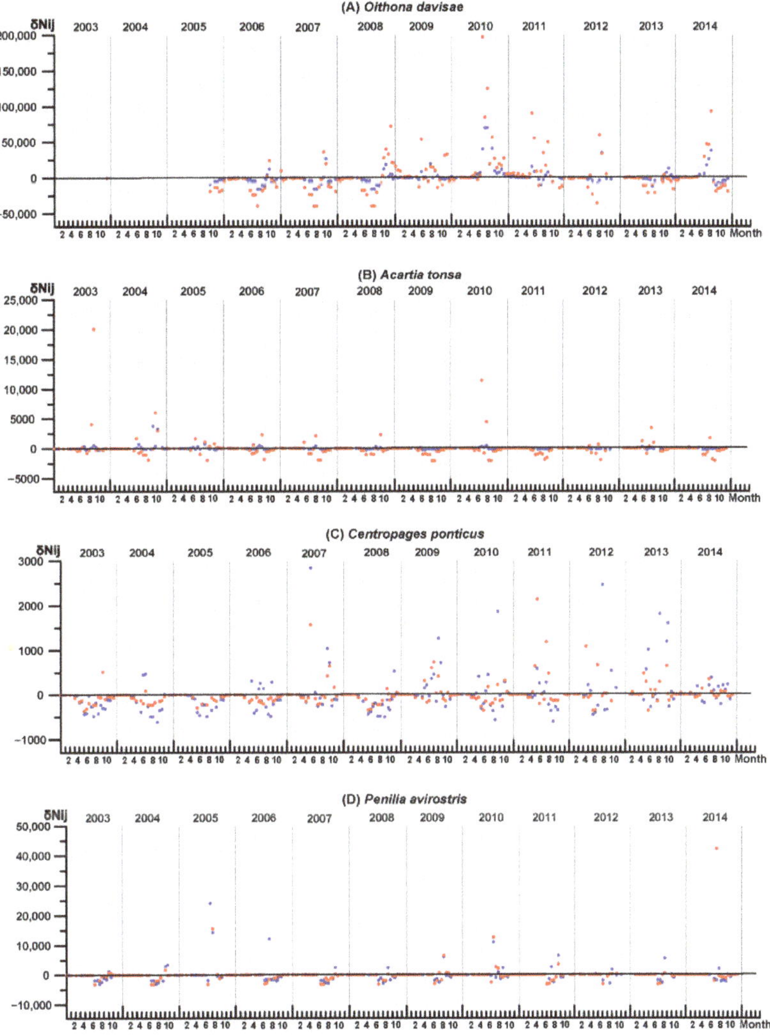

Figure 8. Long-term fluctuations in anomalies of warm-water crustaceans abundance relative to the annual variation (see details of the method in Section 2.4) in Sevastopol Bay at station 1 (blue) and station 2 (red): (**A**)—*Oithona davisae;* (**B**)—*Acartia tonsa;* (**C**)—*Centropages ponticus;* (**D**)—*Penilia avirostris.*

Eurythermal crustacean assemblages were represented by *A. clausi*, *P. parvus* and *P. polyphemoides*. The pattern of interannual variability of *A. clausi* showed a significant increase in population in 2008 and 2013 (Figure 9A) associated with seasonal picks in its abundance in spring and autumn (Figure 10A).

The annual average abundance of *A. clausi* ranged within 125 ± 35 ind. m^{-3} and 922 ± 408 ind. m^{-3} at station 1 and within 83 ± 17 ind. m^{-3} and 485 ± 117 ind. m^{-3} at station 2 (Table 1). In 2010, the abundance was close to the average value (362 ± 162 and 232 ± 62 ind. m^{-3} at stations 1 and 2, respectively). The annual average abundances of *P. parvus* increased at both stations in period 2009–2011 (Figure 9B), with the maximum density in 2010 (1830 ± 567 ind. m^{-3} at station 1 and 731 ± 192 ind. m^{-3} at station 2) (Table 1). The highest abundance of *P. parvus* occurred in November 2010 at the mouth of the bay (9100 ind. m^{-3}). The abundance was also high in April–May and in October–December in 2010.

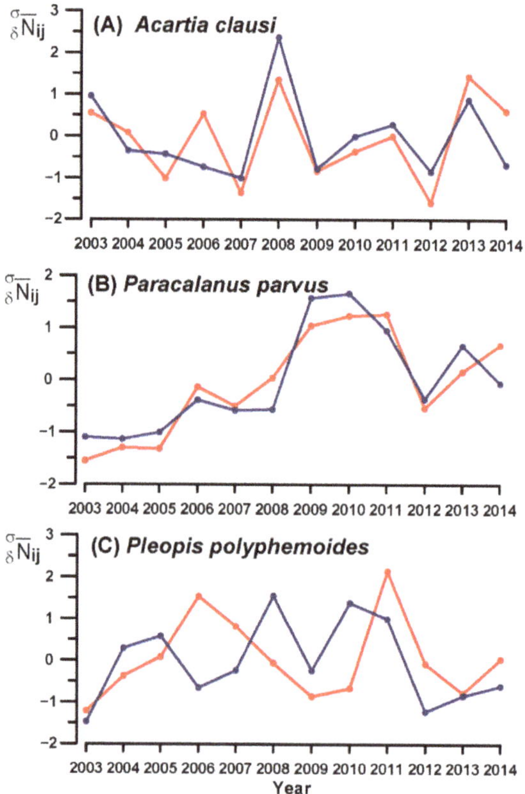

Figure 9. Pattern of interannual variability of abundance of eurythermal crustaceans in Sevastopol Bay at station 1 (blue) and station 2 (red) presented as normalized annual anomalies (see details of the method in Section 2.4): (**A**)—*Acartia clausi*; (**B**)—*Paracalanus parvus*; (**C**)—*Pleopis polyphemoides*.

Positive anomalies of *P. parvus* were observed from January to May 2010 and from October 2010 to February 2011 (Figure 10B). The abundance of *P. parvus* was significantly higher at the mouth of the bay over the study period.

Unlike other species, the extremes of the interannual variability in the abundance of *P. poliphemoides* did not coincide at two stations (Figure 9C). At the mouth of the bay, the maximum of annual average density was in 2010 (865 $\pm$ 374 ind. m^{-3}) followed by 2008 (799 $\pm$ 513 ind. m^{-3}) and 2011 (795 $\pm$ 426 ind. m^{-3}). In the middle of the bay, the most abundant year was 2011 (2027 $\pm$ 1086 ind. m^{-3}), although 2010 was also abundant (1056 $\pm$ 511 ind. m^{-3}) (Table 1). These interannual extremes were largely contributed by strong positive anomalies in abundance observed in May–June at both stations (Figure 10C).

The density of cold-water species *P. elongatus* and O. *similis* was low throughout the study period (Table 1). Their annual average abundance fluctuated from year to year, and the long-term variability of its anomalies had an irregular pattern (Figure 11A,B). A common feature of the long-term variability curves for both species was an increase in the density in 2009 and a decline in 2008 and 2010. (Figure 11A,B). Negative anomalies in the seasonal dynamics of *P. elongatus* and O. *similis* were observed from January to April in 2008 and 2010 (Figure 12A,B).

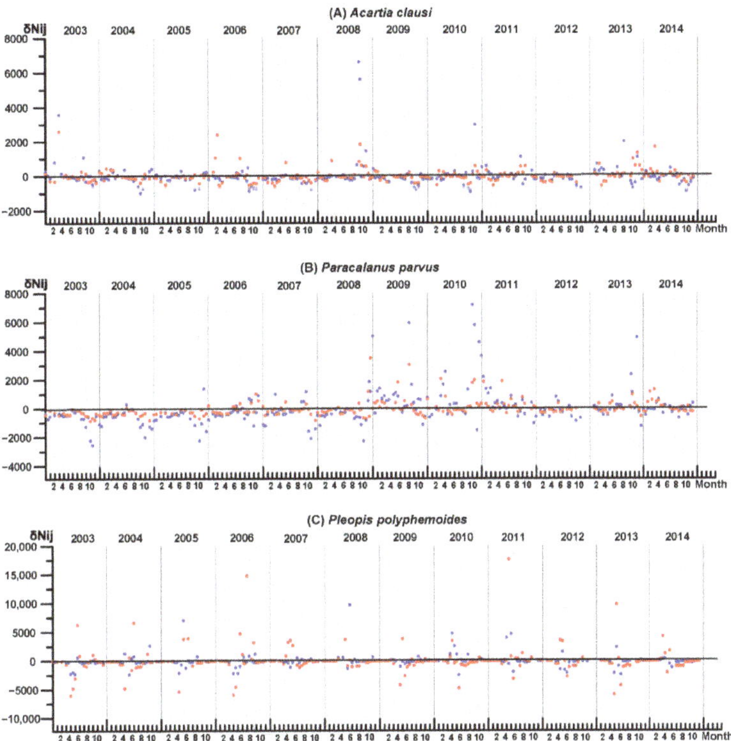

Figure 10. Long-term fluctuations in anomalies of eurythermal crustaceans abundance relative to the annual variation (see details of the method in Section 2.4) in Sevastopol Bay at station 1 (blue) and station 2 (red): (**A**)—*Acartia clausi*; (**B**)—*Paracalanus parvus*; (**C**)—*Pleopis polyphemoides*.

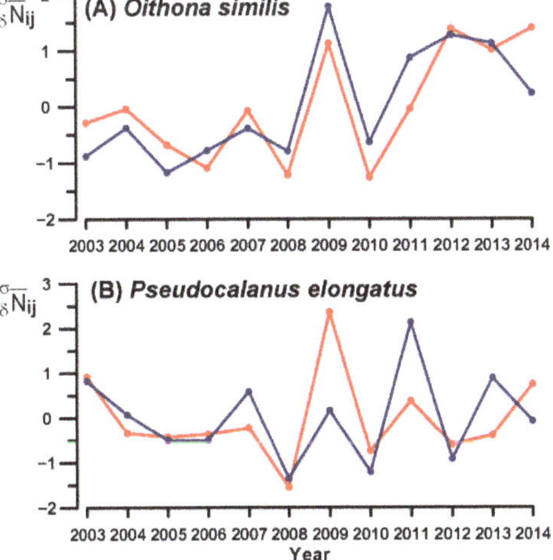

Figure 11. Pattern of interannual variability of abundance of cold-water crustaceans in Sevastopol Bay at station 1 (blue) and station 2 (red) presented as normalized annual anomalies (see details of the method in Section 2.4): (**A**)—*Oithona similis*; (**B**)—*Pseudocalanus elongatus*.

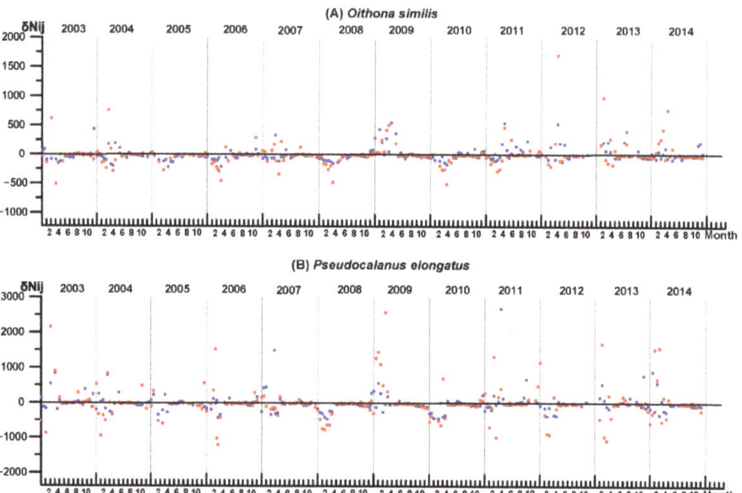

Figure 12. Long-term fluctuations in anomalies of cold-water crustaceans abundance relative to the annual variation (see details of the method in Section 2.4) in Sevastopol Bay at station 1 (blue) and station 2 (red): (**A**)—*Oithona similis*; (**B**)—*Pseudocalanus elongates*.

The annual average abundance of *P. elongatus* ranged between 55 ± 21 ind. m^{-3} and 352 ± 141 ind. m^{-3} at station 1 and 130 ± 40 ind. m^{-3} and 598 ± 228 ind. m^{-3} at station 2. The maximum seasonal abundance of *P. elongatus* occurred in February and March, the strongest peak being observed in March 2009 (about 4000 ind. m^{-3}). The density of *O. similis* varied from 23 ± 9 ind. m^{-3} to 160 ± 46 ind. m^{-3} at station 1 and from 20 ± 7 ind. m^{-3} to 247 ± 151 ind. m^{-3} at station 2. The maximum seasonal abundance of *O. similis* was observed in February and April, the highest value of about 1500 ind. m^{-3} being found in April 2014.

4. Discussion

As a distinct signature of contemporary global warming, the MHWs are increasing in frequency, duration and magnitudes, posing a serious threat for the marine ecosystem. The Black Sea is an example of semi-closed sea experiencing a rapid warming, which is considered an amplified precursor of the changes to expect in the greater oceans [3]. Recent studies have shown that the frequency of MHW in the Black Sea has increased by a factor of five in the last two decades compared with the two previous decades [4].

In this study, we assessed the response of the zooplankton in Sevastopol Bay to the summer 2010 event, which was among the most persistent and intense MHW recorded in the Black Sea. In order to interpret the changes observed in 2010, the patterns of seasonal dynamics and interannual variability in the abundance of crustacean species were analyzed based on a dataset of zooplankton samples collected twice per month between 2003 and 2014.

The analysis of the SST in Sevastopol Bay showed that the study period was the warmest duodecad since at least 1950. Within this period, the SST showed a strong increasing trend between 2003 and 2010. 2010 was the warmest year since 1950 and exhibited positive SST anomalies almost all year round. The summer 2010 MHW starts at the beginning of June and lasts for 3 months, reaching the maximum amplitudes (daily SST anomaly > 4 °C) between the end of July and mid-August. Extreme positive SST anomalies in 2010 led to a sharp increase in the abundance of three warm-water species of crustaceans, namely *O. davisae, A. tonsa and P. avirostris,* and their share in the mesozooplankton community of Sevastopol Bay. An increase in the annual abundance in 2010 was also observed for two eurythermal species *P. parvus* and *P. poliphemoides*, although it is not related to the

MHW event but can be rather explained by warm SST anomalies during the first part of the year. We now discuss in detail the response of each species to the MHV 2010 depending on unique peculiarities of its biology, seasonal dynamics and sensitivity to high temperature.

4.1. Warm-Water Assemblages

The largest contribution to the total abundance of crustaceans in 2010 was made by a non-indigenous species (NIS) of warm-water copepod *Oithona davisae*. This species was discovered in the Black Sea for the first time in October 2005 (it was initially misidentified as *Oithona brevicornis*) [26,27]. Its abundance rapidly increased, and already in 2006, *Oithona davisae* outnumbered other copepods in summer–autumn and on average per year [28]. A logarithmic acceleration phase such as a phase of NIS invasion pattern was observed in 2006–2008, when abundance raised sharply [29,30]. Afterwards, the growth was limited, and density remained at the same level as in 2009–2014, with the exception of 2010. In 2010, the abundance of *O. davisae* was extremely high with a maximum in July–August during the peak of the summer MHW 2010. A general positive trend in *O. davisae* abundance over 2003–2010 coincides with a strong warming SST trend in this period. We suggest that the increase in temperature during this period promoted the rapid growth in the population of *O. davisae*.

The other non-indigenous warm-water copepod *A. tonsa* appeared in the Black Sea in the 1970s [31,32]. It dominated the Sevastopol Bay in summer–autumn to 2006 before colonization of the bay by *O. davisae*. A considerable and statistically significant decline in *A. tonsa* abundance occurred from 2006 to 2014 due to competitive interactions between the two non-indigenous copepods [30]. It is worth noting that the negative effect of the non-indigenous *O. davisae* on populations of *Acartia omori*, *Micosetelle norvegica* and *P. parvus* have been also found in Tokyo Bay [33]. A sharp rise of the *A. tonsa* population in Sevastopol Bay was observed in July and August 2010 as a response to the MHW. The positive correlation of *O. davisae* and *A. tonsa* abundance with temperature was revealed in other areas of the world ocean [34–36].

At a seasonal scale, *P. avirostris* occurred in Sevastopol Bay in May–November and had one pronounced peak in August–September. The same seasonal dynamics was observed in the Mediterranean Vigo and Trieste regions, while in the subtropical highly productive waters of the Arabian Sea (Gulf of Oman), its population persisted all the year round. A regional link between the abundance of this species and temperature was also reflected in Gulf of Oman [37]. In Sevastopol Bay, the *P. avirostris* population was most abundant in 2005 and 2010 when strong positive SST anomalies were reported in August and September. A similar link between the average long-term abundance of *P. avirostris* and SST in August was reported for the coastal waters near Sevastopol [10].

Calanoid copepod *C. ponticus* is endemic to the Mediterranean and Black Sea [38,39]. It is a typical warm-water species that appears in plankton only during the warm season. However, unlike the warm-water species described above, *O. davisae* and *A. tonsa*, its peaks occurred in June and September at 22–23 °C. In July and August, when SST reached its maximum value, *C. ponticus* density declines. The annual average density of *C. ponticus* was relatively low in 2010 in Sevastopol Bay.

4.2. Eurythermal Assemblages

The native eurythermal *Acartia clausi* is one of the most common and numerous copepod in the World Ocean and also in the coastal area of the Black Sea [15,40–42]. In Sevastopol Bay, it was present year-round and reproduced throughout the year. According to long-term routine observations of zooplankton in the coastal area near Sevastopol in 1961–1969, a high abundance of *A. clausi* was observed in the years with negative temperature anomalies [10]. This is rather in line with our data: at the mouth of the bay, *A. clausi* exhibits the positive anomalies in its annual average density in the years when the SST anomalies in the bay was lower than 0.6 °C (2003, 2004, 2006, 2008, 2011 and 2013). In 2010, the annual population of *A. clausi* was close to its long-term average value, showing a seasonal peak in November, when the seasonal temperature drops. Thus, this eurythermal-type species did not show any response to the 2010 MHW.

Our results further suggested that the warm anomalies 2010 affected populations of eurythermal copepods *P. parvus* and *P. polyphemoides*. The highest average annual abundance of these species was reported in 2010 in Sevastopol Bay. The rise in *P. parvus* density was observed from February to June and in September-November 2010. The average annual abundance of *P. polyphemoides* in 2010 was higher than in other years, with positive abundance anomalies in spring and autumn, while in July and August, its abundance was low. Overall, the increase in annual abundance in 2010 of these eurythermal species can be explained by the rise of temperature in the first part of 2010, and it is not related to the summer MHW. Indeed, although both species occurred in the plankton of the bay all year round, their seasonal peaks occur in the spring and autumn and not in July–August when the 2010 MHW occurred. Interestingly, V.N. Grese with co-authors documented a significant summertime abundance of *P. parvus* in the coastal arear near Sevastopol in 1961–1969 [10]. The discrepancy in the seasonal patterns with their study could possibly be explained by the fact that the optimal temperature range for the *P. parvus* population development was reported between 10 and 20 °C, while over the study period, the late summer SST in Sevastopol Bay was near 26 °C, which led to a seasonal population decline.

4.3. Cold-Water Assemblage

Cold-water crustaceans were represented by copepods *P. elongatus* and *O. similis*. *P. elongatus* is common in the temperate eastern North Atlantic Ocean, including the Black Sea and some localities in Mediterranean Sea [43,44]. *O. similis* is cosmopolitan, distributed from tropical to polar waters [36,45]. In the summer, both species stay in the open Black Sea under a thermocline and appear in the surface waters and coastal areas in the cold season. In Sevastopol Bay, these copepods reached their greatest abundance in the first half of the year and were not found in summer. The year 2010 was one of the years characterized by the lowest annual average abundance of the cold-water species within the study period.

4.4. Concluding Remarks

Among all considered species, the most pronounced response to the summer 2010 MHW was observed in the population of non-native warm-water copepods *O. davisae* and *A. tonsa* at both seasonal and interannual scales. These species showed the ability of rapid population growth with rising temperatures. A large number of previous laboratory studies have indicated that increasing temperatures accelerate the development times of eggs and larval stages (nauplii and copepodids) of copepods [46,47]. Apparently, the extreme temperature in 2010 led to a reduction in the generation time of the warm-water species *O. davisae* and *A. tonsa*, resulting in a sharp increase in their abundance.

These NIS have a number of competitive advantages over native species. Their specific biological features ensured its rapid spread across the world ocean, establishment in new habitats, and successful competition with native species [30,48]. Non-native species also exhibit great flexibility as an adaptive response to environmental changes, especially in the case of climate warming.

The current climate changes significantly reduce the resistance of marine ecosystems to disturbance effects, which greatly facilitates the introduction of alien species into new ecosystems, especially into coastal areas [49]. The number of alien species and their abundance has increased in the Black Sea in recent decades. In addition to the described above *A. tonsa* and *O. davisae*, a new non-indigenous copepod *P. marinus* was reported in Sevastopol Bay in September 2016 [16,50]. All these new species are members of the warm-water mesozooplankton assemblage. Observations in other estuaries have also indicated that non-indigenous zooplankton species usually prevail in summer and autumn [34,36,50,51].

Studies of the climate warming impact on the marine ecosystems may be facilitated by the description of the indicator species. Following Reed Noss, the indicator should be (1) sensitive enough to warn of changes in a timely manner; (2) distributed over a wide geographic area or otherwise widely applicable; and (3) able to provide continuous assessment over a wide range of stresses [52]. Our results suggest that future warming may lead to an increase in *O. davisae* dominance in the mesozooplankton community of the Black Sea coastal area and that among crustaceans observed in this study, this species can be considered as an indicator of the environmental conditions associated with the warming of the Black Sea and the Mediterranean basin as a whole.

Author Contributions: Conceptualization, A.G. and K.G.; validation, A.G., K.S. and V.B., formal analysis, A.G., K.G., O.K. and V.B.; investigation, A.G., O.G., E.S. and T.L.; data curation, A.G., O.K.; writing—original draft preparation, A.G., K.G., O.K. and K.S.; writing—review and editing, A.G., K.G., O.K., T.L. and K.S.; supervision, A.G.; project administration, A.G. and O.K.; funding acquisition, A.G. and K.S. All authors have read and agreed to the published version of the manuscript.

Funding: The research was conducted in the frame of the Russian state assignments of Institute of Biology of the Southern Seas No. 121040600178-6; No. 121030100028-0 and of Marine Hydrophysical Institute FNNN-2021-0002 and the National Science Program "Environmental Protection and Reduction of Risks of Adverse Events and Natural Disasters", approved by the Resolution of the Council of Ministers No. 577/17.08.2018 and supported by the Ministry of Education and Science (MES) of Bulgaria (Agreement No. Д01-279/03.12.2021).

Institutional Review Board Statement: Not applicable.

Informed Consent Statement: Not applicable.

Data Availability Statement: Not applicable.

Conflicts of Interest: The authors declare no conflict of interest.

References

1. Oliver, E.C.J.; Donat, M.G.; Burrows, M.T. Longer and more frequent marine heatwaves over the past century. *Nat. Commun.* **2018**, *9*, 1324. [CrossRef] [PubMed]
2. Pörtner, H.O.; Roberts, D.C.; Masson-Delmotte, V.; Zhai, P.; Tignor, M.; Poloczanska, E. Summary for policymakers. In *IPCC Special Report on the Ocean and Cryosphere in a Changing Climate*; IPCC: Geneva, Switzerland, 2019.
3. Stanev, E.V.; Peneva, E.; Chtirkova, B. Climate Change and Regional Ocean Water Mass Disappearance: Case of the Black Sea. *J. Geophys. Res. Ocean.* **2019**, *124*, 4803–4819. [CrossRef]
4. Mohamed, B.; Ibrahim, O.; Nagy, H. Sea Surface Temperature Variability and Marine Heatwaves in the Black Sea. *Remote Sens.* **2022**, *14*, 2383. [CrossRef]
5. Barriopedro, D.; Fischer, E.M.; Luterbacher, J.; Trigo, R.M.; García-Herrera, R. The Hot Summer of 2010: Redrawing the Temperature Record Map of Europe. *Science* **2011**, *332*, 220–224. [CrossRef] [PubMed]
6. Soulié, T.; Vidussi, F.; Mas, S.; Mostajir, B. Functional Stability of a Coastal Mediterranean Plankton Community During an Experimental Marine Heatwave. *Front. Mar. Sci.* **2022**, *9*, 831496. [CrossRef]
7. Richardson, A.J. In hot water: Zooplankton and climate change. *ICES J. Mar. Sci.* **2008**, *65*, 279–295. [CrossRef]
8. Maskas, D.L.; Beaugrand, G. Comparisons of zooplankton time series. *J. Mar. Syst.* **2010**, *79*, 286–304. [CrossRef]
9. Sazhina, L.T. Fertility of the mass pelagic copepoda in the Black Sea. *Zool. Zhurnal* **1971**, *50*, 586–588. (In Russian)
10. Greze, V.N.; Baldina, E.P.; Bileva, O.K. Dynamics of the abundance and production of the principal components of Zooplankton in the Neritic Zone of the Black Sea. *Mar. Biol.* **1971**, *24*, 12–49. (In Russian)
11. Evans, R.; Lea, M.-A.; Hindell, M.A.; Swadling, K.M. Significant shifts in coastal zooplankton populations through the 2015/16 Tasman Sea marine heatwave. *Estuar. Coast. Shelf Sci.* **2020**, *235*, 106538. [CrossRef]
12. McKinstry, C.A.E.; Campbell, R.W.; Holderied, K. Influence of the 2014–2016 marine heatwave on seasonal zooplankton community structure and abundance in the lower Cook Inlet, Alaska. *Deep. Sea Res. Part II Top. Stud. Oceanogr.* **2022**, *195*, 105012. [CrossRef]
13. Zaika, V.E. Marine biodiversity of the Black Sea and Eastern Mediterranean. *Ecol. Sea* **2000**, *51*, 59–62. (in Russian).
14. Benedetti, F.; Ayata, S.D.; Irisson, J.O.; Adloff, F.; Guilhaumon, F. Climate change may have minor impact on zooplankton functional diversity in the Mediterranean Sea. *Divers. Distrib.* **2019**, *25*, 568–581. [CrossRef]
15. Gubanova, A.D.; Altukhov, D.A.; Stefanova, K.; Arashkevich, E.G.; Kamburska, L.; Prusova, I.Y.; Svetlichny, L.S.; Timofte, F.; Uysal, Z. Species composition of Black Sea marine planktonic copepods. *J. Mar. Syst.* **2014**, *135*, 44–52. [CrossRef]
16. Gubanova, A.; Drapun, I.; Garbazey, O.; Krivenko, O.; Vodiasova, E. *Pseudodiaptomus marinus* Sato, 1913 in the Black Sea: Morphology, genetic analysis, and variability in seasonal and interannual abundance. *PeerJ* **2020**, *8*, e10153. [CrossRef]
17. Greze, V.N. Zooplankton. In *Basics of Biological Productivity of the Black Sea*; Greze, V.N., Ed.; Naukova Dumka: Kiev, Ukraine, 1979; pp. 143–164. (In Russian)

18. Finenko, Z. Biodiversity and Bioproductivity. In *The Black Sea Environment. The Handbook of Environmental Chemistry; Water Pollution*; Part, Q., Kostianoy, A.G., Kosarev, A.N., Eds.; Springer: Berlin/Heidelberg, Germany, 2008; Volume 5, pp. 351–374.
19. Garmashov, A. Hydrological characteristics of Sevastopol Bay. *Int. Multidiscip. Sci. Geo Conf. SGEM* **2019**, *19*, 247–252. [CrossRef]
20. Gubanov, V.I.; Gubanova, A.D.; Rodionova, N.Y. Diagnosis of water trophicity in the Sevastopol bay and its offshore. In *Current issues in aquaculture: In Proceedings of the International Scientific Conference, Rostov-on-Don, Russia, 28 September–2 October 2015*; pp. 64–67.
21. Litvinyuk, D.; Mukhanov, V.; Evstigneev, V. The Black Sea Zooplankton Mortality, Decomposition, and Sedimentation Measurements Using Vital Dye and Short-Term Sediment Traps. *J. Mar. Sci. Eng.* **2022**, *10*, 1031. [CrossRef]
22. Donlon, C.J.; Martin, M.; Stark, J.; Roberts-Jones, J.; Fielder, E.; Wimmer, W. The Operational Sea Surface Temperature and Sea Ice Analysis (OSTIA) system. *Remote Sens. Environ.* **2012**, *116*, 140–158. [CrossRef]
23. Postel, L.; Fock, H.; Hagen, W. Biomass and abundance. In *ICES Zooplankton Methodology Manual*; Harris, R.P., Wiebe, P.H., Lenz, J., Skjoldal, H.R., Huntley, M., Eds.; Academic Press: London, UK, 2000; pp. 83–174.
24. Aleksandrov, B.; Arashkevich, E.; Gubanova, A.; Korshenko, A. Black Sea monitoring guidelines mesozooplankton. *Publ. EMBLAS Proj. BSC* 2014. Available online: http://www.blacksea-commission.org/Downloads/Mesozooplankton_Manual_2015_ISBN%20%20978-617-7953-33-2.pdf (accessed on 15 October 2020).
25. Hobday, A.J.; Alexander, L.V.; Perkins, S.E.; Smale, D.A.; Straub, S.C.; Oliver, E.C.J.; Benthuysen, J.A.; Burrows, M.T.; Donat, M.G.; Feng, M.; et al. A hierarchical approach to defining marine heatwaves. *Prog. Oceanogr.* **2016**, *141*, 227–238. [CrossRef]
26. Gubanova, A.; Altukhov, D. Establishment of *Oithona brevicornis* Giesbrecht, 1892 (Copepoda: Cyclopoida) in the Black Sea. *Aquat. Invasions* **2007**, *2*, 407–410. Available online: http://www.aquaticinvasions.net/2007/index4.html (accessed on 15 May 2007). [CrossRef]
27. Temnykh, A.; Nishida, S. New record of copepod *Oithona davisae* Ferrari and Orsi in the Black Sea with notes on the identity of Oithona brevicornis. *Aquat. Invasions* **2012**, *7*, 425–431. [CrossRef]
28. Altukhov, D.; Gubanova, A.; Mukhanov, V. New invasive copepod *Oithona davisae* Ferrari and Orsi, 1984: Seasonal dynamics in Sevastopol Bay and expansion along the Black Sea coasts. *Mar. Ecol.* **2014**, *35*, 28–34. [CrossRef]
29. Odum, E. *Fundamentals of Ecology*; Saunders, W.B., Ed.; Saunders: Philadelphia, PA, USA; Moscow, Russia, 1975.
30. Gubanova, A.D.; Garbazey, O.A.; Popova, E.V.; Altukhov, D.A.; Mukhanov, V.S. *Oithona davisae*: Naturalization in the Black Sea, interannual and Seasonal dynamics, and effect on the structure of the planktonic copepod community. *Oceanology* **2019**, *59*, 912–919. [CrossRef]
31. Belmonte, G.; Mazzocchi, M.G.; Prusova, I.Y.; Shadrin, N.V. *Acartia tonsa*: A species new for the Black Sea fauna. *Hydrobiologiya* **1994**, *292*, 9–15. [CrossRef]
32. Gubanova, A.D. Occurrence of *Acartia tonsa* Dana in the Black Sea. Was it introduced from the Metiterranean? *Mediterr. Mar. Sci.* **2000**, *1*, 105–109. [CrossRef]
33. Itoh, H.; Nishida, S. Spatiotemporal distribution of planktonic copepod communities in Tokyo Bay where Oithona davisae ferrari and orsi dominated in mid-1980s. *J. Nat. Hist.* **2015**, *49*, 2759–2782. [CrossRef]
34. Uriarte, I.; Villate, F.; Iriarte, A. Zooplankton recolonization of the inner estuary of Bilbao: Influence of pollution abatement, climate and non-indigenous species. *J. Plankton Res.* **2016**, *38*, 718–731. [CrossRef]
35. Kimmel, D.G.; Roman, M.R. Long-term trends in mesozooplankton abundance in Chesapeake Bay, USA: Influence of freshwater input. *Mar. Ecol. Prog. Ser.* **2004**, *267*, 71–83. [CrossRef]
36. Besiktepe, S.; Kurt, T.T.; Gubanova, A. Mesozooplankton composition and distribution in Izmir Bay, Aegean Sea: With special emphasis on copepods. *Reg. Stud. Mar. Sci.* **2022**, *55*, 102567. [CrossRef]
37. Piontkovski, S.A.; Fonda-Umani, S.; De Olazabal, A.; Gubanova, A.D. *Penilia avirostris*: Regional and Global Patterns of Seasonal Cycles. *Int. J. Ocean. Oceanogr.* **2012**, *6*, 9–25.
38. Sazhina, L.I.; Kovalev, A.V. About the synonymy of the Black Sea copepods. *Zool. J.* **1971**, *50*, 1099–1101. (In Russian)
39. Papantoniou, G.; Danielidis, D.B.; Spyropoulou, A.; Fragopoulu, N. Spatial and temporal variability of small-sized copepod assemblages in a shallow semi-enclosed embayment (Kalloni Gulf, NE Mediterranean Sea). *J. Mar. Biolog. Assoc. UK* **2015**, *95*, 349–360. [CrossRef]
40. Azeiteiro, U.M.; Marques, S.C.; Vieira, L.M.R.; Pastorinho, M.R.D.; Ré, P.A.B.; Pereira, M.J.; Morgado, F.M.R. Dynamics of the Acartia genus (Calanoida: Copepoda) in a temperate shallow estuary (the Mondego estuary) on the western coast of Portugal. *Acta Adriat.* **2005**, *46*, 7–20.
41. Jeffries, H.P. Succession of two Acartia species in estuaries. *Limnol. Oceanogr.* **1962**, *7*, 354–364. [CrossRef]
42. Lee, W.Y.; McAlice, B.J. Seasonal succession and breeding cycles of three species of *Acartia* (Copepoda: Calanoida) in a Marine estuary. *Estuaries* **1979**, *2*, 228–235. [CrossRef]
43. Frost, B.W. A taxonomy of the marine calanoid copepod genus *Pseudocalanus*. *Can. J. Zool.* **1989**, *67*, 525–551. [CrossRef]
44. Razouls, C.; de Bovée, F.; Kouwenberg, J.; Desreumaux, N. Diversity and Geographic Distribution of Marine Planktonic Copepods, 2005–2022. Available online: http://copepodes.obs-banyuls.fr/en (accessed on 18 November 2022).
45. Nishida, S. Taxonomy and distribution of the family Oithonidae (Copepoda, Cyclopoida) in the Pacific and Indian oceans. *Bull. Ocean Res. Inst. Univ. Tokyo* **1985**, *20*, 1–167.
46. Vijverberg, J. Effect of temperature in laboratory studies on development and growth of Cladocera and Copepoda from Tjeukemeer, The Netherlands. *Freshw. Biol.* **1980**, *10*, 317–340. [CrossRef]

47. Huber, V.; Adrian, R.; Gerten, D. A matter of timing: Heat wave impact on crustacean zooplankton. *Fresh Water Biol.* **2010**, *55*, 1769–1779. [CrossRef]
48. Cornils, A.; Wend-Heckmann, B. First report of the planktonic copepod *Oithona davisae* in the northern Wadden Sea (North Sea): Evidence for recent invasion? *Helgol. Mar. Res.* **2015**, *69*, 243–248. [CrossRef]
49. Rice, E.; Dam, H.G.; Stewart, G. Impact of climate change on estuarine zooplankton: Surface water warming in long Island sound is associated with changes in copepod size and community structure. *Estuaries Coasts* **2015**, *38*, 13–23. [CrossRef]
50. Uttieri, M.; Aguzzi, L.; Aiese Cigliano, R.; Amato, A.; Bojanić, N.; Brunetta, M.; Camatti, E.; Carotenuto, Y.; Damjanović, T.; Delpy, F.; et al. WGEUROBUS—Working Group "Towards a European Observatory of the non-indigenous calanoid copepod *Pseudodiaptomus marinus*". *Biol. Invasions* **2020**, *22*, 885–906. [CrossRef]
51. Bollens, S.M.; Breckenridge, J.; Cordell, J.R.; Simenstad, C.; Kalata, O. Zooplankton of tidal marsh channels in relation to environmental variables in the upper San Francisco Estuary). *Aquat. Biol.* **2014**, *21*, 205–219. [CrossRef]
52. Noss, R.F. Indicators for Monitoring Biodiversity: A Hierarchical Approach. *Conserv. Biol.* **1990**, *4*, 355–364. [CrossRef]

Journal of
Marine Science and Engineering

MDPI

Article

Ecosystem Variability along the Estuarine Salinity Gradient: A Case Study of Hooghly River Estuary, West Bengal, India

Diwakar Prakash [1], Chandra Bhushan Tiwary [2] and Ram Kumar [1,*]

1 Ecosystem Ecology Research Unit, Department of Environmental Science, School of Earth, Biological and Environmental Sciences, Central University of South Bihar, SH-7, Gaya-Panchanpur Rd, Fatehpur, Gaya 824326, Bihar, India
2 Department of Zoology, Vidya Bhavan Mahila Mahavidyalay, Siwan 841226, Bihar, India
* Correspondence: ramkumar@cub.ac.in

Abstract: Hooghly River, a ~460 km long distributary of the Ganga River, passes through a highly industrialized Metropolis-Kolkata in West Bengal, India, and eventually empties into the Bay of Bengal at Gangasagar. To determine the patterns and drivers of planktonic community, spatiotemporal variations in water quality and micronutrient content and planktic prokaryotic and microeukaryotic abundance and diversity across the salinity gradient (0.1 to 24.6 PSU) in the Hooghly River estuary (HRE) were studied. Plankton and water samples were collected at six sites during October 2017, February 2018, and June 2018. The biotic parameters—phytoplankton (Chlorophyll *a*), total bacterial abundance (cfu), and copepods—were significantly higher in the downstream estuarine sites than in the upstream riparian sites; conversely, rotifer and cladoceran abundances were significantly higher at upstream stations. The most culturable bacterial strains were isolated from the two freshwater sites and one at the confluence (estuarine) and are characterized as *Bacillus subtilis*, *Pseudomonas songnenesis*, and *Exiguobacterium aurantiacum*. Among zooplankton, rotifers (0.09 ± 0.14 ind L^{-1}) and cladocerans (5.4 ± 8.87 ind L^{-1}) were recorded in higher abundance and negatively correlated with bacterial concentrations at upstream stations. On the temporal scale, February samples recorded lower proportions of bacterivorous zooplankton at the three upstream stations. Cluster analysis separated samples on the basis of seasons and water mass movement. The February samples showed distinct spatial characteristics, as three freshwater (FW) stations grouped together and segregated at second 2nd hierarchical level, whereas the three estuarine stations formed a separate cluster at the 50% similarity level. Samples collected in October 2017 and June 2018 exhibited mixed attributes. June samples recorded higher influence of freshwater discharge. The zooplankton abundance showed significant negative correlation with Chl *a*. Our results demonstrate the relative role of river continuum, land-driven lateral discharge, and seawater intrusion in shaping community structure, which needs to be considered in management and conservation planning of aquatic ecosystems, especially in highly productive and overexploited HRE.

Keywords: bacteria; estuary; river; plankton; trophic structure

Citation: Prakash, D.; Tiwary, C.B.; Kumar, R. Ecosystem Variability along the Estuarine Salinity Gradient: A Case Study of Hooghly River Estuary, West Bengal, India. *J. Mar. Sci. Eng.* **2023**, *11*, 88. https://doi.org/10.3390/jmse11010088

Academic Editors: Marco Uttieri, Ylenia Carotenuto, Iole Di Capua and Vittoria Roncalli

Received: 24 November 2022
Revised: 20 December 2022
Accepted: 26 December 2022
Published: 3 January 2023

1. Introduction

Estuaries are highly productive and dynamic semi-enclosed waterbodies linked to the sea either permanently or periodically and fed by freshwater from river inputs, resulting in a distinct salinity gradient and characteristics biota [1–5]. The complexity in a river estuary is determined by the variability in river water mixing with sea water, resulting in salinity [6], turbidity, and nutrient gradients [7–10]. There has been a long debate about the functioning of estuaries [11–13]; the community structure of phytoplankton, bacterioplankton, and zooplankton; and their relationship and their co-occurrence pattern [14,15]. The common consensus is that planktic communities play a key role in maintaining the ecological functioning of an estuary [16].

The Hooghly River is one of the most important estuarine systems in India because of the discharge from a vast river basin with substantial monsoonal precipitation (70,500 $m^3 \cdot s^{-1}$ peak flow at Farakka), its origin from the largest montane river (~2600 km), and its long tidal zone (~280 km). Being an active tidal estuary, it has distinct biological and physico-chemical characteristics [17]. The commissioning of Farakka Barrage in 1975 facilitated the adequate quantity of Ganga water in the Bhagirathi-Hooghly River system, improving the ecosystem health and riverine-estuarine biodiversity [18–21], finally manifesting as seaward pushing the salinity zone in the estuary [22]. Fish species such as *Rita rita*, *Sperata seenghala*, *Eutropiicthys vacha*, *Wallago attu*, *Clupisoma garua*, *Labeo calbasu*, and *Catla catla* have made their emergence in the upper zone of the Hooghly River estuary (HRE), namely Tribeni and Banlagarh [23], which were reported from this zone before 1975, i.e., prior to the commissioning of the barrage. HRE provides valuable a nursery and recruitment habitat for commercially important species, such as Hilsa, finfish, and shrimp [24,25].

Any short-term or long-term changes are immediately reflected by the change in planktonic community [26,27], as they are self-sustaining, constituting the important components of the microbial loop while channeling carbon and energy from microbes to higher trophic levels by joining the classical food web [14,28–30] in aquatic ecosystems. The microbial loop explains pathway of carbon flow through nutritional food web that begins with dissolved organic matter (DOM) and reaches to the highest trophic levels bypassing some and passing through various trophic levels. The main stakeholders of the microbial loop include bacteria, zooplankton, phytoplankton, and other nutrient-cycling organisms [14,30,31]. The relative densities of bacterivorous, herbivorous, carnivorous, and omnivorous zooplankton are a reliable indicator of the functioning of the microbial loop and of ecosystem health on the spatial scale [32,33]. The zooplankton community comprises diverse feeding groups, such as bacterivores, detritivores, herbivores, and carnivores [34–36], forming a bridge between the microbial loop and classical food web. Information concerning co-occurrence, distribution, and community composition of the prokaryotic and eukaryotic plankton in the HRE is lacking [37,38]. The spatiotemporal variations of planktonic communities are highly affected by the hydrochemical parameters and physical forces [26,27,39–41]. Therefore, major components of microbial loop, i.e., bacterioplankton, phytoplankton and zooplankton, are likely to be affected by these activities. However, their co-occurrence and distributional patterns have not been studied in the HRE at a spatial scale ranging from fresh water to the estuary mouth. The knowledge of bacterioplankton–zooplankton co-occurrence is very essential, as zooplankton might act as a biotic selector for a specific microbial loop [42]. Bacterioplankton in the present study include culturable strains only because isolation of microbes is still necessary for the extraction of bioactive compounds [43], and this is accomplished by culture-based technique. Descriptions of new taxa of prokaryotes and experimental validation of microbial, ecological, and evolutionary processes are reliably based on culture based techniques. Therefore, this study isolated culturable bacterial strains and concentrated on culture-based methods.

The present study aims to elucidate the heterogeneity and co-occurrence of planktic community along the salinity gradient ranging from freshwater to the estuary mouth and ecological drivers shaping the planktic community structure in the HRE. To achieve these objectives, the study identified the ubiquitous nature and heterogeneity in distribution patterns of aquatic biological communities, including bacterioplankton, phytoplankton, and zooplankton (Rotifera, Cladocera, and Copepoda), at spatial scales during October 2017, February 2018, and June 2018 in the HRE, India. The samples were collected at six sites along the main salinity axis (0.1 to 25 PSU), from Barrackpore before the metropolis Kolkata to the estuary mouth. To elicit whether the co-occurrence or abiotic parameters are responsible for differential distribution patterns, we estimated water quality and micronutrient concentrations at all the six sites. At stations where higher correlation coefficient values for Bacteria *vs.* Rotifera abundance (($R = -0.76$) were recorded, we isolated the bacterial strain with >50% occurrence for further characterization.

2. Materials and Methods

2.1. Study Site

The HRE is a part of Ganga River system that originates from Bhagirathi (upper stretch), flows southwards through the lower Ganga deltaic plane, and merges with the Bay of Bengal in Sundarbans as the Hooghly River in West Bengal, India (Figure 1). Kolkata city, one of the largest metropolises along the Hooghly River, having a population of about 14.5 million, utilizes the river water for drinking and domestic and industrial purposes and also discharges sewage and sludge into the river [44]. Beginning upstream of the metropolitan city of Kolkata and downstream to the confluence, six sampling sites were chosen comprising the agricultural-industrial-anthropogenic and riverine (salinity: 0.1 to 0.45 PSU)-estuarine (salinity: 4.32 to 24.6 PSU) salinity gradient along the HRE (Figure 1).

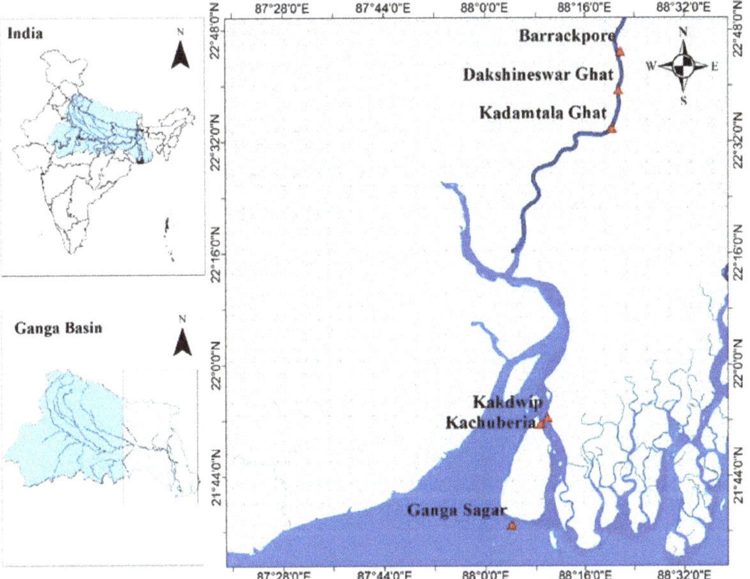

Figure 1. Geographic distribution of sampling sites along the Hooghly River estuary. Red triangles indicate the location of six sampling stations from Barrackpore to the confluence at Gangasagar (Bay of Bengal).

Sampling of surface water was performed at three riverine and three estuarine sites during the post-monsoon season, i.e., October 2017 and spring–February 2018, the and pre-monsoon season, i.e., June 2018, in the HRE (Figure 1). Samples were collected in cleaned polypropylene bottles from the surface to 20 cm depth for physio-chemical, bacteriological, phytoplankton, and zooplankton analysis. The details of sampling sites, abbreviations used hereafter, and common stressors at each site are provided in Table 1.

2.2. Environmental Parameters

In total, we collected 72 surface water samples in pre-cleaned, acid-washed polythene bottles for water quality analyses during the three sampling cruises. The salinity was determined by Hanna Instruments HI98319 marine salinity tester, as salinity has influence on the demography of zooplankton [45]. For the determination of dissolved oxygen, the Winkler titration method was used. Surface water temperature was measured using a mercury glass thermometer. The flow rate was measured by mechanical flow meter (hydrobios model 438110) The dissolved nutrients (nitrate and phosphate) were estimated by colorimetric methods (following [46]) using a spectrophotometer (PerkinElmer UV/VIS Spectrometer Lambda 25, Waltham, MA, USA) after filtering the water through 0.45 μm filter paper (GF/F-Whatman, Maidstone, UK) within 12 h of sampling, and the filtered waters were stored in 100 mL pre-cleaned, acid-washed polythene bottles at 4 °C. All the parameters were analyzed following the standard procedures for water sampling and examination of water quality [47].

Table 1. Ecological stressors, abbreviations used, and coordinates of the six stations in the HRE sampled during October 2017, February 2018, and June 2018 for the present study.

Sampling Stations (Code)	Coordinates	Altitude (ft) ASL	Salinity (PSU) Trophic Status N:P Ratio	Ecological Stresses
Barrackpore (BRK)	22.75272° N 88.36212° E (Gandhighat)	13.12	0.1–0.42 13:0.04	Industrial effluents, domestic sewage disposal, boating, bathing, occasional Immersion of idols
Dakshineshwar (DKS)	22.65643° N 88.35682° E (Dakshineshwarghat)	9.8	0.14–0.45 (18:0.092)	This site is 128 km away from the sea mouth of the river and has an estuarine condition due to significant tidal oscillation of ∼3 m. Here, the river flows through the densely populated region in Kolkata city. Mostly untreated sewage disposes into river water near Dakshineshwar ghat. The river water is also accessed for washing, bathing, and for many religious rituals
Kadamtala (KDM)	22.565° N 88.3387° E (Kadamtalaghat)	3.2	0.13–0.45 (14:0.052)	Bathing, Washing clothes, domestic effluents, ferry service, spiritual rituals, immersion of idols, oil leaching, leakage of oil from mechanized boat.
Kakdwip (KDP)	21.87208° N 88.16383° E (Harwood Point Ferry service)	0	4.32–10.64 (43:0.10)	Frequent dredging, boating, fishing, etc.
Kachuberia (KCB)	21.85903° N 88.14433° E (Kachuberiaghat Near govt. Jetty Gangasagar)	0	6.29–17.79 (60:0.18)	Frequent dredging, boating, fishing, etc.
Gangasagar (GS)	21.63307° N 88.07498° E (Gangasagar Mohana sea Beach)	0	13.09–24.6 (74:0.33)	Boating, tourist activities, dredging, fishing

2.3. Bacteriological Analysis

2.3.1. Enumeration, Isolation, and Characterization of Culturable Bacterial Strains

To study the bioactive potential for further prospecting, we estimated culturable bacterial concentrations and isolated the most culturable strains from three different sites. Five replicates of the water samples were collected in sterile polypropylene bottles from the surface to 15–20 cm deep and transported to the laboratory at 4 °C in an icebox (Table 1; Figure 1) during October 2017, February 2018, and June 2018. In the laboratory, water samples were stored at −21 °C until further processing. In the laboratory, bacterial concentrations were estimated by direct plate count method following [48]. The surface water samples, collected from different sites along an anthropic gradient in the HRE, were spread on media plates. Total bacterial density (colony forming unit: CFU mL^{-1}) was enumerated on nutrient agar plate, which was incubated at 37 °C for 24 ± 1.5 h [48]. The bacterial counts obtained were used to estimate the number of bacteria grown on the media plates used for DNA extraction. Colonies with different morphologies were subcultured into pure cultures by inoculating them into freshly prepared agar plates [49]. The most abundant colony at each of the three sites was recorded for subsequent identification and statistics. At two sites, BRK and DKS were recorded for relatively higher abundance of bacterivorous zooplankton and higher strength of association between bacterioplankton and zooplankton. With an aim to identify highly abundant bacterial colony at these three sites (BRK, DKS, and GS), we isolated the bacterial colony with 50–70% occurrence and further cultured for sequencing and phylogenetic analysis [48,50].

2.3.2. Clustering, Alignment, and Phylogenetic Analysis of 16S rRNA Gene Fragments in Most Culturable BRK2, DKS, and GS1 Bacterial Strains

DNA from the bacterial culture—BRK2, DKS, and GS1 strains—was isolated using the bacterial gDNA isolation kit (XcelGen, Gujarat, India). Isolation of DNA was carried out according to manufacturer's instructions. First, 1.2% agarose gel was used to evaluate the quality of isolated DNA, and a single band of high-molecular band of the PCR amplicon was detected (Figure S1). Amplification of isolated DNA was performed with 16S rRNA-specific primer (8F and 1492R) using Veriti® 96 well thermal cycler (Model No. 9902, Thermo Fisher Scientific, Waltham, MA, USA). Sanger sequencing was performed using BDT v3.1 Cycle sequencing kit with M13F and M13R primers on ABI 3730xl Genetic Analyzer was performed after the PCR amplicon was enzymatically purified. A consensus sequence of 1284, 1465 bp, and 1487 of 16S rRNA was generated by using aligner software from forward- and reverse-sequence data. The consensus sequence of all the three strains, accession numbers, and origin are shown as Table S1. All nucleotide sequences were deposited at the National Center for Biotechnology Information (NCBI) strain library with accession numbers provided in (Table S1).

The evolutionary history was inferred by using the maximum likelihood method and Tamura–Nei model [51]. The tree with the highest log likelihood (−10,338.41, −10,518.13, −10,395.19) is shown, respectively, for BRK, DKS, and GS strain. Initial tree(s) for the heuristic search were obtained automatically by applying neighbor-joining and BioNJ algorithms to a matrix of pairwise distances estimated using the Tamura–Nei model and then selecting the topology with superior log likelihood value. This analysis involved 61 nucleotide sequences. There were a total of 1284, 1465, and 1484 positions in the final dataset of each strain (BRK, DKS, and GS). Evolutionary analyses were conducted in MEGA11 [52].

2.4. *Phytoplankton Analysis*

For qualitative analyses, five replicates of 1 L surface water samples were preserved in neutral Lugol's solution (1% Lugol's solution and 4% formalin) and brought to the laboratory for species identification. In the laboratory, samples were concentrated 10 times by centrifugation, and algal cells were observed on a Sedgewick rafter cell under a compound microscope (10×–400× magnification), and abundant species were identified using the standard key [53–57] and AlgaeBase (www.algaebase.org (accessed on 12 July 2018)) [56]. The amount of primary productivity was estimated in terms of chlorophyll *a* (Chl *a*). Chlorophyll pigment was analyzed through extraction using a mixture of dimethyl sulfoxide and 90% acetone [58] and enumerated by spectrometry using a Turner TD-700 fluorimeter (New York, NY, USA) following the standard method [46,59].

2.5. *Zooplankton Analyses*

Zooplankton samples were collected by making surface tows (0–20 cm) with a customized plankton net with 53 μm mesh size, 0.25 m mouth diameter, and preserved in 4% (*w*/*v*) buffered formalin immediately after collection in a 100 mL transparent bottle. At each site, 100 L water was filtered in five replicates. In the laboratory, the plastic bottles containing preserved zooplankton were thoroughly mixed, and a 1 mL subsample was drawn with a fine pipette to a Sedgwick–Rafter plankton counting cell for enumeration under the compound microscope (model no: Olympus CX21LED, Bartlett, TN, USA). The numbers per liter of each genus was quantified and calculated using the following formula:

$$N = \frac{A \times C}{L}$$

where N denotes the number of plankton per liter, A is the average number of plankton in all counts, C is the volume of original concentrate in ml, and L is the volume of original water filtered, expressed in liters. Zooplankton species were identified to their lowest possible taxon [60–66].

Trophic-Based Zooplankton Community Analysis

Based on the published literature (Supplementary Table S5) and our own observations on propensity of feeding [35,67–71], the zooplankton communities identified at each station were characterized on the basis of functional feeding mode. Different fractions of zooplankton representing bacterivorous, herbivorous, carnivorous, and omnivorous types were segregated following standard literature [34–36,69,71–78] and analyzed separately.

2.6. *Data Analysis*

To elicit variations at spatio-temporal scales, the similarities of community composition among the sampling stations for each sampling date and also among sampling dates were compared. We first determined the centroid vector that represents the average composition of the group/species. Spatial heterogeneity was estimated using the mean and standard deviation of the similarities from the estimated similarity vector. We calculated the Bray–Curtis index to characterize the dissimilarities between samples (β-diversity). Square-root-transformed species abundance data were used for constructing the Bray–Curtis matrix of dissimilarity with average linkages group classification [79]. As the Bray–Curtis similarity mixes the differences due to species losses and species turnover, we also partitioned this index to understand both components of dissimilarity. For abiotic parameters, the distance between two samples was measured by Euclidean distance (ED), as ED is more appropriate for a low-dimensional data set [79].

To characterize the zooplankton diversity present in each sample (α-diversity), we calculated the Shannon diversity index (H′) ($H' = -\text{Sum}(Pi \times \text{Log}(Pi))$) (Shannon, 1948), evenness index (J′) ($J' = H'/\text{Log}(S)$) [80], and species richness (d) ($d = S - 1/\text{Log}(N)$) [81]. To determine the variations among samples, non-metric multidimensional scaling (NMDS) ordination was computed based on Euclidean distance [82]. To identify the drivers of species abundance, pairwise correlation of water quality, and biotic parameters, the degree of a linear association between any two of the parameters was measured using Pearson's correlation coefficient (R). To test the distribution of data, the Shapiro–Wilk test of normality was used, and outliers were detected using scattering plot prior to Pearson's correlation analysis. Highly correlated parameters that may influence the community structure were identified. Indexes of dissimilarity, Shannon's index (α-diversity), Pielou's index (evenness), and Euclidean distance were calculated with PRIMER-version 6.0.

3. Results

3.1. *Spatio-Temporal Patterns*

All the estimated abiotic (Figure 2) and biotic (Figure 3) parameters except DO level, rotifer (Figure 3B), and cladoceran abundance, showed significant seaward increase; in contrast, the rotifer abundance showed a significantly seaward decreasing trend ($R^2 = 0.6$; $p < 0.0001$).

3.2. *Physicochemical Parameters*

The surface water temperature ranged from 26–29 °C with an average of 27 °C and recorded a significant ($R^2 = 0.3$; $p < 0.02$) seaward increase (Figure 2A). The highest average concentration of Ca^{++}, Na^{+}, and K^{+} was found at the mouth (Figure 2A–C; Table S2). The trophic level-related parameters (nitrate, phosphate, bacterial concentration, and Chl *a*) showed higher values in the estuarine stations (Table S2; Figure 2J,K); however, the mean Chl *a* level was the highest at KCB station preceding the GS (Figure 3C). The dissolved oxygen concentration ranged from 6 to 8.2 mg·L^{-1} but had disorderly spatial variation. The flow rate of the Hooghly River was recorded 0.3 m·s^{-1} in February and 1 m·s^{-1} in August.

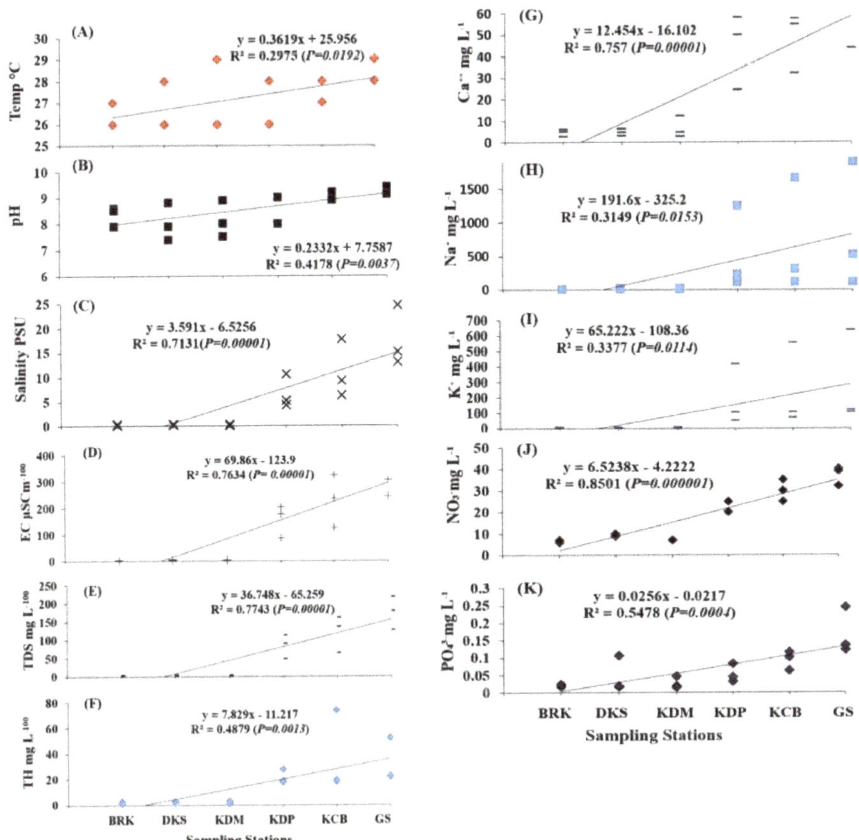

Figure 2. Seaward trends of selected physical parameters estimated for the present study at six sampling sites (Table 1) during October 2017, February 2018, and June 2018, including (**A**) temperature, (**B**) pH, (**C**) salinity, (**D**) electrical conductivity, (**E**) total dissolved solids, (**F**) total hardness, (**G**) calcium, (**H**) sodium, (**I**) potassium, (**J**) nitrate, and (**K**) phosphate.

3.3. Phytoplankton

The phytoplankton species recorded in the sampling stretch were *Pediastrum*, *Spirogyra*, *Coscinodiscus*, *Cyclotella*, *Melosira*, *Ankistrodesmus*, *Aulacoseira*, *Coelastrum*, *Microcystis*, *Oscillatoria*, *Anabaena*, *Aphanocapsa*, *Coscinodiscus radiatus*, *Pleurosigma formosum*, *Coscinodiscus lineatus*, *Biddulphia sinensis*, and *Chaetoceros lorenzianus* (Table S3). The mean Chl *a* concentration varied from 29.1 mg L^{-1} to 219.9 mg L^{-1}, showing significant increase towards the river plume; the mean Chl *a* concentration was the highest at KCB station preceding the confluence GS (Figure 3C). The Chl *a* values showed positive correlation with all the abiotic parameters; however, a significant positive correlation was recorded with nitrate (R = 0.79) and Ca^{++} ion concentration (R = 0.76). With biotic components, the significant negative correlation was recorded between Chl *a* values and zooplankton abundance (Table S6).

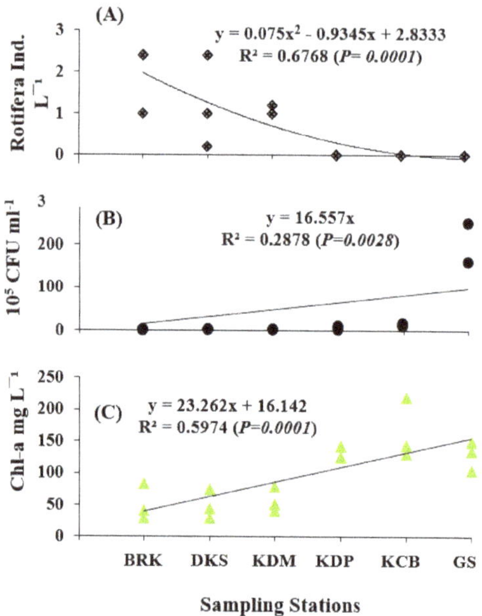

Figure 3. Seaward trends of selected biological parameters estimated for the present study at six sampling sites (Table 1) during October 2017, February 2018, and June 2018 including (**A**) Rotifera, (**B**) bacterial density, and (**C**) chlorophyll-*a*.

3.4. Zooplankton Community Structure

The symmetric map of all the parameters estimated, in rows and columns in principal coordinates, is given in Figure 4, in which the response category points to separate stations on an ordinal scale. Looking at the spatial scale with respect to the horizontal principal axis, all the zooplankton community at all the riverine stations aggregated together on the right side, whereas the last station at the confluence was set aside from other stations and positioned on the left, showing higher variation among sampling seasons (Figure 4).

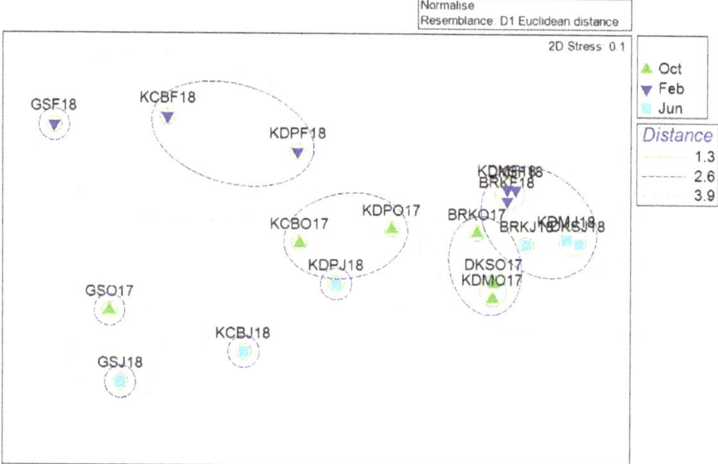

Figure 4. Non-metric multidimensional scaling (NMDS) of all the 18 samples collected from HRE, measured by the Euclidean distance.

The right extreme position of all the riverine samples and their unique position in the middle of the second (vertical) axis indicates that the responses are more in the intermediate categories of the scale rather than a mixture of extreme responses at the temporal scale (Figure 4). The scale values of these samples optimally discriminate between the 18 samples (6 stations × 3 seasons), giving maximum between-sample variance. The second dimension then separates out samples on the basis of seasons, and all the estuarine samples collected in February are polarized towards the top. Upstream estuarine stations (KCB and KDP) aggregated inside, where both extremes of the response scale as well as the missing response are located (Figure 4). As a result, samples were arranged in the ordination of downstream confluence to freshwater stations from left to right. All KDP samples and KCB October samples are in a unique position inside, with relatively high polarization of responses and high missing values. Their position in the middle of both axes reveals more responses in the scale's intermediate categories rather than a mixture of extreme responses at the spatiotemporal dimension. Figure 4 depicts the principal inertias at the positive ends of each axis, which were measured by adding together the percentages of inertia, i.e., 63.1% + 13.3% = 76.4%. This shows a "residual" of 23.6%, which is not shown in the map.

The unique right-side positioning of all the riverine stations may be attributed to the presence of rotifers (0.09 ± 0.14) and cladocerans (5.4 ± 8.87) in dL^{-1}, captured at upstream freshwater stations only, whereas copepods were present at all stations (Figure 5A). Integrating all zooplankton samples (Figure 5A; Table S4) were dominated by the Copepoda (92%) followed by Cladocera (7%) and Rotifera (1%). Total zooplankton abundance was more affected by seasons and showed a disorderly distribution among stations. Zooplankton density showed a peak in February (beginning of spring) at all sampling stations except KDM, where the peak was recorded in June samples (Figure 5A). The peak of the total zooplankton abundance was mainly contributed by the copepods at all stations (Figure 5A). At the three riverine stations, the rotifer densities were significantly lower during the peak of the total zooplankton abundance (Figure 5A). The indices of diversity, richness, and evenness of zooplankton recorded at six selected stations in the HRE are provided in (Figure 5B–D).

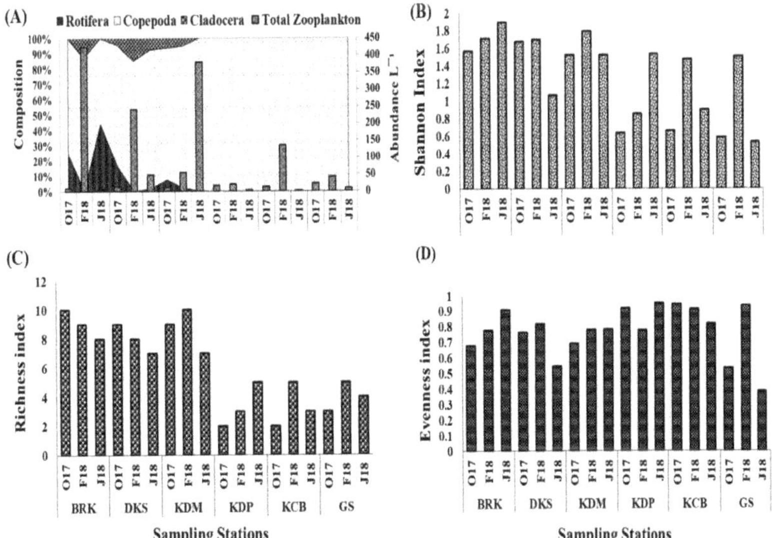

Figure 5. Percent (%) composition of Rotifera, Cladocera, Copepoda, and total zooplankton abundance (L^{-1}) (**A**), Shannon–Wiener index of diversity (**B**), species richness (**C**), and Simpson index for evenness (**D**) of zooplankton, recorded during October 2017, February 2018, and June 2018 at six selected stations in the HRE.

The heterogeneity at the spatial scale was estimated by the Bray–Curtis similarity matrix, which was observed to be higher than the heterogeneity at the temporal scale (Figure 6). It shows that the average similarity between all sampling dates was higher than the average similarity between the stations on a single sampling date. The cluster analysis gives further insights into the relative role of seawater and freshwater in shaping the community structure and segregate stations accordingly (Figure 6). The first hierarchical level separates the June samples of DKS and KDM and February samples of the two uppermost stations (BRK and DKS) from the remaining samples at 90% dissimilarity, from which June samples of DKS were separated at the second hierarchical level at 75% dissimilarity. June samples of DKS mainly represented the Cyclopoida adults, copepodites, and the Cladocera *Moina macrocopa*, whereas February samples of upper two stations and June samples of KDM grouped together and represented copepod nauplii, cyclopoids, and calanoids. The June samples of BRK, KDP, and KCB grouped together and separated at the third hierarchical level, whereas the February samples of all the estuarine and lowermost riverine (KDM) stations grouped together and separated at the fourth hierarchical level. Samples collected in February 2018 at the three estuarine stations and the lower most riverine (KDM) station grouped together and separated from remaining samples at the hierarchical level V (VIIB). The highly indicative zooplankton of cluster VIIIB are copepod nauplii and harpacticoids (Figure 6). This cluster clearly indicates further upward intrusion of the marine community in February. All the October samples aggregated at intermediate position and showed clear separation of riverine and estuarine stations (Figure 6). October samples showed distinct spatial variations, where riverine samples were separated from estuarine samples at the fifth hierarchical level, and the June samples of GS (estuarine mouth) joined this cluster (Figure 6).

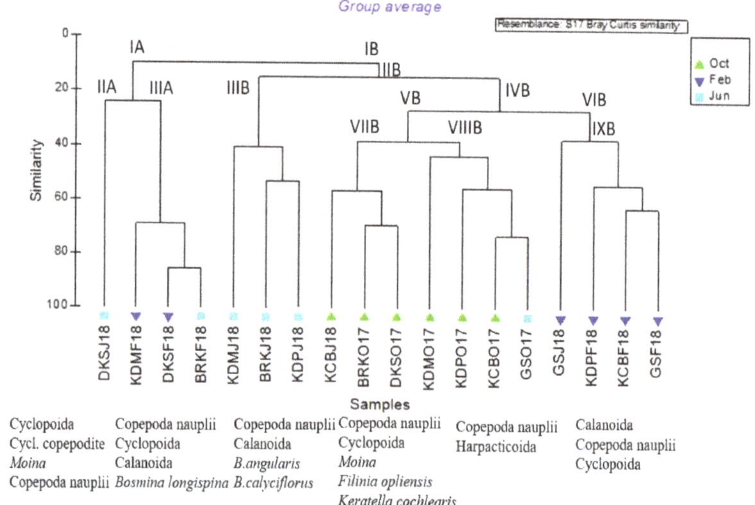

Figure 6. Bray–Curtis similarity of zooplankton and most abundant taxa for each cluster at six sampling stations during (October 2017, February 2018, and June 2018) in the HRE.

The trophic-based structuring of the zooplankton community revealed that the community was dominated by the omnivores followed by the herbivores and the bacterivores, respectively (Figure 7A–C). Variations in trophic-based community structure were more prominent at the spatial scale than the temporal scale. Bacterivorous and detritivorous species were recorded at upper stations and limited to KDP, whereas downstream stations mainly represented the omnivorous copepods (Figure 7A–C). Detritivores were mainly represented by the bdelloid rotifers at BRK and DKS. Differences in bacterivorous fraction of zooplankton community were not significant among the three riverine stations

(Figure 7A–C); however, at temporal scale, February samples recorded significantly lower fractions of bacterivorous zooplankton at all the three riverine stations.

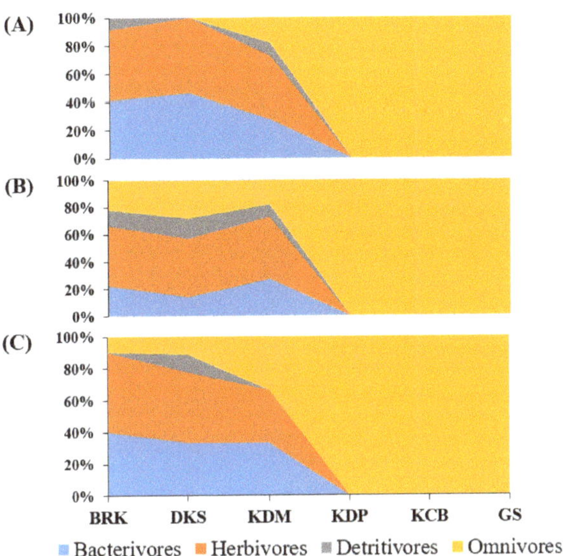

Figure 7. Percent (%) composition of zooplankton trophic guild at six sampling stations during October 2017 (**A**), February 2018 (**B**), and June 2018 (**C**) in the HRE.

3.5. Bacteriological Analyses

Total culturable bacterial density CFU mL^{-1} varied from 0.06 to 300 $\times$ 10^5 CFU mL^{-1} at six selected sampling stations during October 2017, February 2018, and June 2018 in HRE. With the lowest bacterial concentration at Barrackpore and the highest at Gangasagar, unique spatial differences and significant seaward (R = 0.66 $p < 0.0028$) increase in density of culturable bacteria were recorded in the present study (Figure 8).

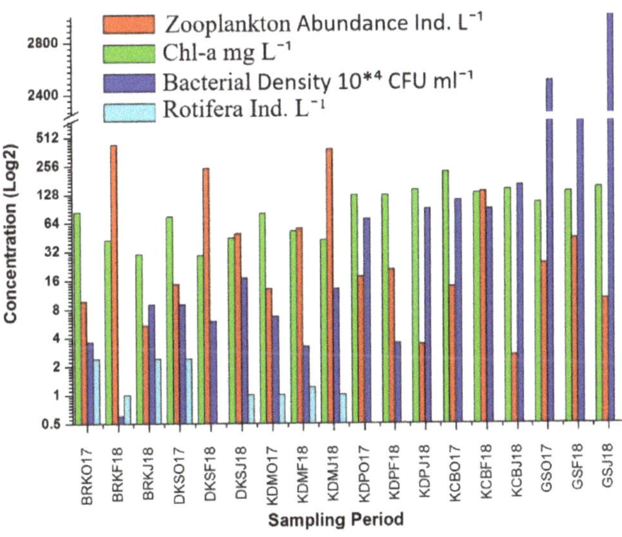

Figure 8. Total Rotifera (Ind L^{-1}), bacterial density (CFU mL^{-1}), total zooplankton abundance, and Chl *a* recorded in October 2017, February 2018, and June 2018 at six sampling stations in the HRE.

Overall bacterial densities were higher at downstream stations (KDP, KCB, and GS) than the upper freshwater stations (BRK, DKS, and KDM). Amongst upstream stations, bacterial densities were negatively correlated with total rotifer density and Chl *a* concentration; however, the correlation was significant for Bacteria *vs.* Rotifera abundance only (Figure 9; Table S6, $p < 0.01$). The most abundant culturable strains were *Bacillus subtilis*, *Pseudomonas songnensis*, and *Exiguobacterium aurantiacum*, respectively, at BRK, DKS, and GS (Figure 10).

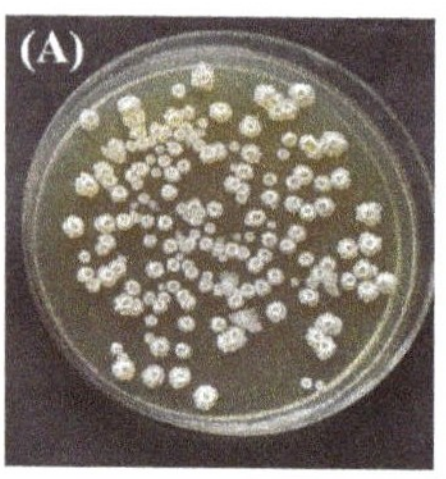

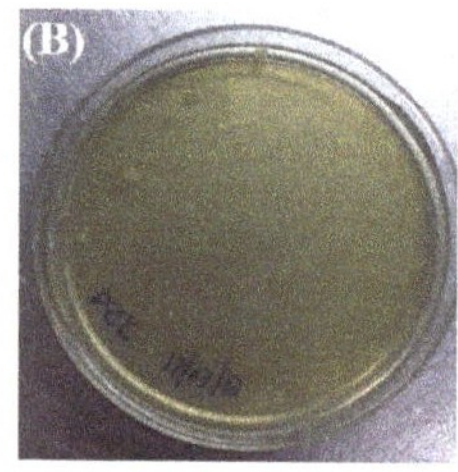

BRK2 (*Bacillus subtilis*) **DKS (*Pseudomonas songnenesis*)** **GS1 (*Exiguobacterium aurantiacum*)**

Figure 9. Most dominant bacterial strains *Bacillus subtilis* (**A**), *Pseudomonas songnenesis* (**B**), and *Exiguobacterium aurantiacum* (**C**) isolated from three selected sampling stations (BRK, DKS, and GS) in the HRE.

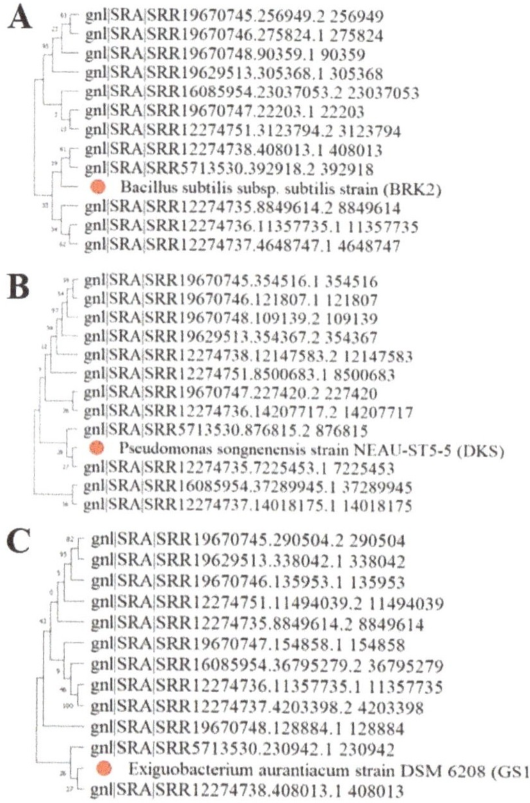

Figure 10. Phylogenetic analysis of *Bacillus subtilis* (**A**), *Pseudomonas songnenensis* (**B**), and *Exiguobacterium aurantiacum* (**C**) with their twelve different orthologues, isolated from three selected sampling stations (BRK, DKS, and GS) in the HRE.

3.6. Characterization of Highly Abundant Bacterial Taxa at BRK, DKS, and GS

The branch lengths and topology of a phylogenetic tree of the isolated strains were attained by the maximum likelihood technique. The phylogenetic trees were formed based on evolutionary distance data of 16S rRNA gene sequences. The tree shows the integrated relationship of the isolated strains with most similar Gangetic bacterial species genes based on highly similarity 95–100% taken from the 12 different Gangetic samples generated by illumine and Nanopore sequencing platforms Figure 10A–C.

3.7. Interrelationship among Taxa

The pairwise multiple correlations among means are summarized in Table S4. The total zooplankton abundance showed negative correlations with all other parameters, recording maximum strength of association with Chl *a*, but the values were not statistically significant. Zooplankton taxon-specific correlation with Chl *a* gives insight, explaining the group-wise differential relationship: Rotifera ($R = -0.67$; $p < 0.01$) and Cladocera ($R = -0.43$; $p < 0.05$) showed a significant negative correlation with Chl *a*, whereas Copepoda ($R = -0.4$; $p < 0.1$) did not correlate significantly with Chl *a*. Among freshwater stations, Cladocera and Copepoda exhibited significant positive correlations in all the three sampling cruises. Rotifera concentration correlated negatively with bacterial concentration, pH, EC, salinity, nitrate, phosphate, and Ca^{++} levels (Table S6). Among biotic components, Chl *a* values were in a core position, showing negative correlation separately with all the three zooplankton groups, whereas the bacterial concentration showed significant negative correlation with rotifer fraction of the zooplankton only (Table S6).

4. Discussion

The observed spatio-temporal variations in zooplankton, phytoplankton, and bacterioplankton concentrations and the interrelationship among taxa and spatial occurrence of the most abundant culturable bacterial strains in the HRE along the salinity level ranging from 0.01 to 25 (PSU) unequivocally confirm the complex nature and dynamicity of the estuary. The local complexity and seaward gradient of nutrient concentrations, salinity, turbidity, and Chl *a* [7,41,83] have been demonstrated in various estuarine ecosystems globally. The strength of association between zooplankton (separately with Rotifera, Cladocera, and Copepoda) with bacteria and Chl *a* in the present study provide additional insights into relative role of bacterial and algal carbon in supporting zooplankton community in highly dynamic, tropical estuary such as the HRE.

4.1. Interrelationship among Taxa

At upper stations, the rotifer community, recording lower abundance with higher diversity, exhibited significant negative correlation with bacterial concentrations, albeit bacteria and rotifer showed an overlapping relationship with abiotic parameters (e.g., pH, EC, salinity, TDS, nitrate, phosphate, and Ca^{++}). Similar physical requirements though significant negative correlation between Bacteria and Rotifera reflect strong grazing pressure by Rotifera on prokaryotes. On the other hand, the negative correlations of Chl *a* concentration with total zooplankton abundance in general and separately with all the zooplankton groups in particular suggest that zooplankton placed major grazing pressure on the phytoplankton community in the whole HRE. The prokaryotic community is controlled by the smaller zooplankton, particularly Rotifera, at the upstream stations. The majority of rotifers are filter-feeding and nanophagous which substantially utilize smaller organisms of the microbial web (bacteria and nano-flagellates and -ciliates) and are able to thrive on bacteria in eutrophic waters [32,84]. Earlier studies have reported the key role of autotrophic protists in shaping the zooplankton community structure by forming the base of food chain in the estuarine ecosystem [85,86].

The observed gradient in taxonomic diversity of zooplankton along the river to the sea continuum has also been reported in other estuaries [1,41,87]. The differences between the bacterial community in freshwater and estuarine systems can be explained by the fact that the freshwater microbial community are directly under grazing pressure by the smaller zooplankton such as Rotifera, whereas at and around the confluence, microbial abundance is mainly controlled by the grazing pressure from developing stages of larger zooplankton or trophic cascade effect, where bacterivorous flagellates are grazed by Copepoda [85,88,89]. In line with this, previous studies have related microbial growth rates with rotifer grazing effects [90–93]. It may be noted that omnivores are a major controlling factor in estuarine stations, and this higher abundance of omnivorous zooplankton in downstream saline stations offsets the trophic cascade effects, resulting in higher bacterial abundance. The weak association of copepods with Chl *a* indicate bottom-up functioning of trophic cascade in estuaries. Omnivore-driven dampening of trophic cascades has been reported in other estuaries also [89,94].

4.2. Ecological Drivers of Planktonic Community Structure

The dissimilarity between sites (β-diversity) may reflect two different ecological situations: (i) It could reflect the loss of species from one site compared to the other. In this situation, the species present at one site are nested inside a larger set of species present at the other site. Alternatively, the following site records none of the species recorded at preceding site. In the latter case, there is a complete turnover of species between the two sites. In the HRE, both events occur concurrently, and the Bray–Curtis index takes into account both the components of dissimilarity. Further, in a distance-based multivariate model, variability among 72 samples based on riverine and estuarine locations was consistent with the main drivers, with salinity itself explaining 39.6%, 40.9%, and 37.4% variability, respectively, for rotifers, bacteria, and Chl *a* concentration. The freshwater-seawater salinity range is not continuous but rather subdivided in distinct stages, which is generally manifested in planktonic abundance [95]. The extensively used salinity classification is the Venice system that is 0–0.5, 0.5–5, 5–18, 18–30, and 30–40 (PSU). Further, based on salinity-range data, ref. [96] explained five salinity zones including 0–4, 2–14, 11–18, 16–27, and 24-marine. The present results based on explicit criteria of planktonic abundance along with the salinity range support the salinity classification by [96].

Surface water temperature increased downward from 26 to 29 °C with the average temperature value of 27.3 °C and showed significant positive correlation with bacteria and Chl *a* concentration. The role of temperature in maintaining planktonic community and invasion of alien species in estuary has been emphasized earlier [95]. Nutrient-related parameters (nitrate and phosphate) and TDS explain the higher abundance of Rotifera in upstream stations, where 56% of the factors determining Chl *a* concentration is the nitrate level and 52% of the factors contributing higher bacterial concentration is explained by the phosphate levels in the surface water. These microbes can use nitrate reductase, nitrite reductase, glutamine synthetase, and other compounds for nitrogen assimilation [97,98].

Ionic concentrations also show a vital role in unravelling the zooplankton, phytoplankton, and bacterioplankton community structure [99–101]. Hence, the discernible spatial distribution of most abundant bacterial strains and higher bacterial concentration at downstream stations can be explained by differential sodium and calcium requirements in bacterioplankton and phytoplankton. Freshwater and halophilic bacteria have overlapping physiological attributes but different sodium requirements. The marine strains require higher sodium and calcium level to grow, while freshwater and terrestrial strains such as *E. coli* can multiply at a higher rate without sodium [102]. In an estuary, the riverine and marine bacterial species, having ecologically similar physiological abilities but different sodium requirements, favor the halophilic strains (e.g., *Exiguobacterium aurantiacum* in present study) when freshwater from rivers enter the sea, as the sodium and calcium dependence does not constitute a fundamental ecological difference; rather, it only regulates the locally adapted strains responsible for them [42].

4.3. Spatio-Temporal Pattern

The lowermost riverine station is KDM, which grouped with upper freshwater stations during June 2018 and, in contrast, grouped together with lower estuarine stations during February 2018 (Figure 6). On the other hand, the clustering of June samples of upper two estuarine stations with freshwater stations is suggestive of a monsoon-based regime shift in the Hooghly estuary. A higher volume of freshwater discharge owing to pre-monsoon rainfall results in seaward extension of riverine biota, whereas in the case of lower discharge and reduced river flow during February, it leads to upward extension of marine biota into a lower freshwater station (KDM). The association of October samples of the KDM station was with the estuarine stations and February and June samples with upstream freshwater stations. This gives an insight in the seawater intrusion process in the HRE. This differential grouping of lower freshwater station with saline stations and with freshwater stations depending upon the season highlights the importance of stratification based on salinity in the suppression of turbulent vertical mixing in the estuary. In February and June, weak stratification and strong vertical mixing prevails during this period, and the riverine discharge counters the seawater intrusion. During the monsoon, the strength of the estuary circulation increases as river discharge rises, while the length of seawater intrusion diminishes. However, in October, during post-monsoon season, the suppression of turbulent mixing and the strength of the estuarine circulation is mostly determined by tidal velocity. Seawater incursion associated with estuary circulation reduces as tidal velocity rises but increases when river discharge rises [103]. In fact, Monsoon flows affect all facets of estuarine hydrobiology and community structure. In the light of monsoonal influence, the HRE may be called a "tropical monsoonal estuary" [104]. The monsoonal precipitation-driven rapid decline of salinity and surface runoffs limits the distribution of marine forms, whereas the intermediate condition favors rapid multiplication of the brackish water forms and re-assemblage of the halotolerant groups, thereby resulting in the shifting regimes of transitional stations. This explains the seasonal shifting of the KDM station from freshwater to salt water, as observed in the present study. Other studies have shown the relevance of inshore water zones as zooplankton sources in large rivers such as the Danube [105,106] and St. Lawrence rivers [107]. Estuarine ecosystems in India show two peak periods, and peak time varies from region to region [108]. The present observation corroborates the previous results, in which two peaks were recorded in different months [18,109,110]. The spatial attributes of rivers are not always continuous [111], as proposed by [112]. Local land-driven discharge, changes in drought and flood regimes, and the establishment of diverse hydrological retention zones [113,114] (due to silt deposition) alter flow and river beds differentially during dry and wet seasons [113,115,116]. Therefore, results suggest that the phytoplankton, zooplankton, and bacterioplankton dynamics in the HRE are controlled by the interplay of hydrological regime, nutrient concentrations, and allochthonous inputs.

4.4. Spatial Occurrence of Highly Abundant Bacterial Strains

Bacterial study includes culture-based evaluation because studying morphology, ecology, bioprospecting for human use, and bioremediation for the purpose of culture-based isolation is practiced globally [117]. Consequently, while many environmental studies focus on large descriptions of microbial diversity through whole genome sequencing (WGS) approaches, other environmental studies rely on culturing approaches to estimate the abundance of a given culturable taxa in the environment. Both approaches are usually performed independently, and this does not allow direct comparison of the benefits and constraints of both methods. However, the culture-based approaches select only a subset of culturable bacteria, and it remains unclear to what extent culture-driven enrichments could be compared between different environmental samples. The present study depends on culture-based analysis, in which the upstream stations BRK and DKS recorded *Bacillus subtilis* and *Pseudomonas songnenesis* strains, and the lowermost station GS recorded *Exiguobacterium aurantiacum* as the highly abundant bacterial strain. The presence of soil bacteria in upstream stations indicate the influence of land driven allochthonous discharge as the major contributory factor of bacterial abundance, which enters the aquatic food chain by the microbial degradation and remineralization and its recycling within the pelagic food web. Bacteria are the only organisms capable of recycling DOM, making them an essential component of ecosystem functioning [85]. The differences in bacterial populations could be due in part to differences we observed in DOM and DOC quantity and quality among the three samples, as it has previously been shown that DOM and DOC strongly structure bacterial communities in aquatic environments [118–120].

The differences in bacterial abundance can also be attributed to differential abundance and community composition of zooplankton, as they utilize bacteria directly or indirectly through many other bacterivorous organisms and establish the link to the traditional aquatic food web [121]. Consequently, ecological productivity in estuaries is affected by both top-down and bottom-up mechanisms. Top-down controls, such as meso-zooplankton grazing, may decrease micro-zooplankton populations, enabling phytoplankton species to bloom and altering the overall structure of the microbial community [87,121–123]. At riverine sites around Kolkata city, strong top-down effects are major regulators of bacterioplankton abundance by bacterivorous organisms [124]. Additionally, the lower rotifer abundance during the peak of copepod abundance at upstream freshwater stations is also suggestive of top-down control of rotifers by copepods [41,67,125,126]. In contrast, the dominance of omnivorous and herbivorous fractions of zooplankton and higher Chl *a* and bacterial concentrations near the estuary mouth reflect strong bottom-up impacts, where the nutrient-loaded environment favors microbial and phytoplankton growth that supports omnivorous species and all the trophic guilds in the absence of a distinct trophic cascade [89,93,127].

The bacterial abundance and physicochemical parameters, particularly nutrient concentrations and differential abundance of rotifers and cladocerans, are indicative of land-driven allochthonous influence from urban discharge of the metropolis city of Kolkata. The highest Chl *a* and Na^+ at KCB preceding the confluence (GS) attest to the established facts that estuaries, particularly the mixing zone, are the most productive ecosystem and constitute an important system that provides valuable nursery and recruitment ground for commercially important species. However, with the development of sequencing technologies, further study is needed to elucidate the potential novel functions and phylogenetic relationships by sequencing the genomes of entire communities to understand relative contribution of bacterial community in HRE.

The instant change in community structure recorded in this study at lower salinity levels suggests the potential for underlying change in the oligohaline or limnetic stretches. In line with successful management of the San Francisco estuary based on isohaline condition [128], the present results indicate management options for the HRE, recommended as limited withdrawals to fixed fraction of total river flow beyond a minimum flow threshold [129]. The concerned administration needs to fix a minimum flow target in accordance with ideal region-specific isohalines in the estuary.

5. Conclusions

The zooplankton, phytoplankton, bacterioplankton, and abiotic parameters reported in this paper elucidate the patterns and drivers of differential community structures across the salinity gradient in the HRE. Among zooplankton, rotifers and cladocerans are numerically dominant and exert strong selection pressure on bacterial community and clear the suspended particles from the water column at upstream stations, whereas copepods play a major role in structuring microbial community at downstream estuarine stations. The negative correlation between Chl *a* and bacterial abundance, though insignificant, points to the competition for inorganic nutrients between phytoplankton and bacteria. Spatial variations in the trophic-based zooplankton community structure also suggest differential effects of direct bacterivore behavior by rotifers and omnivore-driven, suppressed trophic cascade effects through copepods, both of which concurrently play an important role in shaping the HRE community. The abiotic parameters such as surface water temperature; elemental concentrations of Ca^{++}, Na^{+}, and K^{+}; and the trophic level-related parameters (nitrate and phosphate) record significant seaward increase, which in turn reflects the increased concentrations of bacteria and Chl *a* (primary production) at downstream estuarine stations.

The three isolated strains of the most culturable bacteria at Barackpore, Dakshineshwar, and Gangasagar, characterized as *Bacillus subtilis*, *Pseudomonas songnenesis*, and *Exiguobacterium aurantiacum*, indicate differential influences of land-driven discharge and spatial heterogeneity in the prokaryotic community structure. The observed alteration in planktonic community structure in the sampled stretch of the HRE points to larger impacts of water extraction and sewage discharge on salinity level, resulting in changes in the riverine community in the sampled limnetic-to-oligohaline stretches.

At the temporal scale, the increased river discharge during pre-monsoon and monsoon season plays an important role in shaping the community structure by upward extension of marine influence during the waning season but downward extension of river influence during the monsoon. Therefore, the complexity of phytoplankton, mesozoopalnkton, and prokaryote communities responding to variable elemental and nutrient concentrations is driven by the differential mixing of freshwater and marine sources. Both bottom-up and top-down effects play a vital role in shaping the community in the HRE.

Increased urbanization with uncontrolled water extraction, discharge of industrial and domestic wastes in coastal waters near the mouth of the Hooghly River, and shoreline development affect the planktic community, consequentially affecting overall ecosystem health.

Furthermore, this study also highlights the role of land discharge at freshwater stations, season-specific seawater intrusion in the river, and abiotic variables including trophic status as drivers of abundance of the prokaryotic and eukaryotic planktonic community. These results suggest limited wastewater discharge and water withdrawals to a fixed fraction of total river flow beyond a minimum flow threshold maintaining isohaline. The concerned administration needs to fix a minimum flow target in accordance with ideal region-specific isohalines in the estuary. Therefore, for any future planning, the volume of water withdrawals and wastewater discharge, monsoon-driven regime shift, and internal trophic-based regulation mechanisms need to be considered.

Supplementary Materials: The following are available online at https://www.mdpi.com/article/10.3390/jmse11010088/s1, Figure S1: 1.2% Agarose gel showing single 1500 bp of 16S rDNA amplicon. Lane 1: 100bp DNA ladder; Lane 2: 16S rDNA amplicon of (A) BRK2, (B) DKS, and (C) GS1 strains. Table S1: Consensus sequence of the three strains BRK2, DKS, and GS1 characterized and their National Center for Biotechnology Information (NCBI) accession numbers and origin. Table S2: Physiochemical and biological parameters at six selected stations (BRK, DK2S, KDM, KDP, KCB, GS) in Hooghly River estuary. Table S3: Phytoplankton species recorded at six selected sampling stations in HRE during October 2017, February 2018, and June 2018. Table S4: Zooplankton taxa recorded at six selected sampling stations in HRE during October 2017, February 2018, and June 2018. Table S5: Zooplankton species/group identified for the present study, propensity of their feeding, and relevant references. Table S6: Pairwise Pearson correlation matrix between total zooplankton and major groups (Rotifera, Copepoda, Cladocera), with biotic (bacterial density, Chl *a*) and abiotic parameters (pH, EC, temperature, salinity, DO, total hardness, nitrate, phosphate, potassium, sodium, and calcium).

Author Contributions: R.K. conceived the concept; D.P. performed the field sampling data analyses; R.K. worked for interpretation of results and C.B.T. critically provided feedback and edited the manuscript. Both authors approved the submission of the manuscript. All authors have read and agreed to the published version of the manuscript.

Funding: Financial support was provided by the Department of Biotechnology, Government of India under river cleaning project (BT/PR20543/BCE/8/1398/2016).

Institutional Review Board Statement: Not applicable.

Informed Consent Statement: Not applicable.

Data Availability Statement: Data are available as Supplementary Materials. Additional data can be obtained on request to dprakashevs@cub.ac.in.

Acknowledgments: We are thankful to Department of Biotechnology for the project ((BT/PR20543/BCE/8/1398/2016)) to R.K. under the river-cleaning programme. D.P. thanks UGC for a National Fellowship.

Conflicts of Interest: The authors declare no conflict of interest.

References

1. Elliott, M.; McLusky, D.S. The need for definitions in understanding estuaries. *Estuar. Coast. Shelf Sci.* **2002**, *55*, 815–827. [CrossRef]
2. McLusky, D.S.; Elliott, M. Transitional waters: A new approach, semantics or just muddying the waters? *Estuar. Coast. Shelf Sci.* **2007**, *71*, 359–363. [CrossRef]
3. Whitfield, A.K.; Elliott, M. Fishes as indicators of environmental and ecological changes within estuaries: A review of progress and some suggestions for the future. *J. Fish Biol.* **2002**, *61*, 229–250. [CrossRef]
4. Chícharo, L.; Chícharo, M.A.; Ben-Hamadou, R. Use of a hydrotechnical infrastructure (Alqueva Dam) to regulate planktonic assemblages in the Guadiana estuary: Basis for sustainable water and ecosystem services management. *Estuar. Coast. Shelf Sci.* **2006**, *70*, 3–18. [CrossRef]
5. Tweedley, J.R.; Dittmann, S.R.; Whitfield, A.K.; Withers, K.; Hoeksema, S.D.; Potter, I.C. Hypersalinity: Global distribution, causes, and present and future effects on the biota of estuaries and lagoons. In *Coasts Estuaries*; Elsevier: Amsterdam, The Netherlands, 2019; Chapter 30; pp. 523–546. ISBN 9780128140031. [CrossRef]
6. Svetlichny, L.; Hubareva, E.; Khanaychenko, A.; Uttieri, M. Response to salinity and temperature changes in the alien Asian copepod *Pseudodiaptomus marinus* introduced in the Black Sea. *J. Exp. Zool.* **2019**, *331*, 416–426. [CrossRef] [PubMed]
7. Cloern, J.E.; Jassby, A.D. Patterns and scales of phytoplankton variability in estuarine–coastal ecosystems. *Estuaries Coasts* **2010**, *33*, 230–241. [CrossRef]
8. Cloern, J.E.; Foster, S.Q.; Kleckner, A.E. Phytoplankton primary production in the world's estuarine-coastal ecosystems. *Biogeosciences* **2014**, *11*, 2477–2501. [CrossRef]
9. Cloern, J.E.; Jassby, A.D.; Schraga, T.S.; Nejad, E.; Martin, C. Ecosystem variability along the estuarine salinity gradient: Examples from long-term study of San Francisco Bay. *Limnol. Oceanogr.* **2017**, *62*, S272–S291. [CrossRef]
10. Testa, J.M.; Murphy, R.R.; Brady, D.C.; Kemp, W.M. Nutrient-and climate-induced shifts in the phenology of linked biogeochemical cycles in a temperate estuary. *Front. Mar. Sci.* **2018**, *5*, 114. [CrossRef]
11. Hume, T.M.; Snelder, T.; Weatherhead, M.; Liefting, R. A controlling factor approach to estuary classification. *Ocean Coast. Manag.* **2007**, *50*, 905–929. [CrossRef]
12. Dürr, H.H.; Laruelle, G.G.; van Kempen, C.M.; Slomp, C.P.; Meybeck, M.; Middelkoop, H. Worldwide typology of nearshore coastal systems: Defining the estuarine filter of river inputs to the oceans. *Estuaries Coasts* **2011**, *34*, 441–458. [CrossRef]
13. Chilton, D.; Hamilton, D.P.; Nagelkerken, I.; Cook, P.; Hipsey, M.R.; Reid, R.; Sheaves, M.; Waltham, N.J.; Brookes, J. Environmental Flow Requirements of Estuaries: Providing Resilience to Current and Future Climate and Direct Anthropogenic Changes. *Front. Environ. Sci.* **2021**, *9*, 764218. [CrossRef]
14. Azam, F.; Fenchel, T.; Field, J.G.; Gray, J.S.; Meyer-Reil, L.A.; Thingstad, F. The ecological role of water-column microbes in the sea. *Mari. Ecol. Prog. Serie.* **1983**, *10*, 257–263. [CrossRef]
15. Mikhailov, I.S.; Zakharova, Y.R.; Bukin, Y.S.; Galachyants, Y.P.; Petrova, D.P.; Sakirko, M.V.; Likhoshway, Y.V. Co-occurrence networks among bacteria and microbial eukaryotes of Lake Baikal during a spring phytoplankton bloom. *Mic. Eco.* **2019**, *77*, 96–109. [CrossRef]
16. Yang, Y.; Gao, Y.; Chen, Y.; Li, S.; Zhan, A. Interaction-based abiotic and biotic impacts on biodiversity of plankton communities in disturbed wetlands. *Div. Dist.* **2019**, *25*, 1416–1428. [CrossRef]
17. Roy, A.P.; Pandit, A.R.; Sharma, A.P.; Bhaumik, U.T.; Majunder, S.; Biswas, D.K. Socioeconomic status and livelihood of fisher women of hooghly estuary. *Inland Fish. Soc. India* **2015**, *47*, 49–56.
18. Dutta, N.; Malhotia, J.C.; Bose, B.B. Hydrology and seasonal fluctuations of the plankton in the Hooghly estuary. In *Symposium on Marine and Freshwater Plankton in the Indo-Pacific*; Indo-Pacific Fisheries Council: Bangkok, Thailand, 1954; pp. 35–47.
19. Bose, B.B. Observations on the hydrology of the Hooghly Estuary. *Inland J. Fish.* **1956**, *3*, 101–118.
20. Manna, R.K.; Satpathy, B.B.; Roshith, C.M.; Naskar, M.; Bhaumik, U.; Sharma, A.P. Spatio-temporal changes of hydro-chemical parameters in the estuarine part of the river Ganges under altered hydrological regime and its impact on biotic communities. *Aquat. Ecosyst. Health Manag.* **2013**, *16*, 433–444. [CrossRef]
21. Das, B.K.; Ray, A.; Johnson, C.; Verma, S.K.; Alam, A.; Baitha, R.; Sarkar, U.K. The present status of ichthyofaunal diversity of river Ganga India: Synthesis of present v/s past. *Acta Ecol. Sin.* **2021**, in press. [CrossRef]
22. Ramesh, R.; Lakshmi, A.; Sappal, S.M.; Bonthu, S.R.; Suganya, M.D.; Ganguly, D.; Purvaja, R. Integrated management of the Ganges delta, India. *Coas. Estu.* **2019**, 187–211. [CrossRef]
23. ICAR—Central Inland Fisheries Research Institute. *Assessment of Fish and Fisheries of the Ganga River System for Developing Suitable Conservation and Restoration Plan*; Sanctioned under National Mission on Clean Ganga, vide NGRBA Order NO.T-17 /2014 15/526/NMCG-Fish and Fisheries Dated 13 July 2015; CFRI: Barrackpore, India, 2019.

24. Bhaumik, U.; Sharma, A.P. The fishery of Indian Shad (*Tenualosa ilisha*) in the Bhagirathi-Hooghly river system. *Fish. Chimes* **2011**, *31*, 21–27.
25. Sharma, A.P.; Joshi, K.D.; Naskar, M.; Das, M.K. *Inland Fisheries & Climate Change: Vulnerability and Adaptation Options*; NICRA: Phek, India, 2014.
26. Bianchi, F.; Acri, F.; Aubry, F.B.; Berton, A.; Boldrin, A.; Camatti, E.; Cassin, D.; Comaschi, A. Can plankton communities be considered as bioindicators of water quality in the lagoon of Venice? *Mar. Pollut. Bull.* **2003**, *46*, 964–971. [CrossRef] [PubMed]
27. Hsiao, S.H.; Lee, C.Y.; Shih, C.T.; Hwang, J.S. Calanoid copepods of the Kuroshio Current east of Taiwan, with a note on the presence of *Calanus jashnovi* Hulseman, 1994. *Zool. Stud.* **2004**, *43*, 323–331.
28. Pomeroy, L.R.; Leb, W.P.J.; Azam, F.; Hobbie, J.E. The microbial loop. *Oceanography* **2007**, *20*, 28–33. [CrossRef]
29. Orellana, M.V.; Pang, W.L.; Durand, P.M.; Whitehead, K.; Baliga, N.S. A role for programmed cell death in the microbial loop. *PLoS ONE* **2013**, *8*, e62595. [CrossRef]
30. Degerman, R.; Lefébure, R.; Byström, P.; Båmstedt, U.; Larsson, S.; Andersson, A. Food web interactions determine energy transfer efficiency and top consumer responses to inputs of dissolved organic carbon. *Hydrobiologia* **2018**, *805*, 131–146. [CrossRef]
31. Fenchel, T. The microbial loop—25 years later. *J. Exp. Mar. Bio. Ecol.* **2008**, *366*, 99–103. [CrossRef]
32. Prakash, D.; Kumar, R.; Rajan, K.; Patel, A.; Yadav, D.; Dhankar, R.; Khudssar, A.F. Integrated application of macrophytes and zooplankton for wastewater treatment. *Front. Environ. Sci. Water Wastewater Manag.* **2022**, *10*, 941841. [CrossRef]
33. Bachy, C.; Hehenberger, E.; Ling, Y.C.; Needham, D.M.; Strauss, J.; Wilken, S.; Worden, A.Z. Marine Protists: A Hitchhiker's Guide to their Role in the Marine Microbiome. In *The Marine Microbiome*; Springer: Cham, Switzerland, 2022; pp. 159–241. [CrossRef]
34. Kumar, R. Feeding modes and associated mechanisms in Zoolplankton. In *Ecology of Plankton*; Kumar, A., Ed.; Daya Publishing House: Delhi, India, 2004; pp. 220–226.
35. Kumar, R.; Rao, T.R. Demographic responses of adult *Mesocyclops thermocyclopoides* (Copepoda, Cyclopoida) to different plant and animal diets. *Freshw. Biol.* **1999**, *42*, 487–501. [CrossRef]
36. Roy, S.P.; Roy, R.; Prabhakar, A.K.; Pandey, A.; Kumar, R.; Tseng, L.C. Spatio-temporal distribution and community structure of zooplankton in the Gangetic Dolphin Sanctuary, 2009. *Aquat. Ecosyst. Health Manag.* **2013**, *16*, 374–384. [CrossRef]
37. Sarkar, S.K.; Singh, B.N.; Choudhury, A. The ecology of copepods from Hoogly estuary, west Bengal, India. *Mahasagar-Bull. Natl. Inst. Oceanogr.* **1986**, *19*, 103–112.
38. Roshith, C.M.; Meena, D.K.; Manna, R.K.; Sahoo, A.K.; Swain, H.S.; Raman, R.K.; Das, B.K. Phytoplankton community structure of the Gangetic (Hooghly-Matla) estuary: Status and ecological implications in relation to eco-climatic variability. *Flora* **2018**, *240*, 133–143. [CrossRef]
39. Waniek, J.J. The role of physical forcing in initiation of spring blooms in the northeast Atlantic. *J. Mar. Syst.* **2003**, *39*, 57–82. [CrossRef]
40. Sridhar, R.; Thangaradjou, T.; Senthil, K.S.; Kannan, L. Water quality and phytoplankton characteristics in the Palk Bay, southeast coast of India. *J. Environ. Biol.* **2006**, *27*, 561–566.
41. Hwang, J.S.; Kumar, R.; Hsieh, C.W.; Kuo, A.Y.; Souissi, S.; Hsu, M.H.; Chen, Q.C. Patterns of zooplankton distribution along the marine, estuarine and riverine portions of the Danshuei ecosystem in northern Taiwan. *Zool. Stud.* **2010**, *49*, 335–352.
42. Wang, Q.; Hao, Z.; Ding, R.; Li, H.; Tang, X.; Chen, F. Host dependence of zooplankton-associated microbes and their ecological implications in freshwater lakes. *Water* **2021**, *13*, 2949. [CrossRef]
43. Ling, L.L.; Schneider, T.; Peoples, A.J.; Spoering, A.L.; Engels, I.; Conlon, B.P.; Lewis, K. A new antibiotic kills pathogens without detectable resistance. *Nature* **2015**, *517*, 455–459. [CrossRef]
44. Sarkar, S.K.; Saha, M.; Takada, H.; Bhattacharya, A.; Mishra, P.; Bhattacharya, B. Water quality management in the lower stretch of the river Ganges, east coast of India: An approach through environmental education. *J. Clean. Prod.* **2007**, *15*, 1559–1567. [CrossRef]
45. Beyrend-Dur, D.; Kumar, R.; Rao, T.R.; Souissi, S.; Cheng, S.H.; Hwang, J.S. Demographic parameters of adults of *Pseudodiaptomus annandalei* (Copepoda: Calanoida): Temperature–salinity and generation effects. *J. Exp. Mari. Biol. Ecol.* **2011**, *404*, 1–14. [CrossRef]
46. Strickland, J.D.H.; Parsons, T.R. *A Practical Handbook of Seawater Analysis*; Fisheries Research Board of Canada: Ottawa, ON, Canada, 1972.
47. APHA. *Standard Methods for the Examination of Water and Wastewater*, 21st ed.; American Public Health Association (APHA): Washington, DC, USA, 2005.
48. Aneja, K.R. *Experiments in Microbiology, Plant Pathology and Biotechnology*; NewAge International: Lincolnshire, UK, 2007; p. 58.
49. Selvin, J.; Lanong, S.; Syiem, D.; De Mandal, S.; Kayang, H.; Kumar, N.S.; Kiran, G.S. Culture-dependent and metagenomic analysis of lesser horseshoe bats' gut microbiome revealing unique bacterial diversity and signatures of potential human pathogens. *Micr. Patho.* **2019**, *137*, 103675. [CrossRef] [PubMed]
50. Böttger, E.C. Rapid determination of bacterial ribosomal RNA sequences by direct sequencing of enzymatically amplified DNA. *FEMS Microbiol. Lett.* **1989**, *65*, 171–176. [CrossRef]
51. Tamura, K.; Nei, M. Estimation of the number of nucleotide substitutions in the control region of mitochondrial DNA in humans and chimpanzees. *Mol. Bio. Evol.* **1993**, *10*, 512–526. [CrossRef]
52. Tamura, K.; Stecher, G.; Kumar, S. MEGA11, molecular evolutionary genetics analysis version 11. *Mol. Bio. Evol.* **2021**, *38*, 3022–3027. [CrossRef] [PubMed]

53. John, D.M.; Whitton, B.A.; Brook, A.J. *The Freshwater Algal flora of the British Isles: An Identification Guide to Freshwater and Terrestrial Algae*; Cambridge University Press: Cambridge, UK, 2002.
54. Prescott, G.W. *How to Know the Fresh-Water Algae—An Illustrated Key for the Identifying the More Common Freshwater Algae to Genus, with Hundreds of Species Named and Pictured and with Numerous Aids for the Study*; W.C. Brown Co.: Dubuque, IA, USA, 1964; p. 293.
55. Singh, J.; Saxena, R.C. An Introduction to Microalgae: Diversity and Significance. In *Handbook of Marine Microalgae*; Academic Press: Cambridge, MA, USA, 2015; pp. 11–24. [CrossRef]
56. Guiry, M.D.; Guiry, G.M. Algae Base. World-Wide Electronic Publication, National University of Ireland, Galway. 2022. Available online: http://www.algaebase.org/ (accessed on 12 July 2018).
57. Singh, P.; Gupta, S.K.; Guldhe, A.; Rawat, I.; Bux, F. Microalgae Isolation and Basic Culturing Techniques. In *Handbook of Marine Microalgae*; Elsevier Inc.: Amsterdam, The Netherlands, 2015; pp. 43–54. [CrossRef]
58. Shoaf, W.T.; Lium, B.W. Improved extraction of Chlorophyll *a* and *b* from algae using dimethyl sulfoxide. *Limnol. Oceanogr.* **1976**, *21*, 926–928. [CrossRef]
59. Parsons, T.R.; Maita, Y.; Lalli, C.M. *A Manual of Chemical and Biological Methods for Seawater Analysis*; Pergamon Press: Oxford, UK, 1984; p. 173.
60. Sehgal, K.L. *Planktonic Copepods of Freshwater Ecosystems*; Inter Print: New Delhi, India, 1983.
61. Michael, R.G.; Sharma, B.K. *Indian Cladocera (Crustacea, Branchiopoda, Cladocera)*; Zoological Survey of India: Kolkata, India, 1988; p. 262.
62. Dodson, S.I.; Frey, D.G. Cladocera and other branchiopoda. In *Ecology and Classification of North American Freshwater Invertebrates*; Thorp, J.H., Covich, A.P., Eds.; Academic Press: New York, NY, USA, 1991; pp. 764–776.
63. Edmondson, W.T. *Fresh-Water Biology, Wipple and Ward Reprint (Indian Reprint)*; International Books & Periodicals Supply Service: New Delhi, India, 1992; p. 24B/5.
64. Battish, S.K. *Freshwater Zooplankton of India*; Oxford and IBH Publishing Co., Pvt., Ltd.: New Delhi, India, 1992.
65. Sharma, B.K. *Freshwater Rotifers (Rotifers: Eurotatoria) Fauna of West Bengal State Faunal Series*; Zoological Survey of India: Calcutta, India, 1992; Volume 3, pp. 1–121.
66. Dumont, H.J. On the diversity of Cladocera in the tropics. *Hydrobiologia* **1994**, *272*, 27–38. [CrossRef]
67. Kumar, R. Effect of *Mesocyclops thermocyclopoides* (Copepoda, Cyclopoida) predation on population dynamics of different prey: A laboratory study. *J. Freshw. Ecol.* **2003**, *18*, 383–393. [CrossRef]
68. Ramakrishna Rao, T.; Kumar, R. Patterns of prey selectivity in the cyclopoid copepod *Mesocyclops thermocyclopoides*. *Aqua. Ecol.* **2002**, *36*, 411–424. [CrossRef]
69. Fernando, C.H. *A Guide to Tropical Freshwater Zooplankton, Identification, Ecology and Impact on Fisheries*; Backhuys Publishers: Leiden, The Netherlands, 2002.
70. Tseng, L.C.; Kumar, R.; Dahms, H.U.; Chen, Q.C.; Hwang, J.S. Monsoon-driven succession of copepod assemblages in coastal waters of the northeastern Taiwan Strait. *Zool. Stud.* **2008**, *47*, 46.
71. Kumar, R.; Souissi, S.; Hwang, J.S. Vulnerability of carp larvae to copepod predation as a function of larval age and body length. *Aquaculture* **2012**, *338*, 274–283. [CrossRef]
72. Voigt, M.; Koste, W. *Rotatoria, Die Rädertiere Mitteleuropas*, 2nd ed.; Gebrüder Borntraeger: Stuttgart, Germany, 1978; Volume 1, 673p.
73. Brandl, Z. Freshwater copepods and rotifers: Predators and their prey. In *Rotifera X*; Springer: Dordrecht, NJ, USA, 2005; pp. 475–489. [CrossRef]
74. Ooms-Wilms, A.L.; Postema, G.; Gulati, R.D. Evaluation of bacterivory of Rotifera based on measurements of in situ ingestion of fluorescent particles, including some comparisons with Cladocera. *J. Plank. Resea.* **1995**, *17*, 1057–1077. [CrossRef]
75. Starkweather, P.L.; Gilbert, J.J.; Frost, T.M. Bacterial feeding by the rotifer *Brachionus calyciflorus*: Clearance and ingestion rates, behavior and population dynamics. *Oecologia* **1979**, *44*, 26–30. [CrossRef] [PubMed]
76. Kim, H.W.; Hwang, S.J.; Joo, G.J. Zooplankton grazing on bacteria and phytoplankton in a regulated large river (Nakdong River, Korea). *J. Plank. Res.* **2000**, *22*, 1559–1577. [CrossRef]
77. Leasi, F.; Ricci, C. Musculature of two bdelloid rotifers, *Adineta ricciae* and *Macrotrachela quadricornifera*: Organization in a functional and evolutionary perspective. *J. Zool. Syst. Evol. Res.* **2010**, *48*, 33–39. [CrossRef]
78. Kiørboe, T. How zooplankton feed: Mechanisms, traits and trade-offs. *Bio. Rev.* **2011**, *86*, 311–339. [CrossRef]
79. Field, J.G.; Clarke, K.R.; Warwick, R.M. A practical strategy for analysing multispecies distribution patterns. *Mar. Ecol. Prog. Ser.* **1982**, *8*, 37–52. [CrossRef]
80. Pielou, E.C. The measurement of diversity in different types of biological collections. *J. Theor. Biol.* **1966**, *13*, 131–144. [CrossRef]
81. Margalef, R. Some concepts relative to the organization of plankton. *Oceanogr. Mar. Biol. Annu. Rev.* **1967**, *5*, 257–289.
82. Hunt, B.P.V.; Pakhomov, E.A.; Trotsenko, B.G. The macrozooplankton of the Cosmonaut Sea, east Antarctica (30° E–60° E), 1987–1990. *Deep-Sea Res. Pt. I Oceanogr.* **2007**, *54*, 1042–1069. [CrossRef]
83. Azhikodan, G.; Yokoyama, K. Spatio-temporal variability of phytoplankton (Chlorophyll-*a*) in relation to salinity, suspended sediment concentration, and light intensity in a macrotidal estuary. *Cont. Shelf Res.* **2016**, *126*, 15–26. [CrossRef]
84. Gilbert, J.J. Food niches of planktonic rotifers: Diversification and implications. *Limnol. Oceanogr.* **2022**, *67*, 2218–2251. [CrossRef]
85. Zöllner, E.; Hoppe, H.G.; Sommer, U.; Jürgens, K. Effect of zooplankton-mediated trophic cascades on marine microbial food web components (bacteria, nanoflagellates, ciliates). *Limnol. Oceanogr.* **2009**, *54*, 262–275. [CrossRef]

86. Dalu, T.; Froneman, P.W.; Richoux, N.B. Phytoplankton community diversity along a river-estuary continuum. *Trans. R. Soc. South Afr.* **2014**, *69*, 107–116. [CrossRef]
87. Murrell, M.C.; Stanley, R.S.; Lores, E.M.; Di Donato, G.T.; Flemer, D.A. Linkage between microzooplankton grazing and phytoplankton growth in a Gulf of Mexico estuary. *Estuaries* **2002**, *25*, 19–29. [CrossRef]
88. Hobbie, J.E. A comparison of the ecology of planktonic bacteria in fresh and salt water. *Limnol. Oceanogr.* **1988**, *33*, 750–764. [CrossRef]
89. Johnson, K.D.; Grabowski, J.; Smee, D.L. Omnivory dampens trophic cascades in estuarine communities. *Mari. Ecol. Prog. Seri.* **2014**, *507*, 197–206. [CrossRef]
90. McManus, G.B.; Ederington-Cantrell, M.C. Phytoplankton pigments and growth rates, and microzooplankton grazing in a large temperate estuary. *Mar. Eco. Prog. Ser.* **1992**, *87*, 77–85. [CrossRef]
91. Lehrter, J.C.; Pennock, J.R.; McManus, G.B. Microzooplankton grazing and nitrogen excretion across a surface estuarine-coastal interface. *Estuaries* **1999**, *22*, 113–125. [CrossRef]
92. Chen, G.Q. A microbial polyhydroxyalkanoates (PHA) based bio-and materials industry. *Che. Soc. Rev.* **2009**, *38*, 2434–2446. [CrossRef] [PubMed]
93. York, J.K.; McManus, G.B.; Kimmerer, W.J.; Slaughter, A.M.; Ignoffo, T.R. Trophic links in the plankton in the low salinity zone of a large temperate estuary: Top-down effects of introduced copepods. *Estuaries Coasts* **2014**, *37*, 576–588. [CrossRef]
94. Fahimipour, A.K.; Levin, D.A.; Anderson, K.E. Omnivory does not preclude strong trophic cascades. *Ecosphere* **2019**, *10*, e02800. [CrossRef]
95. Svetlichny, L.; Hubareva, E.; Uttieri, M. Ecophysiological and behavioural responses to salinity and temperature stress in cyclopoid copepod *Oithona davisae* with comments on gender differences. *Mediterr. Mar. Sci.* **2021**, *22*, 89–101. [CrossRef]
96. Bulger, A.J.; Hayden, B.P.; Monaco, M.E.; Nelson, D.M.; McCormick-Ray, M.G. Biologically-based estuarine salinity zones derived from a multivariate analysis. *Estuaries* **1993**, *16*, 311–322. [CrossRef]
97. Jetten, M.S.M. The microbial nitrogen cycle. *Environ. Microbiol.* **2008**, *10*, 2903–2909. [CrossRef]
98. Kanamori, K.; Weiss, R.L.; Roberts, J.D. Ammonia assimilation in *Bacillus polymyxa*. 15N NMR and enzymatic studies. *J. Biol. Chem.* **1987**, *262*, 11038–11045. [CrossRef]
99. Zhou, J.; Richlen, M.L.; Sehein, T.R.; Kulis, D.M.; Anderson, D.M.; Cai, Z. Microbial community structure and associations during a marine dinoflagellate bloom. *Front. Microbiol.* **2018**, *9*, 1201. [CrossRef]
100. Bickel, S.L.; Tang, K.W.; Grossart, H.P. Structure and function of zooplankton-associated bacterial communities in a temperate estuary change more with time than with zooplankton species. *Aquat. Microb. Ecol.* **2014**, *72*, 1–15. [CrossRef]
101. Lehman, P.W.; Kurobe, T.; Huynh, K.; Lesmeister, S.; Teh, S.J. Covariance of Phytoplankton, Bacteria, and Zooplankton Communities Within Microcystis Blooms in San Francisco Estuary. *Front. Microbiol.* **2021**, *12*, 1184. [CrossRef]
102. Hrenovic, J.; Ivankovic, T. Survival of *Escherichia coli* and *Acinetobacter junii* at various concentrations of sodium chloride. *EurAsian J. Biosci.* **2009**, *3*, 144–151. [CrossRef]
103. Bricheno, L.M.; Wolf, J.; Sun, Y. Saline intrusion in the Ganges-Brahmaputra-Meghna megadelta. *Estuar. Coast. Shelf Sci.* **2021**, *252*, 107246. [CrossRef]
104. Rath, A.R.; Mitbavkar, S.; Anil, A.C. Response of the phytoplankton community to seasonal and spatial environmental conditions in the Haldia port ecosystem located in the tropical Hooghly River estuary. *Environ. Monit. Assess.* **2021**, *193*, 1–24. [CrossRef] [PubMed]
105. Reckendorfer, W.; Keckeis, H.; Winkler, G.; Schiemer, F. Abundance in the River Danube, Austria: Ce of inshore retention. *Freshw. Biol.* **1999**, *41*, 583–591. [CrossRef]
106. Paul, S.; Karan, S.; Ghosh, S.; Bhattacharya, B.D. Hourly variation of environment and copepod community of the Ganges River Estuary of India: Perspectives on sampling estuarine zooplankton. *Estuar. Coast. Shelf Sci.* **2019**, *230*, 106441. [CrossRef]
107. Casper, A.F.; Thorp, J.H. Diel and lateral patterns of zooplankton distribution in the St. Lawrence River. *River Res. Appl.* **2007**, *23*, 73–85. [CrossRef]
108. Vijith, V.; Sundar, D.; Shetye, S.R. Time-dependence of salinity in monsoonal estuaries. *Estuar. Coast. Shelf Sci.* **2009**, *85*, 601–608. [CrossRef]
109. Shetty, H.P.C.; Saha, S.B.; Ghosh, B.B. Observations on the distribution and fluctuations of plankton in the Hooghly-Matlah estuarine system, with notes on their relation to commercial fish landings. *Indian J. Fish.* **1961**, *8*, 326–363.
110. Baidya, A.U.; Choudhury, A. Distribution and abundance of zooplankton in a tidal creek of Sagar Island, Sundarbans, West Bengal. *Environ. Ecol.* **1984**, *2*, 333–337.
111. Ward, J.W.; Stanford, J.A. *Intermediate-Disturbance Hypothesis: An Explanation for Biotic Diversity Patterns in Lotic Ecosystems*; Dynamics of Lotic Systems, Ann Arbor Science: Ann Arbor, MI, USA, 1983; pp. 347–356.
112. Vannote, R.L.; Minshall, G.W.; Cummins, K.W.; Sedell, J.R.; Cushing, C.E. The river continuum concept. Can. *J. Fish. Aquat. Sci.* **1980**, *37*, 130–137. [CrossRef]
113. Schiemer, F.; Keckeis, H.; Reckendorfer, W.; Winkler, G. The "inshore retention concept" and its significance for large rivers. *Arch. Hydrobiol.* **2001**, *135*, 509–516. [CrossRef]
114. Picapedra, P.H.; Fernandes, C.; Baumgartner, G.; Lansac-Tôha, F.A. Effect of slackwater areas on the establishment of plankton communities (testate amoebae and rotifers) in a large river in the semi-arid region of northeastern Brazil. *Limnetica* **2018**, *37*, 19–31. [CrossRef]

115. Medeiros, E.S.; Arthington, A.H. Allochthonous and autochthonous carbon sources for fish in floodplain lagoons of an Australian dryland river. *Environ. Biol. Fishes* **2011**, *90*, 1–17. [CrossRef]
116. Lucena, L.C.A.; Melo, T.X.D.; Medeiros, E.S.F. Zooplankton community of Parnaíba River, Northeastern Brazil. Acta Limnolo. *Brasilie* **2015**, *27*, 118–129. [CrossRef]
117. Li, Y.; Wu, C.; Zhou, M.; Wang, E.T.; Zhang, Z.; Liu, W.; Xie, Z. Diversity of cultivable protease-producing bacteria in Laizhou Bay sediments, Bohai Sea, China. *Front. Microbiol.* **2017**, *8*, 405. [CrossRef] [PubMed]
118. Judd, K.E.; Crump, B.C.; Kling, G.W. Variation in dissolved organic matter controls bacterial production and community composition. *Ecology* **2006**, *87*, 2068–2079. [CrossRef] [PubMed]
119. Berge, O.; Monteil, C.L.; Bartoli, C.; Chandeysson, C.; Guilbaud, C.; Sands, D.C.; Morris, C.E. A user's guide to a data base of the diversity of *Pseudomonas syringae* and its application to classifying strains in this phylogenetic complex. *PLoS ONE* **2014**, *9*, e105547. [CrossRef]
120. Llirós, M.; Inceoğlu, Ö.; García-Armisen, T.; Anzil, A.; Leporcq, B.; Pigneur, L.-M.; Viroux, L.; Darchambeau, F.; Descy, J.-P.; Servais, P. Bacterial community composition in three freshwater reservoirs of different alkalinity and trophic status. *PLoS ONE* **2014**, *9*, e116145. [CrossRef]
121. Wollrab, S.; Diehl, S.; De Roos, A. MSimple rules describe bottom-up and top-down control in food webs with alternative energy pathways. *Ecol. Lett.* **2012**, *15*, 935–946. [CrossRef]
122. Buskey, E.J.; Montagna, P.A.; Amos, A.F.; Whitledge, T.E. Disruption of grazer populations as a contributing factor to the initiation of the Texas brown tide algal bloom. *Limnol. Oceanogr.* **1997**, *42*, 1215–1222. [CrossRef]
123. Ger, K.A.; Urrutia-Cordero, P.; Frost, P.C.; Hansson, L.A.; Sarnelle, O.; Wilson, A.E.; Lürling, M. The interaction between cyanobacteria and zooplankton in a more eutrophic world. *Harmful Algae* **2016**, *54*, 128–144. [CrossRef]
124. Yoshida, T.; Jones, L.E.; Ellner, S.P.; Fussmann, G.F.; Hairston, N.G. Rapid evolution drives ecological dynamics in a predator–prey system. *Nature* **2003**, *424*, 303–306. [CrossRef]
125. Kumar, R.; Rao, T.R. Effect of the cyclopoid copepod *Mesocyclops thermocyclopoides* on the interactions between the predatory rotifer *Asplanchna intermedia* and its prey *Brachionus calyciflorus* and *B. angularis*. *Hydrobiologia* **2001**, *453/454*, 261–268. [CrossRef]
126. Rosińska, J.; Romanowicz-Brzozowska, W.; Kozak, A.; Gołdyn, R. Zooplankton changes during bottom-up and top-down control due to sustainable restoration in a shallow urban lake. *Environ. Sci. Pollut. Res.* **2019**, *26*, 19575–19587. [CrossRef]
127. Glibert, P.M.; Fullerton, D.; Burkholder, J.M.; Cornwell, J.C.; Kana, T.M. Ecological stoichiometry, biogeochemical cycling, invasive species, and aquatic food webs: San Francisco Estuary and comparative systems. *Revie. Fish. Scie.* **2011**, *19*, 358–417. [CrossRef]
128. Kimmerer, W.J. Physical, biological, and management responses to variable freshwater flow into the San Francisco Estuary. *Estuaries* **2002**, *25*, 1275–1290. [CrossRef]
129. Flannery, M.S.; Peebles, E.B.; Montgomery, R.T. A percent-of-flow approach for managing reductions of freshwater inflows from unimpounded rivers to southwest Florida estuaries. *Estuaries* **2002**, *25*, 1318–1332. [CrossRef]

Journal of
Marine Science and Engineering

Opinion

Can Marine Hydrothermal Vents Be Used as Natural Laboratories to Study Global Change Effects on Zooplankton in a Future Ocean?

Hans-Uwe Dahms [1], Subramani Thirunavukkarasu [2] and Jiang-Shiou Hwang [3,4,5,*]

1 Department of Biomedical Science and Environmental Biology, Kaohsiung Medical University, Kaohsiung 80708, Taiwan
2 Department of Zoology, University of Madras, Guindy Campus Chennai, Tamilnadu 600025, India
3 Institute of Marine Biology, National Taiwan Ocean University, Keelung 20224, Taiwan
4 Center of Excellence for Ocean Engineering, National Taiwan Ocean University, Keelung 20224, Taiwan
5 Center of Excellence for the Oceans, National Taiwan Ocean University, Keelung 20224, Taiwan
* Correspondence: jshwang@mail.ntou.edu.tw; Tel.: +886-2-24622192 (ext. 5304); Fax: 886-2-24629464

Abstract: It is claimed that oceanic hydrothermal vents (HVs), particularly the shallow water ones, offer particular advantages to better understand the effects of future climate and other global change on oceanic biota. Marine hydrothermal vents (HVs) are extreme oceanic environments that are similar to projected climate changes of the earth system ocean (e.g., changes of circulation patterns, elevated temperature, low pH, increased turbidity, increased bioavailability of toxic compounds. Studies on hydrothermal vent organisms may fill knowledge gaps of environmental and evolutionary adaptations to this extreme oceanic environment. In the present contribution we evaluate whether hydrothermal vents can be used as natural laboratories for a better understanding of zooplankton ecology under a global change scenario.

Keywords: Hydrothermal vent; mortality; mesozooplankton; pelagic; climate change; future ocean

Citation: Dahms, H.-U.; Thirunavukkarasu, S.; Hwang, J.-S. Can Marine Hydrothermal Vents Be Used as Natural Laboratories to Study Global Change Effects on Zooplankton in a Future Ocean? *J. Mar. Sci. Eng.* **2023**, *11*, 163. https://doi.org/10.3390/jmse11010163

Academic Editors: Marco Uttieri, Ylenia Carotenuto, Iole Di Capua and Vittoria Roncalli

Received: 30 November 2022
Revised: 28 December 2022
Accepted: 2 January 2023
Published: 9 January 2023

1. Introduction

Zooplankton provides an important functional component of trophic webs and biogeochemical cycling [1]. Zooplankton mediates energy and matter translocation between the pelagic and benthic realm through diurnal migration and passive sinking of particulate organic matter [2]. The spatial distribution and abundance of zooplankton are affected by the transport of water masses as well as by different physical, chemical and biological effects of global change. Global change affects the earth systems including land, oceans, atmosphere, the poles, biogeochemical cycles, biosphere including human populations and society. Global changes of the last 2 centennials caused the change of climate, atmospheric ozone depletion, desertification on land, acidification of aqueous environments including the oceans, pollution in general, species extinctions and distributional range changes, and other large-scale biotic shifts (UN—Oceans, URL). In the oceans, global climate forcing factors provide changes at large spatial scale and regional hydrodynamic circulation patterns across different time scales. In addition, factors such as movement of tectonic plates, volcanic activity causing tsunamis, and biological processes including anthropogenic activities were linked to changed scenarios of the earth system [3–7].

Hydrothermal vents (HVs) caused by suboceanic volcanic activity have several characteristics in common with characteristics summarized as global change (e.g., elevated CO_2 and temperature, low oxygen and pH, elevated trace metal availability, sulfate compounds and turbidity). Ever increased levels of CO_2 and other gases forming acids in the aqueous phase provide ocean acidification with pH reductions in oceanic waters with consequences on oceanic biota [8]. Decrease pH and Eh levels characteristic for CO_2 vents increases the bioavailability and dissolution and/or desorption of metalloids and trace metals [9]. This keeps trace elements (Fe, Cd, Co, Mn, Cu, V and Cr) in solution and bioavailable.

HVs areas are characterized by turbid waters containing elevated trace metal contents. This is comparable to coastal waters with intertidal areas or coral fringe reefs that are at risk by above factors due to anthropogenic activity such as mining, industrial emissions, construction work and natural phenomena, such as heavy rain flushing, landslides which increasingly threaten the coastal waters of a future ocean.

It was suggested to use HVs as templates or natural laboratories that allow research on marine organisms in the highly adverse physicochemical conditions of HVs compared to areas without HVs. HV biota could provide insights in the evolutionary, ecological, genetic, behavioral, physiological and molecular adaptations to extreme marine environments that could be compared with their next phylogenetic relatives away from HV sites [10].

The effects of HVs on marine zooplankton were rarely studied [11]. In the few reports on zooplankton, however, oceanic venting areas have favorable effects on the composition of primary producers (phytoplankton) and the composition of zooplankton that are related to the distribution and abundances of higher trophic levels [12,13]. For their ease of access and allowing revisits, experimental approaches in a cost-saving mode, and linking chemo- and photosynthetic energy pathways, particularly shallow HVs are expected to provide suitable natural laboratories. This holds for studies of environmental extremes, biotic adaptations, and allowing the prediction of responses to a future ocean and its living and non-living resources [10].

We question here whether HVs can provide templates for a 'Future ocean scenario' for zooplankton as well. The goals of our contribution are to: (1) survey existing knowledge about zooplankton at HVs; (2) relate this to global change phenomena; (3) evaluate whether the 'HVs as natural laboratory for global change' concept is suitable for questions related to zooplankton ecology.

2. Zooplankton Research near Hydrothermal Vents

According to [14] studying the shallow HVs at Kueishan Island, Taiwan, taxon diversity and densities of mesozooplankton were increased (for abundance three times higher) at the HV side. This occurred to most zooplankton groups, among others to dinoflagellates, appendicularians, pteropods and copepods (providing the highest number of 34 species). It was reported earlier that HVs increased the assemblage composition and biomass of zooplankton especially at shallow depths in Matupi Harbor (Papua New Guinea) [15,16]. Skebo et al. [13] link the high abundance of copepods in waters adjacent to hydrothermal vents with patchy distributions caused by the avoidance of harsh environments close to the HVs and by the avoidance of jellyfish by swarming behavior. This positive effect may also be caused by hydrothermal fluids that enrich nutrients for algae and increase primary productivity [15] from chemosynthesis and photosynthesis at shallow depths [16]. Such elevated primary productivity would then support elevated densities of zooplankton in HV areas.

Cage experiments resulted in high mortality (>95%) of planktonic copepods that were translocated to HVs at depths of 1–13 m above the seafloor next to HVs of KST island [17]. The mortality value was three times higher than that at distant control sites which were not affected by HV plumes (with 20–30% mortality). There are several reports on the trophic position of HV zooplankton. Hung et al. [18] explained the relatively low C/N ratios of the precipitating particulate organic matter from the HV field of Kueishan Island with a high zooplankton contribution. A food web study by Wu et al. [19] applying $\delta^{13}C$ and $\delta^{15}N$ analysis revealed that the water-column-derived fraction of dead zooplankton provides important energy supplements to carnivores and scavengers like the HV crab *Xenograpsus testudinatus*. Further isotopic niche analysis through this study demonstrated that the contribution of 200 dead zooplankter as a food source to vent crabs living in the center and periphery varied from >34% to $\leq$18%. The results of Chang et al. [2] based on isotope analyses showed that photosynthetic and chemosynthetic producers contributed nearly equally to carbon fixation that is fueling the HV system at Kueishantao. In their study, the authors found both zooplankton and HV crabs acted as important trophic mediators between water column and sea bottom. The results of another isotope study by Wang et al. [20] partially contradicted the above findings in that trophic provisions at the shallow-water HVs of Kueishan Island were mainly provided by phototrophic production (microalgal contribution: 26–54%), then by zooplankton (19–34%) and to a minor extend by chemosynthetic production (14 26%).

3. Variable Hydrographic Effects of HV Effluent Temperature, pH and Chemistry Affecting Zooplankton

The sea floor of HV fields provides a rather heterogeneous environment. HV fluids can reach temperatures of about 116 °C in shallow vents like at Kueishan Island [21] with demarcated thermoclines. Vent fluids and surrounding waters with contrasting chemical and physical characteristics often show strong gradients. HV effluents affect the chemistry to a larger extent at the surface than at the bottom, providing differences of physical and chemical characteristics along the water column axis. Observations that zooplankter at the surface are killed by HV effluents and produce "marine snow" composed of sedimenting plankton carcasses and diverse microbiota including HV bacteria [22].

HV fluid emissions are often unstable and sudden outbreaks of HVs vent fluids are commonly providing vents with large hydrological variability [23]. It is expected that HV biota developed adaptations to tolerate such fluctuating environmental conditions, particularly if they are zooplankter drifting in the water column above HVs. This might provide a limitation to our expectation to use HVs as examples for the more gradual alterations during global changes for decades to come. Variations of other environmentally effective parameters within HV systems are expected as well.

4. Conclusions

We conclude that HVs are particularly useful as "natural laboratories" to approach consequences of global change and global climate change for resident biota that had sufficient time to evolutionary and individually adapt to such extreme environments. However, there is no evidence for an endemic zooplankton assemblage as yet, also not from the better investigated shallow water HV situations. The scarce information available indicates that zooplankton is transported to HV areas and is negatively affected by toxic HV plumes and may die there after such an abrupt environmental transition. This way they provide a high input of allochthonous biomass to the respective HV system. There is a substantial knowledge gap about many issues regarding the fate of zooplankton in HVs as outlined below.

Rather generally, the effects of multiple stressors are difficult to disentangle. This holds for examples for basic phenomena related to simultaneously acidified and warming oceans. The interaction of just these two stressors may differ with taxon, populations, gender and ontogenetic stages. Organisms associated with HVs were shown to have adaptations regarding their reproduction, morphology and behavior. HV biota were also shown to have evolved molecular adaptations to an extreme environment and specialized receptors to find or avoid HVs and their effluents in order to aggregate there or to avoid the HV environment altogether. Such adaptations need to be studied also with zooplankton at HVs since there is a particular knowledge gap here.

The ease of access to shallow-water HV systems -offers scientists the rare opportunity to design meaningful in situ and laboratory experiments. Hydrographic regime changes of the physical and chemical background of zooplankton can instantly be monitored. Fast responses are more difficult to capture in HVs of the deep sea.

The following issues among others related to zooplankton in HV areas are of particular interest: (1) Are there any endemic zooplankton assemblages in HVs? (2) Is zooplankton aggregating or trapped in a toxic environment? (3) Are there taxon-specific differences within patchy distributions? (4) Are the measured higher zooplankton densities near vents caused by dead zooplankton that settles at HV sites? (5) How are the ratios of dead versus alive zooplankton? (5) What are the ultimate mechanisms of toxicity causing mortality among zooplankton at HVs? (6) What are particular mechanisms or adaptations to avoid toxicity effects? (7) To what extent are dead versus alive zooplankter vertically segregated in the water column or are they advectively transported?

Author Contributions: All authors contributed to the study conception and design. All authors have read and agreed to the published version of the manuscript.

Funding: The financial support from the National Science and Technology Council, Taiwan (Grant Nos. MOST 108-2621-M-019-003, MOST 109-2621-M-019-002, MOST 110-2621-M-019-001 and MOST 111-2621-M-019-001) and Center of Excellence for Ocean Engineering, NTOU, Taiwan (Grant No. 109J13801-51, 110J13801-51, 111J13801-51) to J.-S. Hwang is acknowledged here. Grants from MOST to Tan Han Shih (= Hans-Uwe Dahms) are gratefully acknowledged (MOST 107-2621-M-037-001, MOST 108-2621-M-037-001, and MOST 109-2621-M-037-001 to T.H. Shih).

Conflicts of Interest: The authors declare no conflict of interest related to this manuscript.

References

1. Lomartire, S.; Marques, J.C.; Gonçalves, A.M.M. The key role of zooplankton in ecosystem services: A perspective of interaction between zooplankton and fish recruitment. *Ecol. Indic.* **2021**, *129*, 107867. [CrossRef]
2. Lough, A.J.M.; Tagliabue, A.; Demasy, C.; Resing, J.A.; Mellett, T.; Wyatt, N.J.; Lohan, M.C. The Impact of Hydrothermal Vent Geochemistry on the Addition of Iron to the Deep Ocean. *Biogeosciences Discuss.* **2022**, 1–23. [CrossRef]
3. Chang, N.-N.; Lin, L.-H.; Tu, T.-H.; Jeng, M.-S.; Chikaraishi, Y.; Wang, P.-L. Trophic Structure and Energy Flow in a Shallow-Water Hydrothermal Vent: Insights from a Stable Isotope Approach. *PLoS ONE* **2018**, *13*, e0204753. [CrossRef]
4. Chen, S.; Zhang, Y.; Wu, Q.; Liu, S.; Song, C.; Xiao, J.; Band, L.E.; Vose, J.M. Vegetation Structural Change and CO2 Fertilization More than Offset Gross Primary Production Decline Caused by Reduced Solar Radiation in China. *Agric. For. Meteorol.* **2021**, *296*, 108207. [CrossRef]
5. Rampino, M.R.; Caldeira, K.; Zhu, Y. A Pulse of the Earth: A 27.5-Myr Underlying Cycle in Coordinated Geological Events over the Last 260 Myr. *Geosci. Front.* **2021**, *12*, 101245. [CrossRef]
6. Shi, G.; Yan, H.; Zhang, W.; Dodson, J.; Heijnis, H. The Impacts of Volcanic Eruptions and Climate Changes on the Development of Hani Peatland in Northeastern China during the Holocene. *J. Asian Earth Sci.* **2021**, *210*, 104691. [CrossRef]
7. Alhamid, A.K.; Akiyama, M.; Ishibashi, H.; Aoki, K.; Koshimura, S.; Frangopol, D.M. Framework for Probabilistic Tsunami Hazard Assessment Considering the Effects of Sea-Level Rise Due to Climate Change. *Struct. Saf.* **2022**, *94*, 102152. [CrossRef]
8. Bakirci, K.; Kirtiloglu, Y. Effect of Climate Change to Solar Energy Potential: A Case Study in the Eastern Anatolia Region of Turkey. *Environ. Sci. Pollut. Res.* **2022**, *29*, 2839–2852. [CrossRef]
9. Van Dover, C.L. Impacts of Anthropogenic Disturbances at Deep-Sea Hydrothermal Vent Ecosystems: A Review. *Mar. Environ. Res.* **2014**, *102*, 59–72. [CrossRef]
10. Chan, I.; Tseng, L.-C.; Kâ, S.; Chang, C.-F.; Hwang, J.-S. An Experimental Study of the Response of the Gorgonian Coral Subergorgia Suberosa to Polluted Seawater from a Former Coastal Mining Site in Taiwan. *Zool. Stud.* **2012**, *11*, 27–37.
11. Dahms, H.-U.; Schizas, N.V.; James, R.A.; Wang, L.; Hwang, J.-S. Marine Hydrothermal Vents as Templates for Global Change Scenarios. *Hydrobiologia* **2018**, *818*, 1–10. [CrossRef]
12. Skebo, K.; Tunnicliffe, V.; Garcia Berdeal, I.; Johnson, H.P. Spatial Patterns of Zooplankton and Nekton in a Hydrothermally Active Axial Valley on Juan de Fuca Ridge. *Deep. Sea Res. Part I Oceanogr. Res. Pap.* **2006**, *53*, 1044–1060. [CrossRef]
13. Kâ, S.; Hwang, J.-S. Mesozooplankton Distribution and Composition on the Northeastern Coast of Taiwan during Autumn: Effects of the Kuroshio Current and Hydrothermal Vents. *Zool. Stud.* **2011**, *9*, 155–163.
14. Tarasov, V.G.; Gebruk, A.V.; Shulkin, V.M.; Kamenev, G.M.; Fadeev, V.I.; Kosmynin, V.N.; Malakhov, V.V.; Starynin, D.A.; Obzhirov, A.I. Effect of Shallow-Water Hydrothermal Venting on the Biota of Matupi Harbour (Rabaul Caldera, New Britain Island, Papua New Guinea). *Cont. Shelf Res.* **1999**, *19*, 79–116. [CrossRef]
15. Tarasov, V.G.; Gebruk, A.V.; Mironov, A.N.; Moskalev, L.I. Deep-Sea and Shallow-Water Hydrothermal Vent Communities: Two Different Phenomena? *Chem. Geol.* **2005**, *224*, 5–39. [CrossRef]
16. Dahms, H.-U.; Hwang, J.-S. Mortality in the ocean—With lessons from hydrothermal vents off Kueishan Tao, Ne-Taiwan. *J. Mar. Sci. Technol.* **2013**, *21*, 12. [CrossRef]
17. Hung, J.-J.; Peng, S.-H.; Chen, C.-T.; Wei, T.-P.A.; Hwang, J.-S. Reproductive adaptations of the hydrothermal vent crab Xenograpus testudinatus: An isotopic approach. *PLoS ONE* **2019**, *14*, e0211516. [CrossRef]
18. Wu, Y. Occurrence of Ammonium in the Acidic-Circumneutral Coastal Groundwater of Beihai, Southern China: Δ15N, Δ13C, and Hydrochemical Constraints. *J. Hydrol.* **2022**, *615*, 128712. [CrossRef]
19. Wang, T.-W.; Lau, D.C.P.; Chan, T.-Y.; Chan, B.K.K. Autochthony and Isotopic Niches of Benthic Fauna at Shallow-Water Hydrothermal Vents. *Sci. Rep.* **2022**, *12*, 6248. [CrossRef]

20. Kuo, F.-W. *Preliminary Investigation of the Hydrothermal Activities Off Kueishantao Island*; NSYSU: Kaohsiung, Taiwan, 2001.
21. Jeng, M.-S.; Ng, N.K.; Ng, P.K.L. Hydrothermal Vent Crabs Feast on Sea 'Snow'. *Nature* **2004**, *432*, 969. [CrossRef]
22. Hwang, J.S.; Lee, C.S. The mystery of underwater world for tourism of Turtle Island, Taiwan. *Northeast Coast Natl. Scen. Area Adm. Tour Bur. Minist. Transp. Commun. Taiwan* **2003**, 1–103.
23. McDermott, J.M.; Parnell-Turner, R.; Barreyre, T.; Herrera, S.; Downing, C.C.; Pittoors, N.C.; Pehr, K.; Vohsen, S.A.; Dowd, W.S.; Wu, J.-N.; et al. Discovery of Active Off-Axis Hydrothermal Vents at 9° 54′N East Pacific Rise. *Proc. Natl. Acad. Sci. USA* **2022**, *119*, e2205602119. [CrossRef]

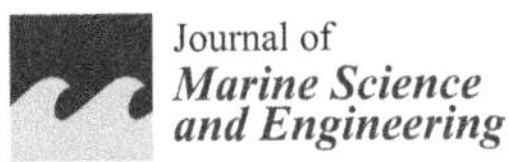
Journal of
Marine Science and Engineering

Article

Cold Dome Affects Mesozooplankton Communities during the Southwest Monsoon Period in the Southeast East China Sea

Yan-Guo Wang [1,†], Li-Chun Tseng [2,†], Xiao-Yin Chen [1], Rou-Xin Sun [1], Peng Xiang [1], Bing-Peng Xing [1], Chun-Guang Wang [1] and Jiang-Shiou Hwang [2,3,4,*]

1 Third Institute of Oceanography, Ministry of Natural Resources, Xiamen 361005, China
2 Institute of Marine Biology, National Taiwan Ocean University, Keelung 202301, Taiwan
3 Center of Excellence for Ocean Engineering, National Taiwan Ocean University, Keelung 202301, Taiwan
4 Center of Excellence for the Oceans, National Taiwan Ocean University, Keelung 202301, Taiwan
* Correspondence: jshwang@mail.ntou.edu.tw; Tel.: +886-9-35289642; Fax: +886-2-24629464
† These authors contributed equally to this work.

Abstract: In order to better understand the cold dome influence on zooplankton community structure, zooplankton samples were collected during the southwest monsoon prevailing period from the southeast waters of the East China Sea. To reduce the bias caused by different sampling months, the samples were collected in June 2018 and in June 2019. An obvious cold dome activity was proven by images of remote sensing satellites during the June 2018 cruise. In contrast, the research area was much affected by open sea high temperature and water masses during the June 2019 cruise. Significant differences in water conditions were demonstrated by surface seawater temperature, salinity, and dissolved oxygen concentrations between the two cruises. Nevertheless, no significant differences were observed concerning mesozooplankton in general, copepods, large crustaceans, other crustaceans, and pelagic molluscs between the June 2018 and June 2019 cruises. However, the mean abundance of gelatinous plankton was significantly different with 1213.08 ± 850.46 (ind./m^3) and 2955.93 ± 1904.42 (ind./m^3) in June 2018 and June 2019, respectively. Noteworthy, a significantly lower mean abundance of meroplankton, with 60.78 ± 47.32 (ind./m^3), was identified in June 2018 compared to 464.45 ± 292.80 (ind./m^3) in June 2019. Pearson's correlation analysis also showed a highly positive correlation of gelatinous plankton and meroplankton with sea surface temperature ($p < 0.01$). The variation of salinity showed a significant negative correlation with gelatinous plankton abundance ($p < 0.05$), and a highly significant negative correlation with the abundance of meroplankton ($p < 0.01$). Only the abundance of meroplankton showed a positive correlation with dissolved oxygen concentrations ($p < 0.05$). The copepod communities were separated in two groups which were consistent with sampling cruises in 2018 and 2019. Based on the specificity and occupancy of copepods, *Macrosetella gracilis*, *Oithona rigida*, *Cosmocalanus darwinii*, *Paracalanus parvus*, and *Calocalanus pavo* were selected as indicator species for the cold dome effect in the study area during June 2018, whereas the indicator species of warm water impact in the open sea were *Calanopia elliptica*, *Subeucalanus pileatus*, *Paracalanus aculeatus*, and *Acrocalanus gibber* during the June 2019 cruise.

Keywords: Copepoda; meroplankton; gelatinous plankton; Kueishan Island; northeast Taiwan

Citation: Wang, Y.-G.; Tseng, L.-C.; Chen, X.-Y.; Sun, R.-X.; Xiang, P.; Xing, B.-P.; Wang, C.-G.; Hwang, J.-S. Cold Dome Affects Mesozooplankton Communities during the Southwest Monsoon Period in the Southeast East China Sea *J. Mar. Sci. Eng.* **2023**, *11*, 508. https://doi.org/10.3390/jmse11030508

Academic Editors: Marco Uttieri, Ylenia Carotenuto, Iole Di Capua and Vittoria Roncalli

Received: 17 January 2023
Revised: 16 February 2023
Accepted: 21 February 2023
Published: 26 February 2023

1. Introduction

Zooplankton, which is considered as a dominant link between primary production and upper trophic levels, plays a pivotal role in shaping marine ecosystems [1,2]. Zooplankton can also be used as bio-indicators of environmental quality and water masses due to their high dependence on environmental conditions and fast responses to environmental variations [3–5]. The abundance and distribution of zooplankton can be affected among other factors by temperature, salinity, and primary production [5–8]. Several studies reported that water masses influenced the zooplankton communities at different spatial and temporal scales in waters of northeast Taiwan [5,9–15]. The distribution, abundance, and species composition of copepods are associated with different water masses in the upwelling waters off northeastern Taiwan [16]. Conversely, increased zooplankton concentration was observed around the upwelling [17]. Tseng et al. [13,18] reported that distribution patterns of mesozooplankton and copepod communities in northern Taiwan varied spatially with distance to land but did not discuss the possibility of the influence of an upwelling cold dome. However, both diatom and larval fish assemblages were strongly affected by monsoon derived water mass succession and topographic upwelling in the offshore of northeast Taiwan [19,20].

The northern shelf of Taiwan is an extremely dynamic oceanic region. The Kuroshio Current (KC), which is the western boundary current of the North Pacific Subtropical Gyre, flows northeastwards along the eastern coast of Taiwan island with occasionally intrusion on the shelf [21–23]. Cold domes, which are formed by the KC intrusion at the edge of the shelf, are observed in the south of the East China Sea [22–26]. Takahashi et al. [27] described a current that flows in the opposite direction to the KC in northeastern Taiwan by high-frequency radar measurements. This current was recently confirmed and named the northeastern Taiwan counter current (NETCC) [28]. In situ observations proved that the main sources and dynamic mechanisms of the NETCC are the counterclockwise flow in the cold dome off northeastern Taiwan and the southward intrusion of the coastal current in northern Taiwan [29,30].

The research area is located in Yilan Bay in northeastern Taiwan, where the water masses are mainly influenced by the interactions of the NETCC, KC, China costal current, and tidal currents [31]. In the present study, a hypothesis was pursued that mesozooplankton communities and the copepod assemblage were influenced by an upwelling cold dome during the southwest monsoon prevailing period in the southeastern East China Sea. The particular aim of the present study was to understand: (1) whether and how the cold dome affected the mesozooplankton communities in the research area; (2) which functional groups of the mesozooplankton were particularly affected from the cold dome phenomenon; and (3) which species could serve as indicator species for the cold dome.

2. Materials and Methods

2.1. Sampling Area and Sampling Strategy

In an attempt to reduce the bias caused by different sampling months, both the zooplankton samples were selected in June in the years 2018 and 2019. The satellite image revealed an upwelling cold dome appearing in the southeast of the East China Sea during the 1 June 2018 cruise (Figure 1A). In order to understand the cold dome effects on the mesozooplankton community in the research area, the samples were recollected from the same stations during the weak phase of the cold dome around the 21 June 2019 (Figure 1B). In total, 9 sampling stations were arranged around Kueishan Island at the edge of the cold dome, Southeast East China Sea (Figure 1C). Samples were collected during the Southwest monsoon prevailing period during the 2018 and 2019 cruises. All samples were collected by horizontal tows from surface waters by a standard north pacific plankton net with a mouth diameter of 45 cm and a mesh size of 200 μm. The tow was kept trawling for 10 min at the 5 m depth layer from near the sea surface with a speed of about 1.0 m/s. The filtered water volume was calculated based on a Hydro-Bios flow-meter, which was mounted in the center of the net mouth. Zooplankton samples gathered at the end of the sampling net were immediately preserved in 5% seawater buffered formaldehyde solution on board for identification and counting in the laboratory. Water temperature, salinity, and dissolved oxygen (DO) were measured prior to the collection of zooplankton samples with a SeaBird CTD sensor (Canada) instrument which was mounted on a rosette sampler.

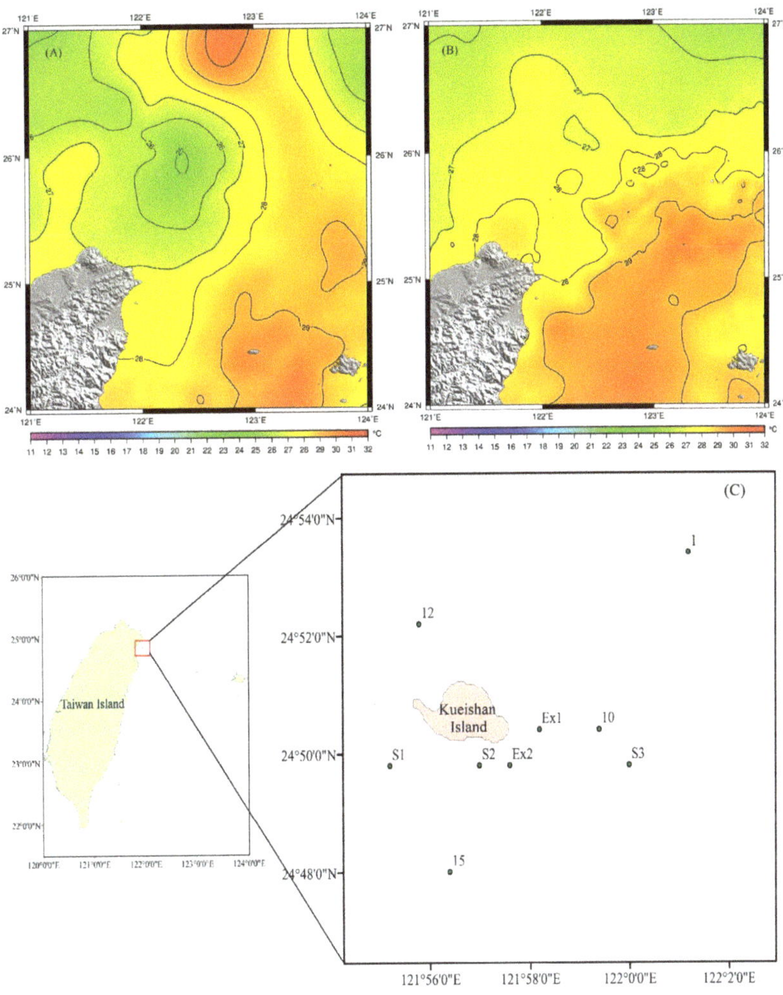

Figure 1. Satellite images of the sea surface temperature on 1 June 2018 (**A**) and 21 June 2019 (**B**), and the maps of the research area and sampling stations (**C**).

2.2. Sample Handling and Identification

In the laboratory, mesozooplankton samples were divided by a Folsom splitter until about 300–500 individuals remained in the subsample. Subsamples were counted and identified by a dissecting microscope (SMZ1500, Nikon, Japan). Adult copepods were counted and identified at species level. Immature copepods were counted at genus level. Other mesozooplankton groups were counted at class or phylum level. Several taxonomic literature and illustrations were used to identify the mesozooplankton species [32–35].

2.3. Data Processing and Statistical Analysis

The mesozooplankton was classified into 6 groups: gelatinous plankton, meroplankton, pelagic Mollusca, Copepoda, large Crustacea, and other Crustacea based on their ecological functioning. Gelatinous plankton included Cnidaria, Urochordata, and Chaetognatha. Mysidacea, Euphausiacea, and Decapoda that were included into the group of large crustaceans. Pteropoda and Heteropoda were considered as taxa of pelagic Mollusca. Taxa of other crustaceans comprised of Amphipoda and Cladocera. The density of individuals belonging to the different groups was calculated by the individuals divided by the filtered water volume and given the unit individuals/m^3 (ind./m^3). Dry weight was estimated referring to the fitted mixed contribution model which was suggested by Tseng [13].

The difference of environmental and biological parameters between different cruises was compared by an independent sample *t*-test. The correlation of mesozooplankton and environmental parameters was calculated by Pearson's correlation analysis. The specificity and occupancy referred to the definition of Dufrene and Legendre [36] and was calculated by the following formula:

$$\text{Specificity} = \frac{N_{individualsi,j}}{N_{individuals_i}};\ \text{Occupancy} = \frac{N_{sites_{i,j}}}{N_{sites_j}}$$

Specificity equaled the ratio of the average abundance of the *i*th species in *j*th community ($N_{individuals_{i,j}}$) and the sum of the average abundance of the *i*th species in different communities ($N_{individuals_i}$). The occupancy was defined as the relative frequency of occurrence of the *i*th in *j*th community. The specificity–occupancy plot was achieved in R language, applying ggplot 2 package.

Group average linkage was used in the cluster analysis with Bray–Curtis similarity. Before cluster analysis, the abundance of copepods species was log (x+1) transformed. Assemblage analysis was performed with Primer 6.

3. Results

3.1. Environmental Characters in the Research Area

The variations in surface sea temperature (Figure 2A), salinity (Figure 2B), and dissolved oxygen (Figure 2C) of each sampling station during two cruises showed a marked difference. The occurrence of the cold dome influenced the hydrological environment of the surrounding waters as revealed by the above three parameters. Furthermore, the results of an independent sample *t*-test showed that the average surface water temperature was significantly different (t = 6.90, $p < 0.001$) during the summer cruises of 2018 and 2019 with an average surface water temperature of 26.16 ± 0.29 °C and 27.20 ± 0.34 °C, respectively (Figure 2A-1). A relatively higher average salinity 34.46 ± 0.02 was recorded during the 2018 cruise. The average salinity was 33.84 ± 0.01 during the 2019 cruise which was significantly lower than during the 2018 cruise (t = 77.67, $p < 0.001$) (Figure 2B-1). Similar to the seawater temperature, the surface water DO was also significantly different (t = 35.94, $p < 0.001$) in 2018 and 2019 with an average of 5.58 ± 0.07 (mg/L) and 6.71 ± 0.06 mg/L), respectively (Figure 2C-1).

3.2. Composition of Mesozooplankton

The mean abundance of mesozooplankton of two sampling cruises (n = 18) was 9786.15 ± 5052.66 (ind./m^3) during the Southwest monsoon prevailing period in the research area. The statistical results showed that there was no significant difference ($p > 0.05$, *t*-test) of the mesozooplankton mean abundance between 2018 and 2019 with mean values of 9792.60 ± 5634.76 (ind./m^3) and 9779.70 ± 4743.34 (ind./m^3), respectively. Copepods were the most dominant taxon in the research area. The mean abundance of copepods was 8278.71 ± 5392.97 (ind./m^3) in 2018 with a percentage 84.25% of the total mesozooplankton (Figure 3A). During 2019, the copepods accounted for 65.20% of the total mesozooplankton mean abundance with 6168.46 ± 3075.17 (ind./m^3) (Figure 3B). The copepod mean abundance between 2018 and 2019 was not significantly different ($p > 0.05$, *t*-test). Both the large Crustacea, other Crustacea, and pelagic Mollusca were not significantly different between these two cruises. However, the mean abundance of gelatinous plankton, which was the second most abundant in the research area, was significantly lower in 2018 (1213.08 ± 850.46 ind./m^3) than in 2019 (2955.93 ± 1904.42 ind./m^3) ($p < 0.05$). The mean abundance of Cnidaria was 75.95 ± 42.64 (ind./m^3) during the 2018 cruise, which was obviously lower than during the 2019 cruise with a mean abundance 133.24 ± 110.78 (ind./m^3). Mean abundance of urochordata was significantly lower during the 2018 cruise than during the 2019 cruise with values of 931.76 ± 788.86 (ind./m^3) and 2652.60 ± 1682.22 (ind./m^3), respectively. There was no obvious difference in the mean abundance of Chaetognatha in the 2018 and 2019 cruise. Due to the obviously higher mean abundance of Bivalvia larva, Brachyura zoea, and Macrura larva during 2019 cruise, the mean abundance of meroplankton was significantly higher in 2019 than in the 2018 cruise with the mean value 464.45 ± 292.80 (ind./m^3) and 60.78 ± 47.32 (ind./m^3), respectively ($p < 0.05$) (Supplementary Materials Table S1).

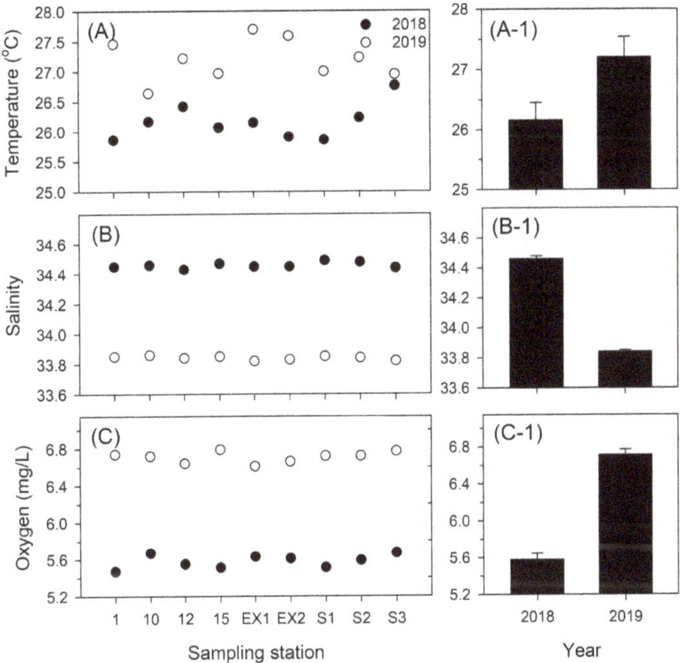

Figure 2. Variation of seawater temperature (**A**); salinity (**B**); and dissolved oxygen (**C**); as well as comparisons of the seawater temperature (**A-1**); salinity (**B-1**); and dissolved oxygen (**C-1**) in average values (mean ± standard deviation) from the cruises in June 2018 and June 2019.

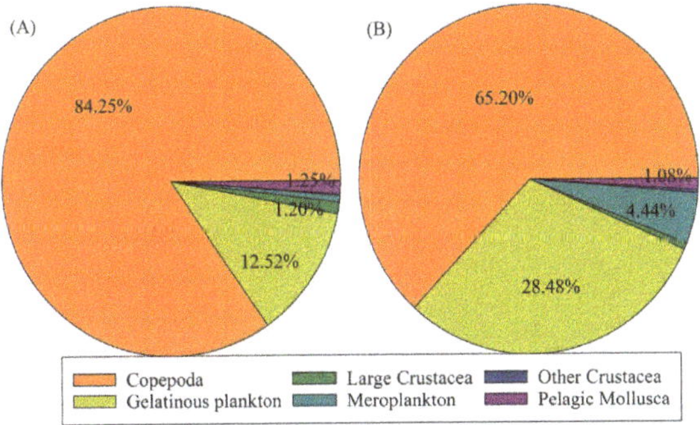

Figure 3. The proportion of functional taxa among mesozooplankton communities collected during the cruises in June 2018 (**A**), and June 2019 (**B**).

In the present study, we could not find a significant correlation between the total mesozooplankton abundance and environment parameters during the research period (Table 1). The variations of Copepoda, large Crustacea, and pelagic Mollusca also showed no correlation with environmental parameters in the research area. In contrast, the variation of gelatinous plankton correlated significantly positively with surface temperature ($p < 0.01$) and negatively with salinity ($p = 0.02$). In contrast, the meroplankton showed a significant negative correlation with salinity ($p < 0.01$), but a positive correlation with surface temperature ($p < 0.01$) and with surface DO ($p = 0.02$) in the research area. The variation of other Crustacea showed only a positive correlation with surface temperature ($p = 0.04$) in the research area.

Table 1. Pearson's correlation of functional groups mean abundance and environmental parameters. The numbers in parentheses are *p*-values.

Functional Groups	Temperature	Salinity	DO
Copepoda	0.02 (0.94)	0.23 (0.36)	−0.14 (0.58)
Gelatinous plankton	0.63 ** (<0.01)	−0.54 * (0.02)	0.43 (0.08)
Large Crustacea	−0.08 (0.77)	0.32 (0.20)	−0.09 (0.73)
Meroplankton	0.75 ** (<0.01)	−0.72 ** (<0.01)	0.53 * (0.02)
Other Crustacea	0.49 * (0.04)	−0.29 (0.25)	0.21 (0.41)
Pelagic Mollusca	−0.06 (0.83)	0.10 (0.69)	0.00 (1.00)
Total abundance	0.27 (0.27)	−0.02 (0.95)	0.05 (0.83)

**. Correlation is significant at the 0.01 level (2-tailed). *. Correlation is significant at the 0.05 level (2-tailed).

The dry weight of mesozooplankton ranged from 244.70 mg/m^3 to 1063.21 mg/m^3 with mean 555.56 ± 255.69 (mg/m^3) in research area during the 2018 cruise. The dry weight was significantly higher during the 2019 cruise with a mean value of 987.78 ± 521.58 (mg/m^3). The dry weight ranged from 458.67 mg/m^3 to 2016.21 mg/m^3 during the 2019 cruise.

3.3. Copepod Assemblages

Totally, 76 copepod species were identified belonging to the Calanoida, Cyclopoida, Harpacticoida, and Poecilostomatoida during the cruises of the present study. In total, 58 and 53 copepod species were found in 2018 and 2019, respectively (Supplementary Materials Table S1). The percentage of Calanoida copepods was dominant and occupying 56.90% and 64.15% of all copepod species in 2018 and 2019, respectively. These were followed by Poecilostomatoida copepods with a percentage of 34.48% in 2018 and 24.53% in 2019.

The results of the cluster analysis indicated that all samples were divided into three groups at a 37.61% similarity level (Figure 4). Samples collected from S1 station in 2019 had a relative lower similarity to other stations, being arranged into a group with the presence of only 19 copepod species. The dominant species was *Temora turbinata* with a relative abundance of 59.04% in this sample. This was followed by *Canthocalanus pauper* and *Paracalanus aculeatus* with a relative abundance of 14.76% and 7.38%, respectively. The remaining eight samples collected in 2019 were gathered into group b with a similarity of 52.68%. Totally, 50 copepods species were identified, and the dominant species was *T. turbinanta* with a relative abundance of 39.86% in this group. The relative abundance of *P. aculeatus* was 12.29%, representing the second most dominant species in this group. The relative abundance was 6.33% and 5.74% for *Acrocalanus gibber* and *C. pauper*, respectively, in group b. In group c, which included all nine samples collected from the 2018 cruise, 58 copepods species were recorded. There were seven species with relative abundances higher than 5% in this group. The mean abundance of *T. turbinata* accounted for 30.88% of the total mean abundance in this group. The following relative abundance was 11.01% for *Oncaea venusta* in group c. The relative abundance of *Macrosetella gracilis*, *Acrocalanus gracilis*, *Paracalanus parvus*, *Farranula gibbula*, and *Oithona rigida* was 8.37%, 7.86%, 7.78%, 6.73%, and 6.30%, respectively. The results of the cluster analysis clearly indicated a significant difference in the composition of the copepod assemblage between the cruises in June of these two years.

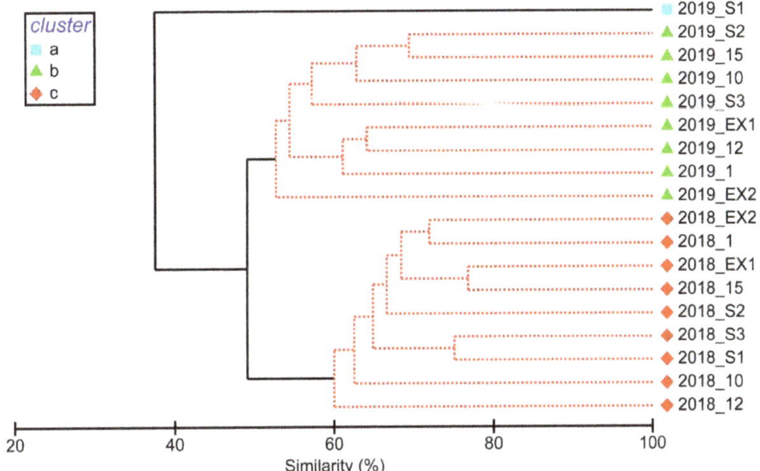

Figure 4. The results of the cluster analysis of the copepod community in each sample, measured by Bray-Curtis similarity distances.

3.4. Specificity and Occupancy of Copepods

The specificity and occupancy of copepods being counted at species level, were calculated and projected to a plot of these two cruises (Figure 5). The results showed that most copepods species were characterized by a relative lower specificity and occupancy during these two cruises. Indicator species are shown in dotted boxes of Figure 5 with specificity and occupancy both higher than 0.8. Indicator species were specific to certain conditions and widely distributed in that environment. There were five species selected as indicator species in 2018. In contrast, four species were selected as indicator species with both specificity and occupancy greater than 0.8 in 2019 (Table 2).

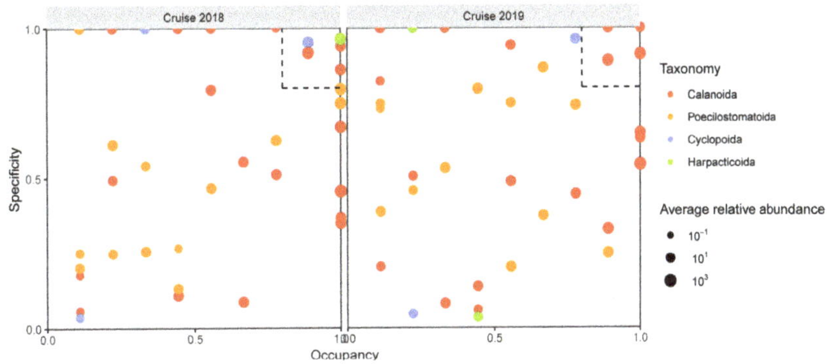

Figure 5. Specificity-occupancy plots of copepods collected from cruises in June 2018 and June 2019.

Table 2. Mean abundance, occupancy, and comparative results of mean abundances using *t*-test for indicator species recorded during the sampling cruises in 2018 and 2019. The numbers in parentheses represent the occupancy (%). * Indicates the indicator species during that cruise. N.A. means not available.

Species Name	2018	2019	*t*-Test
Macrosetella gracilis	384.04 ± 284.23 * (100)	14.45 ± 3.35 (44.44)	$p < 0.01$
Oithona rigida	289.12 ± 624.92 * (88.89)	14.38 ± 30.39 (22.22)	$p < 0.05$
Cosmocalanus darwinii	45.28 ± 54.78 * (100)	2.95 ± 8.85 (44.44)	$p < 0.01$
Paracalanus parvus	356.88 ± 297.73 * (88.89)	32.22 ± 5.52 (33.33)	$p < 0.01$
Calocalanus pavo	108.24 ± 78.30 * (100)	17.47 ± 34.73 (44.44)	$p = 0.085$
Calanopia elliptica	N.A.	21.10 ± 28.36 * (100)	$p < 0.01$
Subeucalanus pileatus	N.A.	36.12 ± 64.73 * (88.89)	$p < 0.01$
Paracalanus aculeatus	48.21 ± 48.62 (66.67)	495.24 ± 401.77 * (100)	$p < 0.01$
Acrocalanus gibber	31.46 ± 55.93 (44.44)	254.70 ± 231.24 * (88.89)	$p < 0.01$

The specificity of *Macrosetella gracilis* was 96.37% with a mean abundance of 384.04 ± 284.22 (ind./m^3) in 2018 (Figure 5). This was followed by *Oithona rigida* and *Cosmocalanus darwinii* with the specificity 95.26% and 93.89%, and mean abundances of 289.12 ± 624.92 (ind./m^3) and 45.28 ± 54.78 (ind./m^3). In 2019, the mean abundance of *M. gracilis*, *O. rigida*, and *C. darwinii* was only 14.45 ± 43.35 (ind./m^3), 14.38 ± 30.39 (ind./m^3), and 2.95 ± 8.85 (ind./m^3), respectively (Table 2). The results of Pearson's correlation analysis showed that there was no significant correlation between the variation of *O. rigida* and environmental parameters in the research area. The variation of *M. gracilis*, however, was significantly positive correlated with salinity, surface temperature ($p = 0.01$) and DO ($p = 0.05$), respectively. The results of the analysis implied that *M. gracilis* should come from high temperature and high salinity Kuroshio water, and its abundance variation could be influenced by the interplay of Kuroshio and East China Sea waters in research area. A positive correlation of the variation of *C. darwinii* and salinity was detected in the study area. The specificity of *Paracalanus parvus* and *Calocalanus pavo* was 91.72% and 86.10% with a mean abundance of 356.88 ± 297.73 (ind./m^3) and 108.24 ± 78.30 (ind./m^3) in 2018. During the 2019 cruise, the mean abundance of *Paracalanus parvus* and *Calocalanus pavo* was only 32.22 ± 50.52 (ind./m^3) and 17.47 ± 34.73 (ind./m^3). Pearson's correlation analysis results showed that both *P. parvus* and *C. pavo* were significantly positively ($p < 0.01$) correlated with surface salinity in the study area.

During the 2019 cruise, the specificity of both *Calanopia elliptica* and *Subeucalanus pileatus* were 100% with a mean abundance of 21.10 ± 28.36 (ind./m^3) and 36.12 ± 64.73 (ind./m^3), respectively (Figure 5). *Paracalanus aculeatus* and *Acrocalanus gibber* were indicator species with specificities of 91.13% and 89.01% and mean abundance of 495.24 ± 401.77 (ind./m^3) and 254.70 ± 231.24 (ind./m^3), respectively. The variation of *S. pileatus* showed no significant correlation with environmental parameters. Whereas the variation of *C. elliptica* showed a significant negative correlation with salinity. *A. gibber* also showed a significantly negative correlation with salinity and a positive correlation with sea surface temperature ($p = 0.01$) and DO ($p = 0.05$). The variation of *P. aculeatus* correlated significantly negatively with salinity and positively with sea surface temperature ($p = 0.01$) and DO ($p = 0.05$).

4. Discussion

Intermittent but common upwelling in the southern East China Sea was associated with local cyclonic circulation and inshore intrusion of Kuroshio waters interacting with ocean topography [22,37,38]. Remote sensing satellite images and in situ measured hydrographic parameters indicated that the study area was obviously affected by cold dome upwelling water in June 2018. Conversely, the study area was mainly dominated by oceanic oligotrophic and warm waters in June 2019. Relatively lower surface water temperature and higher salinity in the present study were observed during the June 2018 cruise. This is mainly explained by the upwelling bringing cold and saline Kuroshio waters to the surface and developing a cold dome region in the area. The cold dome waters were transported to the study area by NETCC during June 2018 [31]. Withal, several studies reported that the cold dome provided a major fishing ground due to the higher concentration of nutrients [16,39,40]. Previous studies noted that physical factors (monsoon, stratification, and upwelling) could reduce the availability of dissolved oxygen [41]. An obviously lower DO concentration was observed in the surface water at the edge of the cold dome during the June 2018 cruise of the present study. DO concentrations were thought to diminish in the upwelling area due to a strong remineralization of sinking organic matter [42]. From a biological perspective, changes in the relative dominance of functional groups in the community could also affect the DO concentrations in the upwelling area [43–47].

Several studies reported that the abundance and distribution of zooplankton were influenced severely by environmental factors such as temperature, salinity, and primary production [5–8,15]. The composition of mesozooplankton was also influenced by environmental parameters in the research area [4,5,11–14]. In the present study, despite no obvious difference of mesozooplankton mean abundance between the cruises in June 2018 and June 2019, the community structure was significantly influenced by the cold dome during June 2018. The relative abundance of copepods was higher in 2018 than in 2019 but without significant differences by statistical comparison between these two study cruises. Madhupratap et al. [48] reported that a higher abundance of copepods was found due to a relatively higher primary production in the upwelling area. In this research, surface water was obviously affected by the cold dome with relative lower temperature and higher salinity during 2018 cruise. Gelatinous plankton and meroplankton showed a significantly correlation with surface water temperature and salinity. The abundance and relative abundance of gelatinous plankton and meroplankton were significantly higher during the June 2019 than during the June 2018 cruise. Thus, we inferred that gelatinous plankton and meroplankton probably had a negative correlation with the cold dome. This phenomenon is explained by the energy contribution from high phytoplankton production followed by a peak production of secondary producers with abundant copepods, as well as many fish species in the upwelling area [49]. In convergent ecosystems, however, food chains are characterized by small flagellate phytoplankton, an abundance of small copepods, and large numbers of gelatinous plankton [50].

Temora turbinata, *C. pauper*, and *O. venusta* were the most common species both during the June 2018 and June 2019 cruise. Due to the effect of cold upwelling waters, the mean abundance of *T. turbinata* and *C. pauper* was slightly lower in June 2018. Several studies proved that *T. turbinata* was commonly found around coastal waters of Taiwan [9,14,51–56]. Previous reports pointed out that *T. turbinata* preferred to occur in waters with seawater temperatures higher than 28 °C, being considered a warm-water indicator species in the northwest of Taiwan [54,57,58]. Consistent with previous research, *T. turbinata* was the most abundant species and also showed an increased tendency to be accompanied with a relative higher water temperature in the present study. This was followed by *Oncaea venusta*, which was one of the easiest oncaeid species to be recognized in mesozooplankton samples, with a mean abundance of 505.35 ind./m^3 in June 2018. Tseng et al. [13] found *O. venusta* as a dominant species in the boundary waters of the East China Sea and Kuroshio Current during the southwest-northeast monsoon transition period. This species commonly provides a substantial fraction of copepod assemblages in coastal and oceanic regions in middle to low latitude epi- and mesopelagic waters worldwide [59–63]. Júnior et al. [64] reported that the abundance and biomass of *O. venusta* positively correlated with seawater temperature in the south Atlantic, ranging from 21 to 27 °C. However, the abundance of *O. venusta* was correlated with relatively lower surface water temperatures of 22–24 °C rather than 26 °C in the northeast of Taiwan [4]. In the present study, the mean abundance of *O. venusta* was significantly higher in June 2018 than during the June 2019 cruise with a relative abundance of 11.01% and 4.07%, respectively, which also showed an inclination to a relatively lower seawater temperature.

The harpacticoid copepod *Macrosetella gracilis*, which was surmised as a most suitable indicator species for upwelling influence in the study area during the June 2018 cruise, is a species that occurs globally in tropical and subtropical oceans and is typically found in association with blooms of *Trichodesmium* spp. [65,66]. In coincidence with the present study results, *M. gracilis* is an indicator species for monsoon derived cold water masses during the northeast monsoon prevailing period in the study area [5]. Species in *Oithona* are described as the most ubiquitous and abundant oceanic copepod species worldwide [67]. The cyclopoid copepod *Oithona rigida*, which makes use of a variety of food items, was less sensitive and more tolerant to extreme environmental conditions, such as high temperature, low nutrients, and low pH [68], and had a higher productivity than Calanoida copepods [69]. In this study, *O. rigida* was also surmised as an indicator species with an occupancy of 88.89% during the June 2018 cruise, showing cold dome upwelling. Our results were supported by the findings of Keister and Tuttle [70], who suggested that species in the genus *Oithona* might migrate to the surface layer due to subsurface hypoxia during upwelling events. The Calanoida copepod *Paracalanus parvus*, playing an important role in ocean fisheries with relatively higher abundance in the western subtropical Pacific, is widely distributed in temperate and tropical regions [71–74]. In addition, *P. parvus* was reported as a coldwater mass indicator species in northeast Taiwan during the northeast monsoon season [5]. The temperate species *Calocalanus pavo* was suggested as an indicator species of the upwelling cold dome influence in June 2018. It was reported as being related to cold-water masses in the northeast of Taiwan with a relative abundance of 0.56% during the monsoonal transition period in 1998 [18]. The omnivorous Calanoida copepod *Calanopia elliptica*, belonging to the Pontellidae family, was recorded in June 2019 with 100% occurrence in the present study. It has been reported as a Lessepsian migration species, indicating the connection between the Red Sea and the southeastern Mediterranean Sea via the Suez Canal [75]. *Subeucalanus pileatus*, a warm coastal and shelf water species, was only recorded during the June 2019 cruise with an occurrence rate of 88.89%. Previous studies emphasized that *P. aculeatus* was widely distributed around Taiwan [5,9,76–78]. In the present study, the abundant *P. aculeatus* was recorded in June 2019, and its abundance was significantly negative correlated with salinity but positively with the change of sea surface temperature. Previous reports found that *A. gibber* was abundant during summer and decreasing during winter and was considered as an indicator species of the Kuroshio Branch Current in the waters of northeast Taiwan [58,78,79]. Abundant *A. gibber* was recorded during June 2019 with the warmest oceanic water affection in the present study. The variations of indicator species in the research area proved that mesozooplankton communities were significantly changing during the southwest monsoon prevailing period from the June 2018 to the June 2019 cruise.

5. Conclusions

Based on two years of zooplankton sampling in the same months, significant differences in the community structure of mesozooplankton and copepod assemblages were revealed in the East China Sea northeast of Taiwan. The hypothesis that the presence of a cold dome was affecting the composition and overall density of mesozooplankton communities could be strengthened. The information obtained from the present study on functional groups and indicator species of copepods could contribute to future studies on the effects of cold domes on secondary producers. Studies on the dynamic composition and functional groups of mesozooplankton and dominant copepod species are needed to better understand the impact of cold domes in the future.

Supplementary Materials: The following supporting information can be downloaded at: https://www.mdpi.com/article/10.3390/jmse11030508/s1, Table S1: Species list with mean abundance and standard error.

Author Contributions: Conceptualization and methodology, Y.-G.W. and J.-S.H.; software, Y.-G.W., C.-G.W., P.X., and L.-C.T.; validation, R.-X.S. and X.-Y.C.; formal analysis, Y.-G.W. and L.-C.T.; investigation, Y.-G.W.; resources, Y.-G.W. and J.-S.H.; data curation, Y.-G.W.; writing—original draft preparation, Y.-G.W.; writing—review and editing, L.-C.T.; visualization, L.-C.T. and B.-P.X.; supervision, B.-P.X.; project administration, J.-S.H.; funding acquisition, J.-S.H. All authors have read and agreed to the published version of the manuscript.

Funding: Financial support from the National Science and Technology Council (NSTC) of Taiwan through grant no. MOST 106-2621-M-019-001, MOST 107-2621-M-019-001, MOST 108-2621-M-019-003, MOST 109-2621-M-019-002, MOST 110-2621-M-019-001 and MOST 111-2621-M-019-001, and Center of Excellence for Ocean Engineering (Grant No. 109J 13801-51 110J13801-51, 111J13801-51) to J.-S.H. This work was financially supported by the Scientific Research Foundation of the Third Institute of Oceanography, MNR, No. 2017010 and No. 2017009, and the Bilateral Cooperation of Maritime Affairs and the Marine Biological Sample Museum (GASI-02-YPK-SW) to C.-G.W. and grant no. MOST 109-2811-M-019-504, MOST 110-2811-M-019-504, and MOST 111-2811-M-019-003 to L.-C.T. The funders had no role on study design, data collection and analysis, decision to publish, or preparation of the manuscript.

Institutional Review Board Statement: Not applicable.

Informed Consent Statement: Not applicable.

Data Availability Statement: Not applicable.

Acknowledgments: The authors are grateful to members of Jiang-Shiou Hwang's laboratory for their assistance during the field works during cruises to the waters off northeastern Taiwan. The authors acknowledge the support from the Ocean Data Bank of the Ministry of Science and Technology, Republic of China for providing temperature and salinity data.

Conflicts of Interest: The authors declare no conflict of interest.

References

1. Sterner, R.W.; Elser, J.J. *Ecological Stoichiometry: The Biology of Elements from Molecules to the Biosphere*; Princeton University Press: Princeton, NJ, USA, 2002; ISSN 0-691-07491-7.
2. Sailley, S.F.; Polimene, L.; Mitra, A.; Atkinson, J.; Allen, J.I. Impact of zooplankton food selectivity on plankton dynamics and nutrient cycling. *J. Plankton Res.* **2015**, *37*, 519–529. [CrossRef]
3. Carter, J.L.; Schindler, D.E.; Francis, T.B. Effects of climate change on zooplankton community interactions in an Alaskan lake. *Clim. Change Responses* **2017**, *4*, 3. [CrossRef]
4. Wang, Y.G.; Zeng, L.C.; Wang, C.G.; Hwang, J.S. The Community and horizontal distribution of Oncaea Philippi, 1843 in the northeast of Taiwan Island. *Mar. Underw. Sci. Technol.* **2020**, *30*, 26–29. (In Chinese)
5. Wang, Y.G.; Tseng, L.C.; Sun, R.X.; Chen, X.Y.; Xiang, P.; Wang, C.G.; Xing, B.P.; Hwang, J.S. Copepods as indicators of different water masses during the Northeast monsoon prevailing period in the northeast Taiwan. *Biology* **2022**, *11*, 1357. [CrossRef] [PubMed]
6. Tan, Y.H.; Huang, L.M.; Chen, Q.C.; Huang, X.P. Seasonal variation in zooplankton composition and grazing impact on phytoplankton standing stock in the Pearl River estuary, China. *Cont. Shelf. Res.* **2004**, *24*, 1949–1968. [CrossRef]
7. Champalbert, G.; Pagano, M.; Sene, P.; Corbin, D. Relationships between meso-and macro-zooplankton communities and hydrology in the Senegal River Estuary. *Estuar. Coast. Shelf S.* **2007**, *74*, 381–394. [CrossRef]
8. Hsiao, S.H.; Kâ, S.; Fang, T.H.; Hwang, J.S. Zooplankton assemblages as indicators of seasonal changes in water masses in the boundary waters between the East China Sea and the Taiwan Strait. *Hydrobiologia* **2011**, *666*, 317–330. [CrossRef]
9. Kâ, S.; Hwang, J.S. Mesozooplankton distribution and composition on the northeastern coast of Taiwan during autumn: Effects of the Kuroshio Current and hydrothermal vents. *Zool. Stud.* **2011**, *50*, 155–163.
10. Kao, S.C.; Lee, M.A.; Iida, K. Diel change in acoustic characteristics and zooplankton composition of the sound scattering layer in I-Lan Bay in northeastern Taiwan. *J. Mar. Sci. Tech.* **2016**, *24*, 22.
11. Tseng, L.C.; Hung, J.J.; Chen, Q.C.; Hwang, J.S. Seasonality of the copepod assemblages associated with interplay waters off northeastern Taiwan. *Helgol. Mar. Res.* **2013**, *67*, 507–520. [CrossRef]
12. Tseng, L.C.; Chou, C.; Chen, Q.C.; Hwang, J.S. Jellyfish assemblages are related to interplay waters in the Southern East China Sea. *Cont. Shelf. Res.* **2015**, *103*, 33–44. [CrossRef]
13. Tseng, L.C.; Molinero, J.C.; Chen, Q.C.; Hwang, J.S. Community structure of zooplankton during the southwest-northeast monsoon transition period in the southeastern East China Sea. *Crustaceana* **2022**, *95*, 667–694. [CrossRef]
14. Wang, Y.G.; Tseng, L.C.; Xing, B.P.; Sun, R.X.; Chen, X.Y.; Wang, C.G.; Hwang, J.S. Seasonal population structure of the copepod Temora turbinata (Dana, 1849) in the Kuroshio Current Edge, Southeastern East China Sea. *Appl. Sci.* **2021**, *11*, 7545. [CrossRef]

15. Hwang, J.S.; Wong, C.K. The China Coastal Current as a driving force for transporting Calanus sinicus (Copepoda: Calanoida) from its population centers to waters off Taiwan and Hong Kong during the winter northeast monsoon period. *J. Plankton RES* **2005**, *27*, 205–210. [CrossRef]
16. Liao, C.H.; Chang, W.J.; Lee, M.A.; Lee, K.T. Summer distribution and diversity of copepods in upwelling waters of the Southeastern East China Sea. *Zool. Stud.* **2006**, *45*, 378–394.
17. Chen, H.Y.; Chen, Y.L.L. Quantity and quality of summer surface net zooplankton in the Kuroshio current-induced upwelling northeast of Taiwan. *Terr. Atmos. Ocean. Sci.* **1992**, *3*, 321–334. [CrossRef]
18. Tseng, L.C.; Dahms, H.U.; Chen, Q.C.; Hwang, J.S. Mesozooplankton and copepod community structures in the southern East China Sea: The status during the monsoonal transition period in September. *Helgol. Mar. Res.* **2012**, *66*, 621–634. [CrossRef]
19. Chiang, K.P.; Shiah, F.K.; Gong, G.C. Distribution of summer diatom assemblages in and around a local upwelling in the East China Sea northeast of Taiwan. *Bot. Bull. Acad. Sin.* **1997**, *38*, 121–129.
20. Hsieh, H.Y.; Lo, W.T.; Wu, L.J.; Liu, D.C.; Su, W.C. Monsoon-driven succession of the larval fish assemblage in the East China Sea shelf waters off northern Taiwan. *J. Oceanogr.* **2011**, *67*, 159–172. [CrossRef]
21. Barkley, R.A. The Kuroshio Current. *Sci. J.* **1970**, *6*, 54–60.
22. Jan, S.; Chen, C.C.; Tsai, Y.L.; Yang, Y.J.; Wang, J.; Chern, C.S.; Gawarkiewicz, G.; Lien, R.C.; Centurioni, L.; Kuo, J.Y. Mean Structure and Variability of the Cold Dome Northeast of Taiwan. *Oceanography* **2011**, *24*, 100–109. [CrossRef]
23. Shen, M.L.; Tseng, Y.H.; Jan, S. The formation and dynamics of the cold-dome off northeastern Taiwan. *J. Marine Syst.* **2011**, *86*, 10–27. [CrossRef]
24. Chern, C.S.; Wang, J. On the water masses at northern offshore area of Taiwan. *Acta Oceanogr.* **1989**, *22*, 14–32.
25. Liu, K.K.; Gong, G.C.; Shyu, C.Z.; Pai, S.C.; Wei, C.L.; Chao, S.Y. Response of Kuroshio upwelling to the onset of the northeast monsoon in the sea north of Taiwan: Observations and a numerical simulation. *J. Geophys. Res.* **1992**, *97*, 12511–12526. [CrossRef]
26. Chung, H.W.; Liu, C.C. Spatiotemporal variation of cold eddies in the upwelling zone off northeastern Taiwan revealed by the geostationary satellite imagery of ocean color and sea surface temperature. *Sustainability* **2019**, *11*, 6979. [CrossRef]
27. Takahashi, D.; Guo, X.; Morimoto, A.; Kojima, S. Biweekly periodic variation of the Kuroshio axis northeast of Taiwan as revealed by ocean high-frequency radar. *Cont. Shelf Res.* **2009**, *29*, 1896–1907. [CrossRef]
28. Hsu, P.C.; Zheng, Q.; Lu, C.Y.; Cheng, K.H.; Lee, H.J.; Ho, C.R. Interaction of coastal countercurrent in I-Lan Bay with the Kuroshio northeast of Taiwan. *Cont. Shelf Res.* **2018**, *171*, 30–41. [CrossRef]
29. He, Y.; Hu, P.; Yin, Y.; Liu, Z.; Liu, Y.; Hou, Y.; Zhang, Y. Vertical migration of the along-slope counter-flow and its relation with the Kuroshio intrusion off northeastern Taiwan. *Remote Sens.* **2019**, *11*, 2624. [CrossRef]
30. Yin, Y.; Liu, Z.; Hu, P.; Hou, Y.; Lu, J.; He, Y. Impact of mesoscale eddies on the southwestward countercurrent northeast of Taiwan revealed by ADCP mooring observations. *Cont. Shelf Res.* **2020**, *195*, 104063. [CrossRef]
31. Hsu, P.C.; Lee, H.J.; Lu, C.Y. Impacts of the Kuroshio and Tidal Currents on the Hydrological Characteristics of Yilan Bay, Northeastern Taiwan. *Remote Sens.* **2021**, *13*, 4340. [CrossRef]
32. Chen, Q.C.; Zhang, S.Z. The planktonic copepods of the Yellow Sea and the East China Sea. I. Calanoida. *Studia Mar. Sin.* **1965**, *7*, 20–131. (In Chinese with English Summary)
33. Chen, Q.C. *Zooplankton of China Seas*; Science Press: Beijing, China, 1992; Volume 1, pp. 1–87, ISBN 7-03-002599-7.
34. Chihara, M.; Murano, M. *An Illustrated Guide to Marine Plankton in Japan*; Tokai University Press: Tokyo, Japan, 1997.
35. Lian, G.S.; Wang, Y.G.; Sun, R.X.; Hwang, J.S. *Species Diversity of Marine Plankton Copepods in China's Seas*; China Ocean Press: Beijing China, 2018; p. 868.
36. Dufrêne, M.; Legendre, P. Species assemblages and indicator species: The need for a flexible asymmetrical approach. *Ecol. Monogr.* **1997**, *67*, 345–366. [CrossRef]
37. Yin, W.; Huang, D. Short-term variations in the surface upwelling off northeastern Taiwan observed via satellite data. *J. Geophys. Res.* **2019**, *124*, 939–954. [CrossRef]
38. Kuo, Y.C.; Lee, M.A.; Chang, Y. Satellite observations of typhoon-induced sea surface temperature variability in the upwelling region off northeastern Taiwan. *Remote Sens.* **2020**, *12*, 3321. [CrossRef]
39. Wang, Y.C.; Lee, M.A. Composition and Distribution of Fish Larvae Surrounding the Upwelling Zone in the Waters of Northeastern Taiwan in Summer. *J. Mar. Sci. Tech.* **2019**, *27*, 8.
40. Hsiao, P.Y.; Shimada, T.; Lan, K.W.; Lee, M.A.; Liao, C.H. Assessing Summertime Primary Production Required in Changed Marine Environments in Upwelling Ecosystems Around the Taiwan Bank. *Remote Sens.* **2021**, *13*, 765. [CrossRef]
41. Adams, K.A.; Barth, J.A.; Chan, F. Temporal variability of near-bottom dissolved oxygen during upwelling off central Oregon. *J. Geophys. Res.* **2013**, *118*, 4839–4854. [CrossRef]
42. Castro, C.G.; Nieto-Cid, M.; Álvarez-Salgado, X.A.; Pérez, F.F. Local remineralization patterns in the mesopelagic zone of the Eastern North Atlantic, off the NW Iberian Peninsula. *Deep-Sea Res. Part I Oceanogr. Res. Pap.* **2006**, *53*, 1925–1940. [CrossRef]
43. Moncoiffé, G.; Alvarez-Salgado, X.A.; Figueiras, F.G.; Savidge, G. Seasonal and short-time-scale dynamics of microplankton community production and respiration in an inshore upwelling system. *Mar. Ecol-Prog. Ser.* **2000**, *196*, 111–126. [CrossRef]
44. Sarthou, G.; Timmermans, K.R.; Blain, S.; Tréguer, P. Growth physiology and fate of diatoms in the ocean: A review. *J. Sea Res.* **2005**, *53*, 25–42. [CrossRef]
45. Rossi, V.; Garçon, V.; Tassel, J.; Romagnan, J.B.; Stemmann, L.; Jourdin, F.; Morin, P.; Morel, Y. Cross-shelf variability in the Iberian Peninsula Upwelling System: Impact of a mesoscale filament. *Cont. Shelf. Res.* **2013**, *59*, 97–114. [CrossRef]

46. López-Sandoval, D.C.; Rodríguez-Ramos, T.; Cermeño, P.; Sobrino, C.; Marañón, E. Photosynthesis and respiration in marine phytoplankton: Relationship with cell size, taxonomic affiliation, and growth phase. *J. Exp. Mar. Biol. Ecol.* **2014**, *457*, 151–159. [CrossRef]
47. Bettencourt, J.H.; Rossi, V.; Renault, L.; Haynes, P.; Morel, Y.; Garçon, V. Effects of upwelling duration and phytoplankton growth regime on dissolved-oxygen levels in an idealized Iberian Peninsula upwelling system. *Nonlinear Proc. Geoph.* **2020**, *27*, 277–294. [CrossRef]
48. Madhupratap, M.; Nair, S.R.S.; Haridas, P.; Padmavati, G. Response of zooplankton to physical changes in the environment: Coastal upwelling along the central west coast of India. *J. Coastal Res.* **1990**, *6*, 413–426.
49. Parsons, T.R. Some ecological, experimental and evolutionary aspects of the upwelling ecosystem. *S. Afr. J. Sci.* **1979**, *75*, 536–540.
50. Mills, C.E. Medusae, siphonophores, and ctenophores as planktivorous predators in changing global ecosystems. *Ices J. Mar. Sci.* **1995**, *52*, 575–581. [CrossRef]
51. Lo, W.T.; Hwang, J.S.; Chen, Q.C. Identity and abundance of surface-dwelling, coastal copepods of southwestern Taiwan. *Crustaceana* **2001**, *74*, 1139–1157. [CrossRef]
52. Lan, Y.C.; Shih, C.T.; Lee, M.A.; Shieh, H.Z. Spring distribution of copepods in relation to water masses in the northern Taiwan Strait. *Zool. Stud.* **2004**, *43*, 332–343.
53. Tseng, L.C.; Kumar, R.; Dahms, H.U.; Chen, Q.C.; Hwang, J.S. Monsoon driven seasonal succession of copepod assemblages in the coastal waters of the northeastern Taiwan Strait. *Zool. Stud.* **2008**, *47*, 46–60.
54. Tseng, L.C.; Kumar, R.; Chen, Q.C.; Hwang, J.S. Faunal shift between two copepod congeners (Temora discaudata and T. turbinata) in the vicinity of two nuclear power plants in southern East China Sea: Spatiotemporal patterns of population trajectories over a decade. *Hydrobiologia* **2011**, *666*, 301–315. [CrossRef]
55. Tseng, L.C.; Wang, Y.G.; Lian, G.S.; Hwang, J.S. A multi-year investigation of the Temoridae (Copepoda, Calanoida) assemblage succession within the interplay waters of the northern South China Sea. *Crustaceana* **2020**, *93*, 519–540. [CrossRef]
56. Lee, P.W.; Hsiao, S.H.; Chou, C.; Tseng, L.C.; Hwang, J.S. Zooplankton fluctuations in the surface waters of the estuary of a Large Subtropical Urban River. *Front. Ecol. Evol.* **2021**, *9*, 598274. [CrossRef]
57. Hwang, J.S.; Souissi, S.; Tseng, L.C.; Seurong, L.; Schmitt, F.G.; Fang, L.S.; Peng, S.H.; Wu, C.W.; Hsiao, S.H.; Twan, W.H.; et al. A 5-year study of the influence of the northeast and southwest monsoons on copepod assemblages in the boundary coastal waters between the East China Sea and the Taiwan Strait. *J. Plankton Res.* **2006**, *28*, 943–985. [CrossRef]
58. Dur, G.; Hwang, J.S.; Souissi, S.; Tseng, L.C.; Wu, C.H.; Hsiao, S.H.; Chen, Q.C. An overview of the influence of hydrodynamics on the spatial and temporal patterns of calanoid copepod communities around Taiwan. *J. Plank. Res.* **2007**, *29*, 97–116. [CrossRef]
59. Fernandes, V.; Ramaiah, N. Mesozooplankton community in the Bay of Bengal (India): Spatial variability during the summer monsoon. *Aquat. Ecol.* **2009**, *43*, 951–963. [CrossRef]
60. De Oliveira-Díaz, C.; De Araujo, A.V.; Paranhos, R.; Bonecker, S.L.C. Vertical copepod assemblages (0–2300 m) off southern Brazil. *Zool. Stud.* **2010**, *49*, 230–242.
61. Lo, W.T.; Dahms, H.U.; Hwang, J.S. Water mass transport through the northern Bashi Channel in the northeastern South China Sea affects copepod assemblages of the Luzon Strait. *Zool. Stud.* **2014**, *53*, 1–9. [CrossRef]
62. Becker, É.C.; Garcia, C.A.E.; Freire, A.S. Mesozooplankton distribution, especially copepods, according to water masses dynamics in the upper layer of the Southwestern Atlantic shelf (26 S to 29 S). *Cont. Shelf. Res.* **2018**, *166*, 10–21. [CrossRef]
63. Fernández de Puelles, M.L.; Gazá, M.; Cabanellas-Reboredo, M.; Santandreu, M.M.; Irogoien, X.; González-Gordillo, J.L.; Duarte, C.M.; Hernánder-León, S. Zooplankton abundance and diversity in the tropical and subtropical ocean. *Diversity* **2019**, *11*, 203. [CrossRef]
64. Júnior, M.D.M.; Miyashita, L.K.; Lopes, R.M. A 3-year study of the seasonal variability of abundance, biomass and reproductive traits of Oncaea venusta (Copepoda, Oncaeidae) in a subtropical coastal area. *J. Plankton Res.* **2021**, *43*, 751–761. [CrossRef]
65. Eberl, R.; Carpenter, E.J. Association of the copepod Macrosetella gracilis with the cyanobacterium Trichodesmium spp. in the North Pacific Gyre. *Mar. Ecol-Prog. Ser.* **2007**, *333*, 205–212. [CrossRef]
66. Vineetha, G.; Karati, K.K.; Raveendran, T.V.; Babu, K.K.I.; Riyas, C.; Muhsin, M.I.; Shihab, B.K.; Simson, C.; Anil, P. Responses of the zooplankton community to peak and waning periods of El Niño 2015–2016 in Kavaratti reef ecosystem, northern Indian Ocean. *Environ. Monit. Assess.* **2018**, *190*, 465. [CrossRef] [PubMed]
67. Gallienne, C.P.; Robins, D.B. Is Oithona the most important copepod in the world's oceans? *J. Plankton Res.* **2001**, *23*, 1421–1432. [CrossRef]
68. Khalifa, U.; Ebenezer, V.; Pierson, J.J. Elevated temperature and low pH affect the development, reproduction, and feeding preference of the tropical cyclopoid copepod Oithona rigida. *Int. J. Environ. Stud.* **2022**, 1–17. [CrossRef]
69. Bhattacharya, B.D.; Hwang, J.S.; Sarkar, S.K.; Rakhsit, D.; Murugan, K.; Tseng, L.C. Community structure of mesozooplankton in coastal waters of Sundarban mangrove wetland, India: A multivariate approach. *J. Marine Syst.* **2015**, *141*, 112–121. [CrossRef]
70. Keister, J.E.; Tuttle, L.B. Effects of bottom-layer hypoxia on spatial distributions and community structure of mesozooplankton in a sub-estuary of Puget Sound, Washington, USA. *Limnol. Oceanogr.* **2013**, *58*, 667–680. [CrossRef]
71. Bowman, T.E. The distribution of Calanoid copepods off the southeastern United States between Cape Hatteras and southern Florida. *Smithson. Contrib. Zool.* **1971**, *96*, 1–58. [CrossRef]
72. Liang, D.; Uye, S. Population dynamics and production of the planktonic copepods in a eutrophic inlet of the Inland Sea of Japan. *III. Paracalanus sp. Mar. Biol.* **1996**, *127*, 219–227. [CrossRef]

73. Kesarkar, K.S.; Anil, A.C. New species of Paracalanidae along the west coast of India: Paracalanus arabiensis. *J. Mar. Biol. Assoc. UK* **2010**, *90*, 399–408. [CrossRef]
74. Hidaka, K.; Itoh, H.; Hirai, J.; Tsuda, A. Occurrence of the Paracalanus parvus species complex in offshore waters south of Japan and their genetic and morphological identification to species. *Plankton Benthos Res.* **2016**, *11*, 131–143. [CrossRef]
75. Aliçli, B.T.; Sarihan, E. Seasonal changes of zooplankton species and groups composition in Iskenderun Bay (North East Levantine, Mediterranean Sea). *Pak. J. Zool.* **2016**, *48*, 1395–1405.
76. Lo, W.T.; Shih, C.T.; Hwang, J.S. Diel vertical migration of the planktonic copepods at an upwelling station north of Taiwan, western North Pacific. *J. Plankton Res.* **2004**, *26*, 89–97. [CrossRef]
77. Hwang, J.S.; Tu, Y.Y.; Tseng, L.C.; Fang, L.S.; Souissi, S.; Fang, T.H.; Lo, W.T.; Twan, W.H.; Hsiao, S.H.; Wu, C.H.; et al. Taxonomic composition and seasonal distribution of copepod assemblages from waters adjacent to nuclear power plant I and II in northern Taiwan. *J. Mar. Sci. Technol.* **2004**, *12*, 4. [CrossRef]
78. Lan, Y.C.; Lee, M.A.; Liao, C.H.; Lee, K.T. Copepod community structure of the winter frontal zone induced by the Kuroshio Branch Current and the China Coastal Current in the Taiwan Strait. *J. Mar. Sci. Technol.* **2009**, *17*, 1–6. [CrossRef]
79. Hsieh, C.; Chen, C.S.; Chiu, T.S. Composition and abundance of copepods and ichthyoplankton in Taiwan Strait (western North Pacific) are influenced by seasonal monsoons. *Mar. Freshw. Res.* **2005**, *56*, 153–161. [CrossRef]

Journal of
Marine Science and Engineering

Article

Physicochemical Drivers of Zooplankton Seasonal Variability in a West African Lagoon (Nokoué Lagoon, Benin)

Alexis Chaigneau [1,2,3,*], François Talomonwo Ouinsou [4], Hervé Hołèkpo Akodogbo [4], Gauthier Dobigny [4,5], Thalasse Tchémangnihodé Avocegan [4], Fridolin Ubald Dossou-Sognon [4], Victor Olaègbè Okpeitcha [2,3,6], Metogbe Belfrid Djihouessi [7] and Frédéric Azémar [8]

1 Laboratoire d'Études en Géophysique et Océanographie Spatiale (LEGOS), Université de Toulouse, CNES, CNRD, IRD, UPS, 31555 Toulouse, France
2 Institut de Recherches Halieutiques et Océanologiques du Bénin (IRHOB), Cotonou 03 BP 1665, Benin
3 International Chair in Mathematical Physics and Applications (ICMPA–UNESCO Chair), University of Abomey-Calavi, Cotonou 01 BP 526, Benin
4 Unité de Recherche sur les Invasions Biologiques, Laboratoire de Recherche en Biologie Appliquée (LARBA), Université d'Abomey-Calavi, Cotonou 01 BP 4521, Benin
5 UMR CBGP, IRD, INRAE, Cirad, Institut Agro Montpellier, MUSE, 34000 Montpellier, France
6 Laboratoire d'Hydrologie Appliquée (LHA), Institut National de l'Eau (INE), African Centre of Excellence for Water and Sanitation (C2EA), Université d'Abomey-Calavi, Abomey-Calavi 01 BP 4521, Benin
7 Laboratoire des Sciences et Techniques de l'Eau (LSTE), Université d'Abomey-Calavi, Abomey-Calavi 01 BP 4521, Benin
8 Laboratoire Écologie Fonctionnelle et Environnement, Université Toulouse3–Paul Sabatier (UPS), CNRS, Toulouse INP, 31555 Toulouse, France
* Correspondence: alexis.chaigneau@ird.fr

Abstract: This study aimed to investigate the seasonal variation of zooplankton diversity and abundance in the Nokoué Lagoon in southern Benin. Through extensive sampling, a total of 109 zooplanktonic taxa were identified and quantified. The average zooplankton abundance was found to be 60 individuals per liter, with copepods and rotifers being the most dominant groups, comprising 68.1% and 29.1% of the total abundance, respectively. The key factor identified as driving the structure of the zooplanktonic assemblages was salinity, which showed significant seasonal variation. The results revealed that during the high water period, when the lagoon was filled with fresh water, rotifers were dominant, zooplanktonic diversity was highest, and abundances were quite high. Conversely, during the low water period, when the lagoon was characterized by brackish water, diversity was minimal, and abundance decreased slightly. The study also found that some areas of the lagoon showed high abundances independent of salinity levels, suggesting that other factors such as riverine inputs or the presence of *acadjas* (home-made brush parks used as fish traps) may also have notable effects on the zooplankton community. Overall, the findings of this study provide valuable insights into the functioning of one of the most biologically productive lagoons in West Africa.

Keywords: zooplankton; diversity and abundance; environmental parameters; seasonal variation; Nokoué Lagoon

Citation: Chaigneau, A.; Ouinsou, F.T.; Akodogbo, H.H.; Dobigny, G.; Avocegan, T.T.; Dossou-Sognon, F.U.; Okpeitcha, V.O.; Djihouessi, M.B.; Azémar, F. Physicochemical Drivers of Zooplankton Seasonal Variability in a West African Lagoon (Nokoué Lagoon, Benin) *J. Mar. Sci. Eng.* **2023**, *11*, 556. https://doi.org/10.3390/jmse11030556

Academic Editors: Marco Uttieri, Ylenia Carotenuto, Iole Di Capua and Vittoria Roncalli

Received: 2 February 2023
Revised: 1 March 2023
Accepted: 3 March 2023
Published: 6 March 2023

1. Introduction

Located in southeastern Benin, the Nokoué Lagoon has been recognized by the Ramsar Convention as a wetland of international importance (Ramsar Site no. 1018). This lagoon represents the largest continental water body in Benin and one of the most biologically productive in West Africa in terms of annual fish catch yields [1]. It contributes to 70% of the national fisheries production [2] and is home to the largest lacustrine villages in West Africa populated by 50,000 people including 12,000 fishermen. Although the Nokoué Lagoon forms a vast natural space that sustains a rich ecosystem, it is surrounded by numerous urbanizations that total more than 1.5 million inhabitants and exert a strong anthropic pressure [3]. Despite its importance for the socio-economic development of Benin and its vulnerability to urban development and global changes, Nokoué Lagoon has been relatively poorly studied. Therefore, it is essential to better characterize and understand the functioning of its ecosystem in order to implement future sustainable management plans.

One of the particularities of Nokoué Lagoon is the strong seasonal variation of its salinity, which varies on average from less than 1 during the heavy rainy season in northern Benin (September–October) and progressively increases during the dry season (December–April) to reach a mean value of ~25 in April [4,5]. These strong seasonal changes in salinity have been shown to structure and strongly impact certain trophic levels of the Nokoué Lagoon ecosystem, such as macroinvertebrates, mangrove oysters, or ichthyofauna [6–8]. However, zooplankton is a key organism in the aquatic food chain as it serves as an intermediary species that allows energy transfer between phytoplankton and higher trophic levels such as zooplanktivorous fish. As we know that salinity can control the response of phytoplankton to nutrients and significantly alter zooplankton dynamics in different lagoon ecosystems (e.g., [9–12]), we therefore hypothesize that strong seasonal variations in salinity in the Nokoué Lagoon generate important changes in zooplankton diversity and abundance.

Although salinity is the parameter that exhibits the greatest seasonal variation and is presumed to be the major environmental parameter impacting the Nokoué ecosystem, other physico-chemical parameters (temperature, depth, pH, turbidity, dissolved oxygen concentration, chlorophyll) also exhibit significant seasonal variations (e.g., [3]). Given their short life cycles, zooplankton organisms can respond rapidly and sensitively to many physical, chemical, and biological changes in aquatic ecosystems [13–15]. Zooplankton are therefore highly sensitive to environmental changes, and changes in zooplankton community composition or abundance are often considered indicative of environmental disturbances in coastal lagoons (e.g., [16–19]). The second hypothesis of this study is therefore that the zooplankton community of Nokoué Lagoon may respond to other physico-chemical parameters than salinity and that the presence and dominance of certain zooplanktonic species could reflect particular environmental conditions.

Furthermore, given that relatively strong spatial gradients in salinity and other physicochemical parameters may exist in the lagoon [3–5], we believe that the zooplankton community may respond to these environmental gradients and show spatial differences across the lagoon. In particular, the zooplanktonic community near the river mouths (impacted by freshwater flows) might probably differ from that near the connection between Nokoué Lagoon and the Atlantic Ocean (impacted by saltwater flows).

Despite the importance of zooplankton in the trophic chain and for maintaining the high biological productivity of Nokoué Lagoon, knowledge on the zooplankton of this lagoon is very limited. Indeed, to the best of our knowledge, the only study that has explored the zooplankton community of Nokoué and its relationship with environmental parameters is [20]. In particular, these authors showed that (i) total zooplankton species richness was 31 taxa, (ii) zooplanktonic abundance was relatively low and dominated by copepods, and (iii) the zooplanktonic community responded mainly to nitrate and ammonium concentrations. Unfortunately, as also mentioned by these authors, the sampling was only conducted during a 4-month period (June–September 2015) in the rainy season (low salinity) and did not allow them to investigate the seasonal variation over a full hydrological cycle. Consequently, our study, based on bimonthly sampling carried out during a complete year, aims to fill this gap, and has the following objectives:

- To inventory the zooplanktonic fauna of Nokoué Lagoon,
- To describe the spatio-temporal variations of zooplankton on a seasonal scale, in terms of diversity and abundance,
- To determine the main physicochemical drivers of the zooplankton community and to verify if salinity is the determining parameter.

2. Materials and Methods

2.1. Study Area and Annual Hydrological Cycle

Nokoué Lagoon is a shallow lagoon (~1.3 m depth on average during the dry season) that extends from 2°20′ E to 2°35′ E in longitude and 6°20′ N to 6°30′ N in latitude. This lagoon is bounded to the west by the Abomey-Calavi plateau, to the east by the Porto-Novo Lagoon, to the north by the deltaic floodplain of the Ouémé and Sô Rivers, and to the south by the city of Cotonou (Figure 1). It extends approximately over 20 km from west to east and a maximum of 11 km from south to north, covering an area of ~150 km^2 at low water period [3,21,22]. The Cotonou Channel, which is 4 km long and 300 m wide, connects Nokoué Lagoon to the Atlantic Ocean and thus allows freshwater and saltwater exchanges between these two environments (Figure 1) [5]. Nokoué Lagoon also communicates in its western part with the small Djonou River (a few meters wide and a few kilometers long) and in the east with the Porto-Novo lagoon via the Totchè canal, but they have little effect on the dynamics of Nokoué Lagoon [21–24].

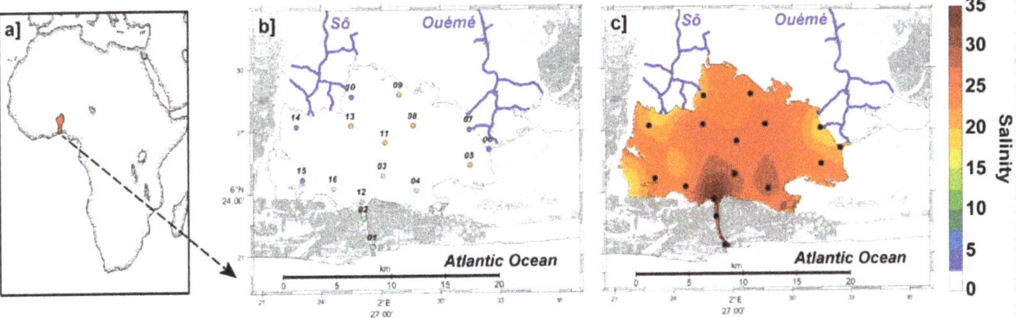

Figure 1. Location of Nokoué Lagoon (Benin) and hydrographic stations carried out between November 2019 and September 2020. (**a**) Benin location in West-Africa. (**b**) Mean salinity (SAL) during high water period in October (SAL ~ 0) and (**c**) during low water period in April (SAL ~ 25). Station numbers are shown in b]; green dots correspond to stations influenced by the Cotonou Canal and the Atlantic Ocean, blue dots by rivers, and red dots are intermediate stations more representative of the central zone of the lagoon.

The annual hydrological regime of Nokoué Lagoon can be characterized by three main periods [3,23,25]. First, during the heavy rainy season in northern Benin (September–November), the lagoon is filled of freshwater and its water-level strongly increases [5,23]. During this period, the highest species richness of ichthyofauna is observed and the lagoon is populated by freshwater fish species [8]. Second, during the dry season (December–April), the lagoon reaches its low water level and is filled with brackish water with an average salinity of ~25 at the end of the dry season [5,23]. During this period, the species richness of the main taxonomic families of the ichthyofauna decreases and the lagoon is depopulated of some freshwater species not tolerant to these high salinity levels [8]. In contrast, some marine fish species enter the lagoon during this season. Third, during the main rainy season in southern Benin (May–July), the lagoon level rises slightly, and its salinity gradually decreases to an average value of less than 10 [5,23].

2.2. Station Location and Sampling Strategy

This study relies on the analysis of six bimonthly campaigns conducted on Nokoué Lagoon and the Cotonou Channel between November 2019 and September 2020. Each of the field campaigns took place during two consecutive days around the 15th of the corresponding month. They consisted of 16 physicochemical and biological sampling stations

(13 in Nokoué Lagoon and 3 in the Cotonou Channel) distributed rather homogeneously in the study area in order to monitor various environmental and biological conditions (Figure 1a). Sampling stations were performed during daytime between ~8 am to ~5 pm. A total of 96 samples were thus taken over the 6 bimonthly campaigns.

2.3. Environmental Parameters

For each of the 96 samples (6 campaigns of 16 stations), the following environmental parameters were acquired at the surface (within the top 20 cm) using both a Conductivity-Temperature-Depth (CTD) probe (Valeport-CTD+) and a multi-parameter probe (WTW-3630IDS): temperature (TEMP, in °C), salinity (SAL), turbidity (TURB, in FTU), dissolved oxygen (DO, in mg L^{-1}). As the water-level of the Nokoué Lagoon strongly varies both spatially and temporally [23], the water-depth (DEPTH, in m) at each station was measured using a Garmin GPSmap 421S echosounder.

Water samples were collected at the surface using a 10 L bucket for measurements of total (organic and inorganic) suspended solids (TSS, in mg L^{-1}), particulate organic matter (POM, in mg L^{-1}), and Chlorophyll-a (CHL-a, in mg m^{-3}). For each water sample, 500 mL were immediately filtered onboard using Whatman GF/F glass fiber filters with a 0.7 µm pore size. The CHL-a concentrations were determined by spectrophotometer after extraction from the filters using 90% acetone [26,27]. For TSS and POM, between 100 and 500 mL (depending on the water turbidity) of water samples were filtered onto both reweighed GF/F filters and Nucleopore membrane filters (0.4 µm pore size). After filtration, filters were dried at 55 °C for 24 h and reweighed to determine TSS concentrations. The GF/F filters were then combusted at 550 °C for 2 h and reweighed to estimate ash weight, POM, and percentage of organic matter (%POM). Based on a statistical analysis, we determined that TSS concentration and turbidity were highly correlated ($r^2 = 0.97$, $n = 742$, $p < 0.05$) and collinear (variance inflation factor, VIF = 14.96). We therefore only retained turbidity and discarded TSS from subsequent analyses. Similarly, as organic and inorganic fractions are complementary, we only retained %POM.

2.4. Zooplankton Collection and Sample Analysis

Zooplankton community composition was determined from 96 samples collected at the 16 stations using a plankton net having a mouth opening diameter of 40 cm and a mesh size of 50 µm. A mechanical flowmeter (General Oceanics-2030R6) was placed on the opening of the net to estimate the volume of water filtered. At each site, the net was rinsed with water from the station, and then towed horizontally at a depth of ~50 cm for 30 to 45 s, which allows an average filtration of ~3–4 m^3 of water. The collected filtrate was then transferred to a pillbox and preserved in 70 mL of 70% ethanol.

The identification and enumeration of zooplankton was done using a digital microscope (Optika B-290TB) and a panel of taxonomic keys [28–47]. For each station, 1 mL of sample was diluted 5 times with distilled water. Then, 1 mL of this subsample (corresponding to a replicate) was observed under the microscope. The identification and enumeration of zooplankton was carried out on at least 5 independent replicas per station, until a species richness plateau was reached, i.e., when no new zooplankton species were observed in three successive replicas. The zooplankton composition and relative abundance for each station was then determined from all pooled replicas. For each sample, counts of zooplankton taxa were converted to abundance (ind. L^{-1}) taking the number of investigated replicas, the subsample ratio, and the total volume of the sample.

In the present study, copepod nauplii could not be diagnosed at the species level and were thus pooled into one single taxon. Some species could not be formally identified (Table 1) but they were distinguished between them taking into account several morphological characteristics. Each differentiated morphotype was assigned to a distinct (unnamed) species. Differentiation of indeterminate copepod species was performed based on (i) the number of antenna segments, (ii) the cephalosome shape, (iii) the number and appearance of metasoma segments, (iv) the number and size of urosoma segments, and (v) the number and size of setae on each furca. As for the undetermined species of rotifers, they could be distinguished by (i) the body shape and (ii) the appearance and position of certain organs (see references quoted above for taxonomic keys).

Table 1. List of zooplankton taxa, their relative abundance (in %) and frequency of occurrence (in %). The inventoried taxa were sampled in Nokoué Lagoon during 6 bimonthly surveys carried out between November 2019 and September 2020. The last column indicates the codes of the most frequent (F_{occ} > 0.25) and abundant (relative abundance higher than 0.5%) taxa, used to study their relationship with environmental variables.

Group	Family	Taxa	Relative Abundance (%)	Frequency of Occurrence F_{occ} (%)	Code
Rotifera	Asplanchnidae	*Asplanchna girodi* de Guerne, 1888	1.55	15.63	
		Asplanchna priodonta Gosse, 1850	0.00	1.04	
		Asplanchna sp. 1	0.32	6.25	
		Asplanchna sp. 2	0.07	8.33	
		Asplanchna sp. 3	0.01	2.08	
		Asplanchna sp. 4	0.06	2.08	
	Brachionidae	*Anuraeopsis navicula* Rousselet, 1911	0.00	2.08	
		Anuraeopsis fissa Gosse, 1851	0.00	1.04	
		Anuraeopsis sp.	0.00	1.04	
		Brachionus angularis Gosse, 1851	2.83	37.50	R03
		Brachionus bidentatus Anderson, 1889	0.28	15.63	
		Brachionus calyciflorus Pallas, 1766	0.54	23.96	
		Brachionus caudatus Barrois & Daday, 1894	1.63	30.21	R08
		Brachionus falcatus Zacharias, 1898	0.73	32.29	R06
		Brachionus mirabilis Daday, 1897	0.01	5.21	
		Brachionus plicatilis Müller, 1786	7.25	57.29	R01
		Brachionus quadridentatus Hermann, 1783	0.00	3.13	
		Epiphanes macroura (Barrois & Daday, 1894)	0.022	7.29	
		Epiphanes sp.	0.03	8.33	
		Keratella cochlearis (Gosse, 1851)	0.09	9.38	
		Keratella lenzi Hauer, 1953	0.35	15.63	
		Keratella sp.	0.51	12.50	
		Keratella tropica (Apstein, 1907)	0.94	34.38	R05
		Plationus patulus (Müller, 1786)	0.06	18.75	
		Platyias quadricornis (Ehrenberg, 1832)	0.06	10.42	
		Trichotria sp.	0.00	2.08	
	Conochilidae	*Conochilus* sp.	0.05	10.42	
	Euchlanidae	*Euchlanis triquetra* Ehrenberg, 1838	0.17	8.33	
	Lecanidae	*Lecane bulla* (Gosse, 1851)	0.27	25.00	
		Lecane crepida Harring, 1914	0.02	3.13	
		Lecane leontina (Turner, 1892)	1.46	26.04	R10
		Lecane ludwigii (Eckstein, 1883)	0.01	6.25	
		Lecane quadridentata (Ehrenberg, 1830)	0.08	6.25	
		Lecane sp.	0.06	15.63	
		Lecane closterocerca (Schmarda, 1859)	0.06	14.58	
		Lecane stenroosi (Meissner, 1908)	0.07	4.17	
	Lepadellidae	*Colurella adriatica* Ehrenberg, 1831	0.09	11.46	
		Colurella hindenburgi Steinecke, 1916	0.00	3.13	
		Lepadella (Lepadella) patella (Müller, 1773)	0.01	2.08	
		Lepadella sp.	0.00	1.04	
		Squatinella lamellaris (Müller, 1786)	0.34	14.58	
	Mytilinidae	*Mytilina mucronata* (Müller, 1773)	0.00	3.13	
	Notommatidae	*Cephalodella gibba* (Ehrenberg, 1830)	0.00	5.21	
		Cephalodella gracilis (Ehrenberg, 1830)	0.00	1.04	
		Cephalodella lipara Myers, 1924	0.00	2.08	
		Cephalodella mira Myers, 1934	0.00	2.08	
		Cephalodella sp. 1	0.02	4.17	
		Cephalodella sp. 2	0.00	1.04	
		Cephalodella sp. 3	0.00	1.04	
		Cephalodella sp. 4	0.02	8.33	
		Cephalodella sp. 5	0.01	1.04	
		Eothinia elongata (Ehrenberg, 1832)	0.00	1.04	
		Notommata pachyura (Gosse, 1886)	0.01	3.13	
		Resticula melandocus (Gosse, 1887)	0.01	8.33	
		Taphrocampa annulosa Gosse, 1851	0.12	7.29	

Table 1. *Cont.*

Group	Family	Taxa	Relative Abundance (%)	Frequency of Occurrence F_{occ} (%)	Code
	Philodinidae	*Philodina* sp. 1	0.20	29.17	
		Philodina sp. 2	0.01	5.21	
		Rotaria neptunia (Ehrenberg, 1830)	0.06	16.67	
	Proalidae	*Proales* sp.	0.03	2.08	
	Scaridiidae	*Scaridium longicaudum* (Müller, 1786)	0.58	8.33	
	Synchaetidae	*Polyarthra* sp.	0.95	21.88	
		Synchaeta bicornis Smith, 1904	1.83	32.29	R07
		Synchaeta pectinata Ehrenberg, 1832	1.73	37.50	R04
		Synchaeta grandis Zacharias, 1893	1.10	23.96	
		Synchaeta sp.	0.01	8.33	
	Testudinellidae	*Testudinella patina* (Hermann, 1783)	0.07	14.58	
		Tetrasiphon sp.	0.02	6.25	
		Trichocerca brachyura (Gosse, 1851)	0.12	15.63	
		Trichocerca lata (Jennings, 1894)	0.00	1.04	
		Trichocerca longiseta (Schrank, 1802)	0.00	1.04	
		Trichocerca platessa Myers, 1934	0.02	1.04	
		Trichocerca rattus (Müller, 1776)	0.01	6.25	
		Trichocerca similis (Wierzejski, 1893)	0.05	12.50	
		Trichocerca sp. 1	0.01	4.17	
		Trichocerca sp. 2	0.01	6.25	
		Trichocerca sp. 3	0.00	1.04	
		Trichocerca tenuior (Gosse, 1886)	0.00	1.04	
	Trochosphaeridae	*Filinia longiseta* (Ehrenberg, 1834)	1.48	38.54	R02
		Filinia opoliensis (Zacharias, 1898)	0.54	28.13	R09
	Trichotriidae	*Macrochaetus* sp. 1	0.01	2.08	
		Macrochaetus sp. 2	0.11	2.08	
Copepoda	unidentified Cyclopoida	Cyclopoid sp. 1	4.92	89.58	C02
		Cyclopoid sp. 2	0.01	6.25	
		Cyclopoid sp. 3	0.04	9.38	
		Cyclopoid sp. 4	0.03	5.21	
		Cyclopoid sp. 5	0.10	38.54	
		Cyclops strenuus strenuus Fischer, 1851	0.03	10.42	
		Ectocyclops sp.	0.27	51.04	
	Corycaeidae	*Corycaeus* sp.	0.00	1.04	
	Oithonidae	*Oithona* sp.	0.72	36.46	C04
		Oithona plumifera Baird, 1843	0.00	1.04	
	Oncaeidae	*Oncaea clevei* Früchtl, 1923	0.22	8.33	
	unidentified Calanoida	Calanoid spp.	1.26	35.42	C05
		Calanoid sp. 1	3.69	75.00	C03
		Calanoid sp. 2	0.01	3.13	
	Temoridae	*Temora turbinata* (Dana, 1849)	0.00	2.08	
	Ectinosomatidae	*Microsetella* sp.	0.41	58.33	
		Microsetella norvegica (Boeck, 1865)	0.00	3.13	
		Microsetella rosea (Dana, 1847)	0.00	1.04	
	Miraciidae	*Macrosetella gracilis* (Dana, 1846)	0.01	9.38	
	Tachidiidae	*Euterpina acutifrons* (Dana, 1847)	0.02	3.13	
		Nauplius	56.27	97.92	C01
Cladocera	Chydoridae	*Alona* sp.	0.00	2.08	
	Daphniidae	*Ceriodaphnia* sp.	0.03	17.71	
	Macrothricidae	*Macrothrix* sp.	0.01	2.08	
	Moinidae	*Moina micrura* Kurz, 1875	0.19	29.17	
	Sididae	*Penilia avirostris* Dana, 1849	0.00	1.04	
Eumalacostraca	Mysidae	*Mysis* sp.	0.02	5.21	
Mollusca	Undetermined	Mollusca spp.	2.52	73.96	M01
Ostracoda	Undetermined	Ostracod sp.	0.04	15.63	

The zooplanktonic specific diversity (H') was estimated by the Shannon index [48,49]:

$$H' = -\sum_{i=1}^{S} \frac{n_i}{N} log_2\left(\frac{n_i}{N}\right)$$

where i is a specific species, S is the total number of species (or species richness), n_i is the number of individuals for species i, and N is the total number of individuals considering all species.

Species evenness was determined by the Pielou's equitability index (J) [50–52]:

$$J = \frac{H'}{log_2 S}$$

This index, which represents the distribution of individuals over species, varies between 0 when a single species dominates, and 1 if all species have an identical abundance.

In addition, the frequency of occurrence (F_{occ}) of each species in all 96 samples was computed (Table 1). A species is considered as frequent if $F_{occ} \geq 0.5$, occasional if $0.25 \leq F_{occ} < 0.5$, infrequent, accidental if $0.05 \leq F_{occ} < 0.25$, and rare if $F_{occ} < 0.05$.

2.5. Data Interpolation and Statistics

For each field campaign, all the results (species richness, abundance, H', J) were spatially interpolated onto a regular grid of ~100 m × 100 m resolution, using an objective interpolation scheme implemented in MATLAB® [53–55].

To test for general significant spatial or temporal variability in the environmental parameters (TEMP, SAL, DEPTH, TURB, %POM, DO, CHL-a) and zooplankton groups, several statistical tests were used. The non-parametric Wilcoxon–Mann–Whitney test (WMW-test) was used to test the significance of variations between the minimum and maximum mean values observed between 2 stations or between 2 campaigns. Similarly, the non-parametric Kruskal–Wallis test (H-test) was applied to test the significance of overall variations observed between the 6 campaigns or between the 16 stations.

In order to test for the effects of the seasons (bi-monthly surveys) and geographic locations (and their interactions) on the environmental variables and zooplankton data, we also used two-way analyses of variance (ANOVA). Two-way ANOVA were performed on Box-Cox transformed data (to approach the hypotheses of normality, independence and homogeneity), and the data were grouped into 3 sub-regions (Figure 1b): the first group includes stations under the direct influence of the Cotonou Channel (green dots in Figure 1b); the second group includes stations under the influence of rivers (blue dots in Figure 1b), while the third group includes stations more representative of the lagoon environment (orange dots in Figure 1b).

General relationships between the diversity and abundance of major zooplankton groups or taxa and environmental variables were investigated using redundancy analyses (RDA) [56]. To do so, we considered only zooplankton groups representing at least 3% of the total diversity and abundance. A Monte Carlo permutation test was used to check the significance ($p < 0.05$) of the relationships between the zooplankton and each environmental variable, using 10,000 randomizations. Environmental variables were ranked according to their quantitative importance through manual selection based on the Monte Carlo permutation test. Environmental variables that do not significantly ($p < 0.05$) increase the explained variance were removed from RDA analyses. Prior to RDA analyses, zooplankton abundances were Hellinger-transformed to down-weight the influence of rare species having low counts and many zeros [57]. Note that RDA analyses were used since detrended canonical correspondence analysis (DCCA) revealed maximum gradient lengths of the response data lower than 2 [58].

We also related abundance and species richness of the main zooplankton groups and taxa to individual environmental variables using generalized additive models (GAMs) [59]. The GAMs allow for nonlinearity in the relationships between the predictor variable (environmental parameter) and the response variable (zooplankton abundance or diversity). Low-rank thin plate splines were applied, and the smoothing parameters were determined through restricted maximum likelihood (REML). The best models, associated with the lowest values of Akaike's information criterion (AIC) and the highest explained deviances, included all 7 environmental variables and 5 degrees of freedom for the smoothing curve functions. In these models, response variables were linked to the additive predictors using log-link functions and Tweedie family distributions. Explained deviances of the GAMs and p-values were examined to retain and describe only those environmental variables that significantly ($p < 0.05$) explained zooplankton variations. The GAMs were carried out using untransformed abundance data.

All statistical analyses, the RDA, and the GAMs were performed using the free R Statistical Software (version 4.0.2; R Foundation for Statistical Computing, Vienna, Austria) and CANOCO 4.5 software [60]. In particular, we used the R package *mgcv* [61,62].

3. Results

3.1. Seasonal Variations of Environmental Parameters

Figure 2 shows the seasonal variations of the physicochemical parameters in Nokoué Lagoon from November 2019 to September 2020. H-tests revealed that all physicochemical factors exhibited significant ($p < 0.05$) bimonthly differences and some of these variables (SAL, DEPTH, DO) also showed significant mean differences between the overall stations (Table 2).

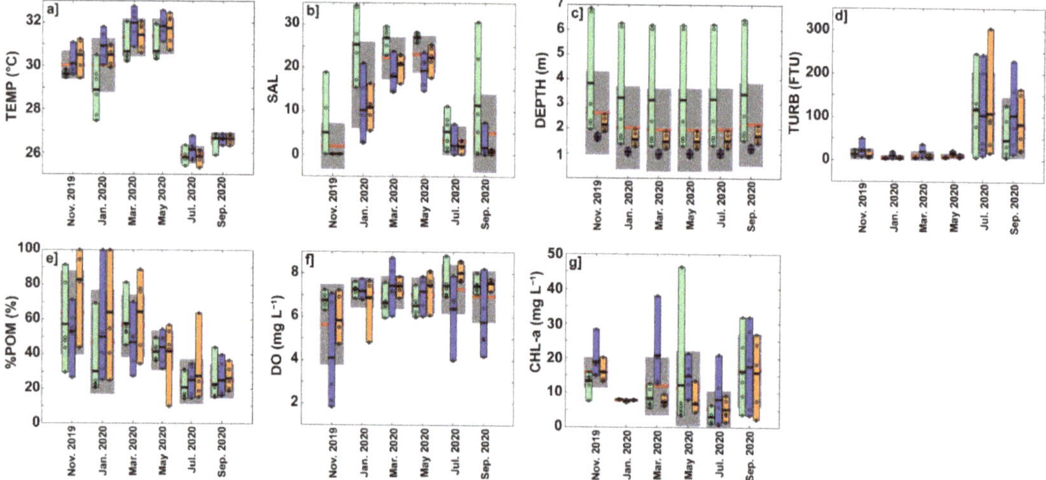

Figure 2. Seasonal variation of environmental parameters for the six bimonthly surveys carried out between November 2019 and September 2020. (**a**) Temperature (TEMP, in °C). (**b**) Salinity (SAL). (**c**) Station depth (DEPTH, in m). (**d**) Turbidity (TURB, in FTU). (**e**) Fraction of particulate organic matter in total suspended material (%POM, in %). (**f**) Dissolved oxygen concentration (DO, in mg L^{-1}). (**g**) Chlorophyll-a concentration (CHL-a, in mg L^{-1}). On each subplot, the gray bars show the mean value (red horizontal line) +/− one standard deviation for all 16 stations, while the green, blue, and orange vertical bars show the range of data in each sub-region defined in Figure 1b. These colored bars thus extend from the minimum to the maximum value observed in each sub-region, and the black horizontal lines show the mean values of the data indicated by the black circles.

Table 2. Results of statistical analyses on environmental and zooplankton variables. Wilcoxon–Mann–Whitney test (WMW-test) was used to test the variations between the minimum (Min.) and maximum (Max.) mean values observed between two campaigns (spatial average) or between two stations (temporal average). Kruskal–Wallis test (H-test) was applied to test the significance of overall variations observed between the six campaigns (temporal analysis, df = 5) or between the 16 stations (spatial analysis, df = 15). Significance levels (S.L.): * $p < 0.05$, ** $p < 0.01$, *** $p < 0.001$, n.s. not significant.

	WMW-Test						H-Test	
	Spatial Average			Temporal Average			Temporal Analysis χ^2 (df = 5)	Spatial Analysis χ^2 (df = 15)
Variable	Min.	Max.	S.L.	Min.	Max.	S.L.		
				Environmental variables				
TEMP (°C)	25.9	31.4	***	28.4	30.1	n.s.	75.8 ***	n.s.
SAL	1.9	23.2	***	5.4	24.9	***	62.5 ***	25.4 *
DEPTH (m)	2.0	2.6	**	1.0	6.4	***	14.8 *	77.0 ***
TURB (FTU)	5.6	110.1	***	7.3	92.2	***	52.7 ***	n.s.
%POM (%)	24.3	63.9	**	26.1	61.9	***	45.2 ***	n.s.
DO (mg L^{-1})	5.6	7.3	**	5.4	7.7	***	15.3 **	37.6 **
CHL-a (mg m^{-3})	5.3	16.5	**	5.9	21.7	*	29.6 ***	n.s.
				Zooplankton				
Species Richness	9	30	***	13	18	n.s.	65.5 ***	n.s.
H′	1.2	3.5	***	1.7	2.7	n.s.	55.4 ***	n.s.
J	0.4	0.7	***	0.4	0.7	**	31.6 ***	n.s.
Zooplankton abundance (ind L^{-1})	36.5	74.9	n.s.	20.0	193.9	*	n.s.	n.s.
Copepod relative abundance (%)	27.0	93.4	***	46.2	87.7	n.s.	47.6 ***	n.s.
Rotifer relative abundance (%)	6.3	71.5	***	11.1	50.9	*	53.8 ***	n.s.

A significant water temperature decrease was observed between maximum values of 30–31 °C from November 2019 to May 2020, and minimum values of 26–27 °C in July and September 2020 during the rainy and flood season (Figure 2a and Table 2). The lagoon is rather spatially homogeneous in temperature, with no significant difference in mean temperature between the stations (TEMP = 28.4–30.1 °C on average, see Table 2). However, the two-way ANOVA analysis revealed significant spatial and temporal variations (without interaction) in the three sub-regions considered (Table 3 and Figure 2a): temperatures tended to be cooler in the Cotonou Channel close to the Atlantic Ocean (green bars in Figure 1a) and warmer in the lagoon and towards the river mouths (orange and blue bars in Figure 2a), more particularly during dry season in January-May.

Mean salinity values strongly increased from November (SAL ~ 0) to May (SAL > 23), before decreasing to less than 5 in July and September (Figure 2b). The two salinity outliers of November 2019 (SAL of 10–20) and September 2020 (SAL of 20–30) were observed in the Cotonou Channel during flood tide. Indeed, during September–November, strong river discharges were observed and seawater can only marginally penetrate the Cotonou Channel under the effect of the tide (e.g., [5,23]). During most of the bimonthly surveys, the minimum salinity values were observed near the river mouths while the maximum values were observed in the Cotonou Channel sub-region (see also [5]). Both the nonparametric tests and the two-way ANOVA revealed significant spatial and temporal salinity variations (Tables 2 and 3).

Figure 2c shows that November 2019 corresponded to the end of the high water period whereas May 2020 corresponded to the end of the low water period. Statistical tests suggested that seasonal and spatial DEPTH variations were significant (Tables 2 and 3). Bathymetry is significantly deeper in the Cotonou Channel (DEPTH > 6 m for Stations 1 and 2) than in the lagoon or close to the river mouths (Figure 2c). This large spatial variability partly masks the non-negligible water-level decrease of 0.60–0.70 m observed between high and low water seasons (see also [23]).

Table 3. Significant F-values ($p < 0.05$) derived from two-way ANOVA on the influence of seasons (bi-monthly campaigns), sub-regions and interaction between seasons and sub-regions on environmental and zooplankton variables. Significance levels: * $p < 0.05$, ** $p < 0.01$, *** $p < 0.001$, n.s. not significant.

Variable	Seasons F-Value (df = 5)	Sub-Regions F-Value (df = 2)	Seasons × Sub-Regions F-Value (df = 10)
		Environmental variables	
TEMP (°C)	192.5 ***	10.7 ***	n.s.
SAL	47.8 ***	8.7 ***	n.s.
DEPTH (m)	6.3 ***	58.6 ***	n.s.
TURB (FTU)	35.0 ***	13.1 ***	2.7 **
%POM (%)	14.4 ***	n.s.	n.s.
DO (mg L^{-1})	4.8 ***	4.0 *	n.s.
CHL-a (mg m^{-3})	9.9 ***	4.1 *	n.s.
		Zooplankton	
Species richness	42.2 ***	n.s.	n.s.
H′	26.5 ***	n.s.	n.s.
J	9.5 ***	n.s.	n.s.
Zooplankton abundance (ind L^{-1})	n.s.	n.s.	n.s.
Copepod relative abundance (%)	18.4 ***	n.s.	n.s.
Rotifer relative abundance (%)	21.4 ***	n.s.	n.s.

Turbidity was relatively low (<15 FTU) from November 2019 to May 2020, and strongly increased to reach mean values of ~100 FTU at the beginning of the rainy season in July 2020 (Figure 2d and Table 2). The TURB showed significant temporal and spatial variability in the three sub-regions, with higher values close to the river mouths (Figure 2d and Table 3). A closer inspection of the turbidity data revealed the presence of a turbid plume that extends to the south of the Ouémé (east of the lagoon), particularly during the rainy season. Figure 2e showed that %POM, which represents the organic fraction of TSS, was maximum in November 2019 (%POM ~ 60%) and decreased to a minimum in July-September 2020 (%POM ~ 25%). Although statistical analyses did not reveal significant spatial variations in %POM ($p > 0.05$; Table 2), the %POM was generally slightly higher in the lagoon environment, than close to the river mouth or in the Cotonou Channel (Figure 2e). Similarly, the %POM was on average significantly (WKW-test, $p < 0.05$) lower at Station 7 (%POM ~ 24%) than at Station 13 (%POM ~ 64%) (Table 2). High TURB values (Figure 2d) were associated with low %POM values (Figure 2e), suggesting that river plumes observed at the beginning of the wet season transport mineral sediments towards the lagoon, thus decreasing the proportion of organic matter.

Mean DO concentration varied from a minimum of less than 6 mg L^{-1} in November 2019 (Figure 2f), during a flood period when the lagoon was filled of freshwater. In contrast, higher DO concentrations (DO > 7 mg L^{-1}) were observed during low water season between January and July of 2020. Both spatial and temporal variations were significant ($p < 0.05$, Tables 2 and 3). During high water period, the water near the river mouths tended to be much less oxygenated than in the rest of the lagoon (Figure 2f).

The CHL-a showed significant temporal variability ($p < 0.05$ Tables 2 and 3) with stronger values (>15 mg m^{-3}) during high-water periods in November 2019 and September 2020, and lower values (<10 mg m^{-3}) in low-water season between January and July 2020 (Figure 2g). Significantly higher CHL-a concentrations were also generally observed near the river mouths (Figure 2g and Table 3). In contrast to the other months, CHL-a in January 2020 was very homogeneous throughout the lagoon and had relatively low values (5–10 mg m^{-3}).

3.2. General Distribution of the Zooplankton Community

A total of 109 zooplanktonic species were inventoried, including 81 taxa of rotifers, 20 of copepods (without considering nauplii), five of cladocerans and three organisms that belonged to three other zooplankton groups (Eumalacostraca, Mollusca, and Ostracoda) (Table 1 and Figure 3). Among all the organisms identified, one rotifer taxon (*Brachionus plicatilis*), four copepod taxa (cyclopoid sp. 1, *Ectocyclops* sp., calanoid sp. 1, and *Microsetella* sp.), copepod nauplii and unidentified mollusks were considered frequent species, with F_{occ} ranging from 51% to ~98% (Table 1). The other species identified were either occasional (15 taxa), accidental (46 taxa) or rare (42 taxa) species (Table 1).

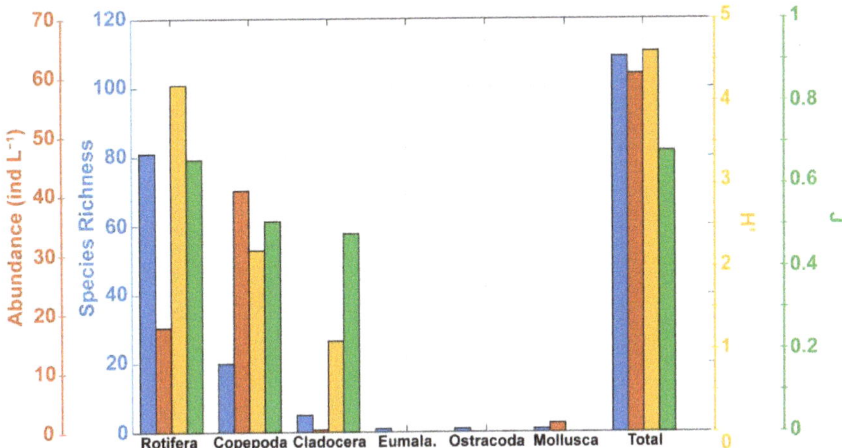

Figure 3. Overall species richness (blue), mean abundance (red), Shannon species diversity index (H′, yellow) and Piélou equitability index (J, green) for each of the zooplankton groups (Rotifera, Copepoda, Cladocera, Eumalacostraca, Ostracoda, Mollusca) as well as for the entire zooplankton community (Total).

Considering the six field campaigns, the average abundance of zooplankton organisms was ~60 ind L^{-1}. This abundance was dominated by copepods (68.1% of the individuals) and rotifers (29.1%) (Figure 3). The average relative abundance of Mollusca was weak (2.5%), whereas cladocerans, Eumalacostraca, and Ostracoda were negligible (Figure 3). For the rest of the manuscript, we focused mainly on the two main groups of zooplankton, namely copepods and rotifers, which represented 92.7% of the total species richness and 97.2% of the zooplankton abundance.

The Shannon diversity index showed an overall value of 4.6 (Figure 3). This index, computed for each group, indicated that rotifers were the most diverse group ($H' = 4.2$) followed by copepods ($H' = 2.2$) (Figure 3). On average, the relative abundance of each species was rather homogeneous within the copepods, rotifers and cladocerans ($J \sim 0.5$–0.7) (Figure 3).

On average over the six surveys, the number of zooplanktonic species observed varied non-significantly ($p > 0.05$) between 13 and 18 species depending on the stations (Table 2). Similarly, no significant variation in species richness was observed between the three sub-regions considered (Table 3). Despite a significant decrease in mean zooplankton abundance from a maximum of ~190 ind L^{-1} in the northwest of the Lagoon to a minimum of ~20 ind L^{-1} in the Cotonou Channel ($p < 0.05$), no significant spatial difference ($p > 0.05$) was observed when considering the abundance distributions at each of the 16 stations or between the three sub-regions (Table 2). This may be due to the presence of multiple intermediate values that reduced noise and increased the statistical power of the analysis, thereby masking the dissimilarity that existed between the two extreme abundance distributions observed between the Channel and northwest of the Lagoon.

On average, copepods were the dominant group, representing 50–80% of the total abundance in the lagoon and reaching almost 90% at the connection of the Cotonou Channel with the Atlantic Ocean (Station 1) (Table 2). In contrast, the relative abundance of rotifers was relatively low (10–50%) over the whole lagoon (Table 2). Spatial variations in relative fractions of copepods or rotifers were not significant ($p > 0.05$; Tables 2 and 3).

The averaged Shannon species diversity index values varied from 1.7 to 2.7 but these weak spatial variations were not significant ($p > 0.05$; Table 2). Piélou's equitability index values broadly followed the *H'* distribution, suggesting that where specific diversity was higher in the lagoon, abundance was more evenly distributed among the different zooplankton species. The Piélou's equitability index values significantly varied between 0.4 and 0.7 ($p < 0.01$; Table 2).

3.3. Seasonal Variation of the Zooplankton Community

Zooplankton species richness peaked during the flood period in November 2019, with 30–40 distinct zooplanktonic taxa at each station (Figure 4a). During the low water period, from January to May 2020, the species richness decreased sharply down to 10–15 taxa at each station. During the rainy period (July 2020) and the beginning of the high water period (September 2020), species richness gradually increased and reached 20–30 species per station in September 2020. These seasonal variations in species richness were highly significant ($p < 0.001$) unlike spatial variations (Tables 2 and 3).

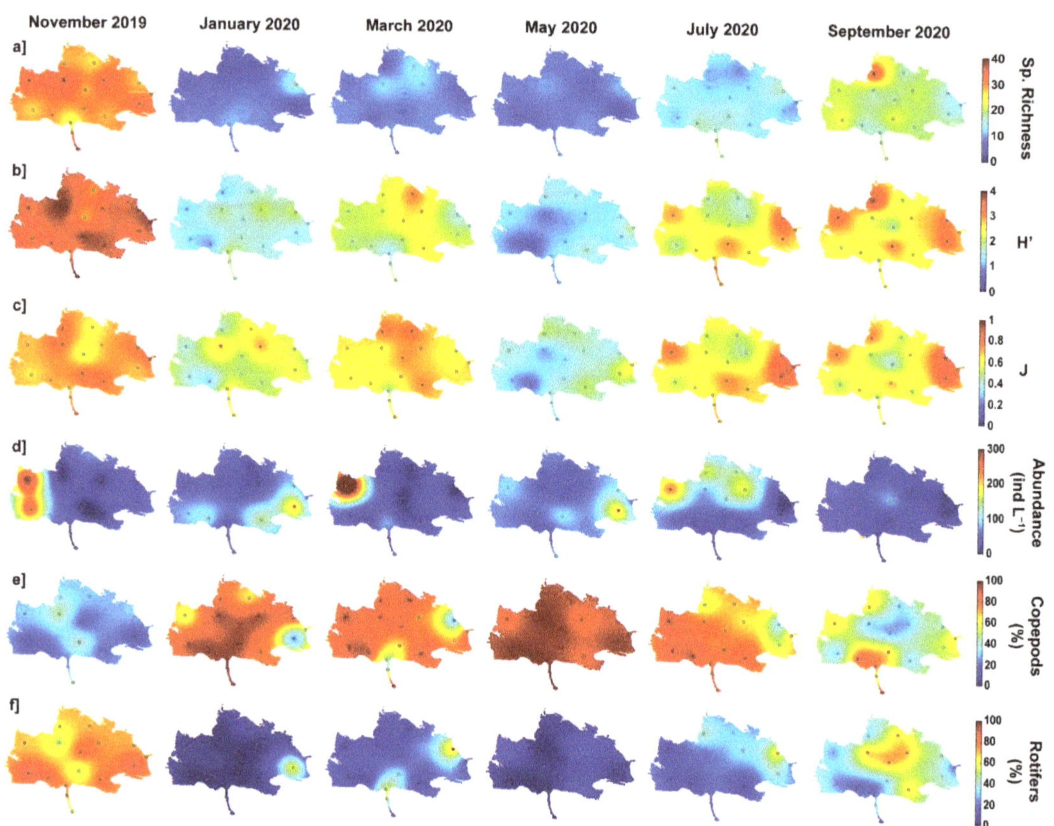

Figure 4. Spatial distribution of bimonthly variation of different zooplankton parameters. (**a**) Species richness (1st row), (**b**) Shannon species diversity index (H', 2nd row), (**c**) Piélou equitability index (J, 3rd row), (**d**) Total abundance (in ind L^{-1}, 4th row), (**e**) Copepod relative abundance (in %, 5th row), and (**f**) Rotifer relative abundance (in %, 6th row).

Spatio-temporal variation in zooplankton abundance was more complex (Figure 4d). Throughout the year, abundance was relatively low in the lagoon (typically 15–40 ind L^{-1}) except in specific locations, usually to the east or west, where values could locally increase to 200–400 ind L^{-1}. The observed spatio-temporal variations, considering all stations/campaigns or the three sub-regions, are not statistically significant (Tables 2 and 3). However, abundance distributions differed significantly ($p < 0.05$) between Station 2 located in the Cotonou Channel (mean abundance of 20 ind L^{-1}) and Station 14 located in the northwestern part of the lagoon (>190 ind L^{-1}) (Table 2).

The relative proportion of copepods varied inversely with abundance (Figure 4e). In November 2019 and September 2020, the percentage of copepods was relatively low in most of the lagoon, with relative abundance values below 60% and 40% in September and November, respectively. During this period, the zooplanktonic assemblage was therefore dominated in abundance by rotifers (Figure 4f). However, during the low water period, from January to May 2020, the proportion of copepods significantly increased ($p < 0.001$; Tables 2 and 3) and exceeded 80% in most areas of the lagoon (Figure 4e). In contrast to temporal variations, spatial variations were not significant except for the proportion of rotifers which showed, on average, significantly ($p < 0.05$) lower values in the Cotonou Channel (Station 1~11%) than near the Ouémé River (Station 7~51%) (Table 2).

The combination of species richness and abundance led to diversity index values that varied similarly to species richness (Figure 4b). The Piélou equitability index values varied similarly, suggesting that abundance was relatively well distributed among species when diversity was high, while some species dominated when abundance was low (Figure 4c). The observed bi-monthly variations of both H′ and J were significant ($p < 0.001$; Tables 2 and 3).

In general, spatial variations in zooplankton were not statistically significant, either between all stations or between the three sub-regions more influenced by rivers or the ocean (Tables 2 and 3). However, we showed that the zooplankton community of the Nokoué Lagoon tended to significantly vary on a seasonal scale, and that it was likely to be strongly impacted by environmental parameters that also showed significant seasonal variability (Figure 2 and Tables 2 and 3). Therefore, we analyzed more thoroughly the relationships between physicochemical factors and zooplankton composition.

3.4. Relationship between Rotifer and Copepod Diversity and Environmental Factors

Zooplankton diversity of the two most abundant groups (rotifers and copepods) was combined with environmental variables to identify the main drivers of the general zooplankton community structure using a RDA (Figure 5a). Four environmental variables significantly constrained ($p < 0.05$) the overall variance of the zooplankton diversity, explaining 54.8% of the total variability. The environmental variable with the greatest explanatory power for zooplankton species richness was salinity ($F = 59.6$, $p < 0.001$) which explained 38.8% of the total variability, followed by dissolved oxygen ($F = 15.5$, $p < 0.001$) which explained 8.8%. Turbidity ($F = 7.9$, $p < 0.01$) and station depth ($F = 5.5$, $p < 0.01$) were of secondary importance, explaining 4.4% and 2.8% of the total variance, respectively. The first RDA axis was mainly scored by salinity ($r = -0.64$) followed by oxygen ($r = -0.43$), whereas axis 2 was mainly scored by water-depth ($r = 0.26$) followed by the three other parameters ($r = \pm0.12$–0.15) (Figure 5a). The RDA analysis revealed that higher copepod diversity was primarily associated with higher salinities (Figure 5a). In contrast, higher rotifer species richness was related to weaker SAL and DO values.

To further investigate the impact of key environmental factors on the diversity of each of the dominant groups (rotifers and copepods), the GAMs were fitted on the response of rotifer and copepod species richness to environmental parameters. Considering all seven environmental variables, the GAMs had the highest explained deviance (76% for rotifers and 57.8% for copepods) and adjusted R^2 (0.70 for rotifers and 0.53 for copepods). The response plots were shown in Figure 5b–e, only for variables with a significant effect ($p < 0.05$).

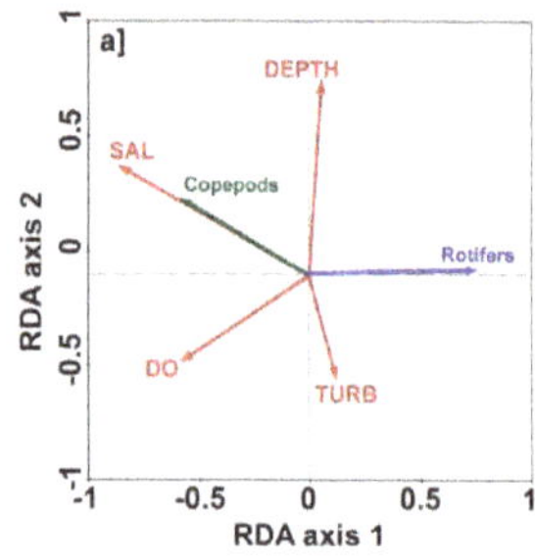

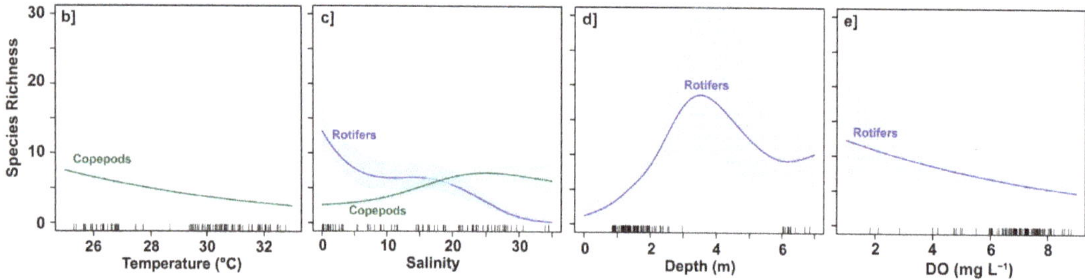

Figure 5. Redundancy analyses (RDA) and generalized additive models (GAMs) relating diversity of copepods and rotifers to environmental variables. (**a**) RDA diagram of species richness and environmental parameters. (**b–e**) GAMs of species richness versus temperature (TEMP), salinity (SAL), depth (DEPTH) and dissolved oxygen (DO). Only significant ($p < 0.05$) relationships are shown for copepods and rotifers. Shaded areas indicate the 95% confidence intervals.

The most significant parameter for copepod diversity was SAL followed by TEMP ($p < 0.001$). Copepod diversity decreased almost linearly with temperature, slightly varying from 7–8 species for TEMP < 26 °C to 3–4 species for TEMP > 32 °C (Figure 5b). In contrast copepod species richness increased with salinity, from ~3–4 species in freshwater to 6–7 species for SAL > 20 (Figure 5c). Copepod species richness was not significantly ($p > 0.05$) influenced by other environmental variables and was therefore not shown in Figure 5d,e.

In a similar way, the most significant parameter for rotifer diversity was SAL ($p < 0.001$) followed by DEPTH ($p < 0.001$) and DO ($p < 0.01$). The other environmental parameters do not significantly ($p > 0.05$) influence rotifer species richness. The evolution of rotifer species richness as a function of salinity showed opposite variations to that of copepods, with a much higher diversity in low-salinity water, and only very few species persisted when SAL > 20 (Figure 5c). The diversity of rotifers tended to be higher at the deepest (Figure 5d) or least oxygenated (Figure 5e) stations. Note, however, the large errors for depths between 3–6 m (Figure 5d) or DO values less than 5 mg L^{-1}, due to the lack of data in these ranges.

3.5. Relationship between Rotifer and Copepod Abundance and Environmental Factors

We investigated the main drivers of the general zooplankton abundance, applying a RDA to the abundance of the two main zooplankton groups (i.e., rotifers and copepods) combined with environmental variables (Figure 6a). The RDA used three variables that significantly constrained ($p < 0.05$) the overall variance of the zooplankton abundance, explaining 47.8% of the total variability: the environmental variable with the greatest explanatory power for zooplankton abundance was salinity (F = 55.9, $p < 0.001$) which explained 37.3% of the total variability. Water temperature (F = 9.6, $p < 0.01$) and station depth (F = 8.1, $p < 0.01$) were of secondary importance, explaining 5.5% and 5% of the total variance, respectively. The first RDA axis was mainly scored by salinity (r = 0.62) followed by temperature (r = 0.19), whereas axis 2 was mainly scored by temperature (r = −0.29) followed by water depth and salinity (r = ±0.07) (Figure 6a). The RDA analysis revealed that higher abundances of copepods and lower abundances of rotifers were mainly associated with higher salinities.

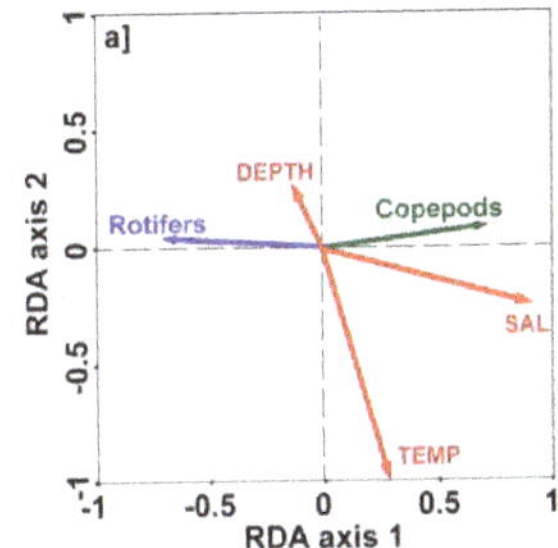

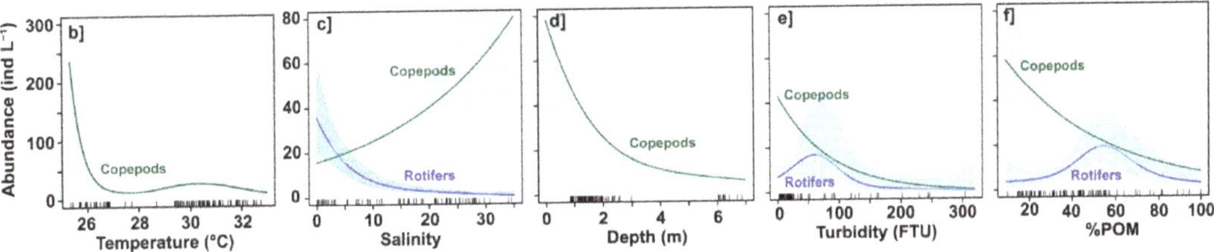

Figure 6. Redundancy analyses (RDA) and generalized additive models (GAMs) relating abundance of copepods and rotifers to environmental variables. (**a**) RDA diagram of abundances and environmental parameters. (**b–f**) GAMs of abundances versus temperature (TEMP), salinity (SAL), depth (DEPTH), turbidity (TURB) and dissolved oxygen (DO). Only significant ($p < 0.05$) relationships are shown for copepods and rotifers. Note the change in scale between subplots (**b**) and (**c–e**). Shaded areas indicate the 95% confidence intervals.

Nonlinear relationships between the abundances of each of the dominant zooplankton groups (rotifers and copepods) and environmental variables were also explored through GAMs (Figure 6b–f). The GAMs including the seven environmental variables have explained deviances of 50.6% for rotifers and 55.2% for copepods, and adjusted R^2 of 0.14 for rotifers and 0.41 for copepods. As previously, response plots were presented in Figure 6b–e, only for variables with a significant effect ($p < 0.05$).

The copepod abundance variability is significantly related to five environmental parameters ($p < 0.001$; Figure 6b–f). Copepod abundance strongly decreased for temperature between 25.5 °C and 26 °C, and then remained around 15–25 ind L^{-1} for TEMP > 26 °C (Figure 6b). Note, however, that the high abundance values for TEMP < 26 °C were highly uncertain because of the lack of data in these temperature ranges. Copepod abundance increased progressively from ~20 ind L^{-1} in freshwater to more than 60 ind L^{-1} for SAL > 30 (Figure 6c). For the other environmental variables (DEPTH, TURB, %POM), the abundance of copepods was found to be higher the lower the depth of the stations, the turbidity of the water and the fraction of organic matter (Figure 6d–f).

The rotifer abundance variability was significantly related to SAL ($p < 0.001$) followed by %POM ($p < 0.01$) and TURB ($p < 0.05$). Rotifers were more abundant for lower salinities and even dominant (>20 ind L^{-1}) for SAL < 3. Rotifer abundance rapidly decreased with salinity and was less than 10 ind L^{-1} for SAL > 10 (Figure 6c). Variations in rotifer abundance as a function of TURB and %POM were more complex and showed a bell shape with relatively high values (~20 ind L^{-1}) for intermediate values of turbidity (~50 FTU) and organic matter (~50%) (Figure 6e,f). However, associated errors were also important for these fitted GAMs. In contrast to copepods, station depths and water temperature did not have a significant impact ($p > 0.05$) on variations in rotifer abundance.

3.6. Relationship between Abundance of the Most Frequent Taxa and Environmental Factors

In this section, we first examined the general relationship between the abundances of the most frequent ($F_{occ} > 25\%$) and most abundant (>0.5%) zooplanktonic taxa and environmental parameters, using an RDA. Therefore, 16 taxa were considered, including five copepod taxa (including Nauplius), 10 rotifer taxa and Mollusca sp. The list of these taxa and their frequency of occurrence were presented in Table 1 and Figure 7. The RDA used four variables that significantly constrained ($p < 0.05$) the overall variance of the species-specific abundances, explaining 23% of the total variability (Figure 7a). The environmental variable with the greatest explanatory power was again salinity ($F = 12.0$, $p < 0.001$), which explained 11.3% of the total variability. The DO concentration ($F = 5.1$, $p < 0.01$), TEMP ($F = 4.2$, $p < 0.01$) and CHL-a ($F = 3.9$, $p < 0.01$) concentration, were of secondary importance, explaining 4.7%, 3.6%, and 3.4% of the variance, respectively. The first RDA axis was mainly scored by SAL ($r = 0.67$) followed by CHL-a ($r = -0.40$), whereas axis 2 was mainly scored by DO ($r = 0.54$) and TEMP ($r = -0.40$) (Figure 6a). Higher abundances of the most frequent copepod taxa were in general associated with higher salinities, whereas the abundance of the most frequent rotifer taxa likely increased with low salinity (or high CHL-a) values.

To further investigate the impact of environmental variables on the most frequently observed zooplankton taxa, we used GAMs to relate the seven environmental variables to these different taxa. Depending on the taxon considered, the explained deviance varied between a minimum value of 33.4% for Calanoid spp. (coded C05) and a maximum value of 94.5% for *Brachionus caudatus* (coded R08) (Table 4). The taxa with the strongest relationship with environmental variables (highest explained deviance) belonged to the rotifer group (e.g., *Brachionus caudatus, Brachionus angularis, Filinia opoliensis, Brachionus falcatus, Lecane leontina*) (Table 4). The most important variable was SAL, followed by CHL-a, as also noted on the RDA (Table 4 and Figure 7a).

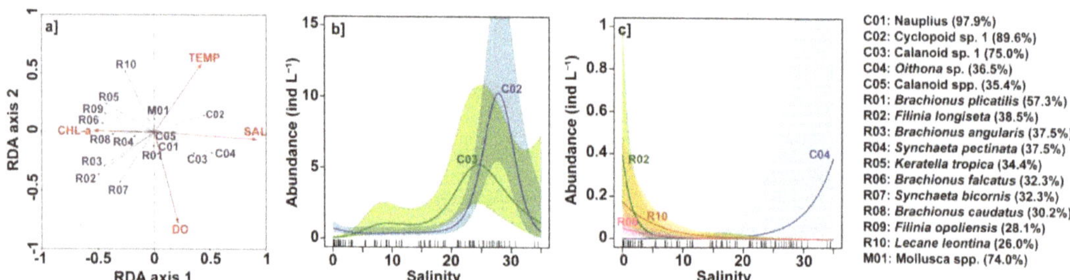

Figure 7. Redundancy analyses (RDA) and generalized additive models (GAMs) relating abundance of the most frequent zooplankton taxa to environmental variables. (**a**) RDA diagram of taxa abundances and environmental parameters. (**b**,**c**) GAMs of taxa abundances versus salinity (SAL). Only the three most significant ($p < 0.05$) relationships are shown for copepods and rotifers. Note the change in scale between subplots (**b**) and (**c**). Taxa were coded according to the legend shown on the right (see also Table 1). Shaded areas indicate the 95% confidence intervals.

Table 4. Each row summarizes the GAM for that particular taxa: percent deviance explained (%Deviance) and significance of each environmental variable. Black cells represent p-values < 0.001, dark gray cells represent p-values < 0.01, light gray cells represent p-values < 0.05, and white cells represent non-significant p-values.

Zooplankton Taxa	Code	%Deviance	SAL	TURB	TEMP	OXY	WL	CHL-a	POM
Nauplius	C01	57.2							
Cyclopoid sp. 1	C02	71.2							
Calanoid sp. 1	C03	58.5							
Oithona sp.	C04	82.1							
Calanoid spp.	C05	33.4							
Brachionus plicatilis	R01	72.2							
Filinia longiseta	R02	79.8							
Brachionus angularis	R03	91.1							
Synchaeta pectinata	R04	63.9							
Keratella tropica	R05	80.1							
Brachionus falcatus	R06	89							
Synchaeta bicornis	R07	81.2							
Brachionus caudatus	R08	94.5							
Filinia opoliensis	R09	89.4							
Lecane leontina	R10	82.3							
Mollusca spp.	M01	67.6							

Salinity was significantly related ($p < 0.05$) to the abundance of 13 of the 16 taxa most frequently and abundantly observed in Nokoué Lagoon (Table 4), and very significantly related ($p < 0.001$) to nine of them. We were therefore studying more specifically the relationship between zooplankton taxa and salinity. However, for the sake of clarity and conciseness, we presented in Figure 7b,c only the three taxa of copepods (Cyclopoid sp. 1, Calanoid sp. 1, Oithona sp.) and rotifers (*Filinia longiseta, Lecane leontina, Brachionus falcatus*) for which the partial deviances explained by salinity were the most important. Concerning the copepods, abundances of Cyclopoid sp. 1 (coded C02) and Calanoid sp. 1 (coded C03) showed almost unimodal distributions, with maximum abundances for salinities between 20 and 30 (Figure 7b). The abundance of *Oithona* sp. (coded C04) also increased for salinities above 20, but its abundance is an order of magnitude lower than that of the other two copepod taxa (Figure 7c). Concerning the rotifers, abundances of *Filinia longiseta* (coded R02) *Brachionus falcatus* (coded R06) strongly increased for SAL < 5, whereas the abundance of *Lecane leontina* (coded R10) started to increase for SAL < 15 (Figure 7c). In very low-salinity waters, *Filinia longiseta* dominated the two other copepod taxa. Although not shown, the abundance of other rotifer taxa showed similar relationships with salinity, with the exception of *Brachionus plicatilis* (coded R01) which showed a roughly unimodal distribution with a marked maximum for a SAL ~ 8. This planktonic species is known to be a euryhaline rotifer that tolerates a wide range of salinities.

4. Discussion

4.1. Zooplankton Composition and Diversity

This study enabled us to identify 109 distinct zooplanktonic taxa in the Nokoué Lagoon, including 81 taxa of rotifers, 20 of copepods, five of cladocerans and three other organisms. Representing ~75% of the species richness, rotifers were thus the most diversified group, as also observed in other regional tropical lagoons, from Ivory Coast to Nigeria [63–66]. The total species richness of 109 zooplanktonic taxa recorded in our study far exceeded that previously reported in Nokoué Lagoon (31 taxa) [20]. This difference could be explained by the sampling that covered here the whole hydrological cycle and a wide range of salinity, while in the previous study zooplankton were analyzed between June and September when salinity was low [20]. In addition, the methodology implemented in our study (see Section 2.1.) was different from that used previously (mesh size, volumes filtered, location of sampling stations, number of replicates, etc.). Note that taxa not identified to specific levels (see Table 1) have been conscientiously distinguished from each other on the basis of clear morphological features (see Section 2.3). As a consequence, the taxonomic list obtained here (Table 1), and the resulting greater diversity cannot be attributed to misclassification and erroneous separation of individuals belonging to the same species.

Relationships between zooplankton diversity and environmental variables, as investigated through RDA and GAMs, showed that salinity was the key parameter for species richness, explaining a large part of the total variance. Consequently, seasonal salinity variations in Nokoué Lagoon, which strongly varied from 0 in the flood period to ~24 in low-water period (Figure 2b; see also [5]), were associated with strong changes in the structure and diversity of zooplankton assemblages. During low-salinity (flood) periods, a higher diversity of rotifers, and thus zooplankton, was observed (Figures 4–6), consistently with other regional studies e.g., [65,67]. During floods, a relatively high freshwater inflow (up to 1200 m^3 s^{-1}) from the Sô and Ouémé rivers was observed [23,68]. These river discharges could transport into the lagoon the numerous zooplankton species (~100 zooplankton taxa, including ~90 rotifers) observed in the rivers [69]. Furthermore, the high dominance of rotifers during the flood period in November 2019 (Figure 4f) was associated with relatively low DO values (Figure 2f), as also suggested by the GAMs (Figure 5e). These low DO values are likely related to the degradation of a significant amount of POM in the lagoon (Figure 2e) that can originate, during the flood season, from several sources, such as the Ouémé deltaic plain, urban water drainage, or *acadjas* (i.e., traditionally made bamboo parks used by local fishermen to grow fish) [19,69]. The degradation of this POM, which consumes oxygen, was generally the main source of nutrients for rotifers [10,63,70,71] which may have increased rotifer diversity during the period of very low salinity and low DO concentrations during flooding. The composition and diversity of rotifers can also reflect a certain level of eutrophication in Nokoué Lagoon. Indeed, rotifers are among the only zooplanktonic organisms that are resistant to high organic matter enrichment and dissolved oxygen depletion [72,73]. The high diversity of Brachionidae and the presence of *Brachionus falcatus* (Figure 7), could be an indicator of eutrophication in Nokoué Lagoon [74,75], as well as the species *Cephalodella gibba* which was often associated with high POM concentrations. In the future, additional measurements of eutrophication, especially nitrogen and phosphorus compounds, would be needed to better interpret zooplankton structure and their use as a bioindicator of environmental conditions in Nokoué Lagoon.

During the low water period, Nokoué Lagoon was very salty, and we observed a strong decrease in the diversity of rotifers associated with an increase in the diversity of copepods (Figures 4 and 5), once again corroborating other regional studies [11,66]. Salinity tends to cause stressful ecological conditions and decreases biodiversity in lagoon ecosystems, particularly rotifers and cladocerans [12]. Rotifers are typically dulcaquicolous organisms (1488 taxa out of 1570 described in this phylum [76]). Only halotolerant rotifers could withstand the haline stress observed in Nokoué Lagoon, such as *Synchaeta bicornis*, or more particularly *Brachionus plicatilis* as observed in Figure 7 [77–79]. These species could tolerate wide variations in salinity [79–82].

In Nokoué Lagoon, total species richness decreased progressively with increasing salinity. These results may seem to contradict the well-accepted Remane's (1934) theory [83], according to which taxonomic diversity would reach a minimum, called the Artenminimum, for salinities of 5 to 8 [83–85]. However, this concept of Artenminimum was mainly based on observations of the diversity of benthic invertebrates in the Baltic Sea, and not of the diversity of pelagic organisms such as zooplankton. Although Remane's species-minimum model has been widely used to explain biodiversity changes along haline gradients, it has also been shown to be inadequate to explain phyto- and zooplanktonic diversities in estuarine habitats (e.g., [86–89]). Indeed, in these regions, and as observed in the

Nokoué Lagoon, planktonic organisms often did not show minimal diversity at salinities intermediate between marine and freshwater (e.g., [89–95]).

4.2. Zooplankton Abundance

Zooplankton abundance was, on average, ~60 ind L^{-1} across the Nokoué Lagoon, but locally as high as 100–300 ind L^{-1} (Figures 3 and 4). These values were of the same order of magnitude as those obtained in other regional lagoons: 100–500 ind L^{-1} in Fresco Lagoon, 120–160 ind L^{-1} in Aby-Tendo-Ehy lagoons, and 50–250 ind L^{-1} in Ebrié Lagoon [65–67]. However, they were slightly lower than the average abundances of 100–400 ind L^{-1} obtained previously in Nokoué Lagoon [20]. This difference may be explained by inter-annual variations as well as by the use of a net with much finer mesh by these authors (30 μm instead of 50 μm), which was likely to collect smaller zooplankton organisms, thus increasing zooplankton abundance.

One of our hypotheses was that the Nokoué Lagoon could show significant differences between the Cotonou Channel located near the Atlantic Ocean and the areas near the river mouths. The separation of the stations into three distinct groups showed indeed that some environmental variables, and in particular salinity, had significant variations ($p < 0.05$) between the three sub-regions (Figure 2 and Table 3) but that this was not the case for zooplankton diversity and abundance (Table 3). However, significant differences on zooplankton may appear between stations with extreme values (see MWM-test in Table 2). In particular, local increase in zooplankton abundance was noted at some stations located at the west and east of Nokoué Lagoon (Figure 4). First, this could be related to the freshwater inflows from the So and Ouémé rivers, which resulted in local desalination processes in these particular areas of the lagoon [5]. These events may be favorable to freshwater zooplankton communities, especially rotifers as well as nauplii, which could maintain high abundances in less saline environments (Figures 6 and 7). Second, the relatively high zooplankton abundance in the western part of Nokoué Lagoon could also be related to the increased presence of brush park fisheries (*acadjas*), used in these areas for trapping and artisanal fishing e.g., [96,97]. Indeed, these artificial parks lead to the local development and increase of biological productivity due to the contribution of nutrients through the decomposition of organic matter from woody materials [1,96,98]. Moreover, *acadjas* parks, which are made up of more than 15 branches per square meter and extend over several hectares [96,97], tend to modify local hydrodynamics by decreasing the intensity of winds and currents [99]. This reduced vertical mixing and turbulence may also enhance the development of zooplankton (e.g., [100]). Thus, higher nutrient concentration and lower turbulence in these zones could explain the higher local zooplankton abundance. Accordingly, it has already been shown that phytoplankton and zooplankton abundances could be four times higher, but that their diversity was lower, in the *acadjas* than in the surrounding areas [99,101]. However, our current dataset did not allow us to highlight a possible link between the proximity of the *acadjas* to our stations (all carried out outside of these brush parks) and the structure of the zooplankton community. For instance, we did not observe any increased presence of species with benthic tendencies (e.g., Lecanidae, Lepadellidae, Mytilinidae, Notommatidae, Chydoridae, Macrothricidae) or bacterivores (e.g., Brachionidae) at these particular stations. More specific studies on the impact of *acadjas* on the structure of planktonic communities will therefore be necessary to reach any definitive conclusions.

The RDA and GAMs analyses revealed a strong relationship between salinity and zooplankton abundance. Indeed, salinity was the primary driver of zooplankton abundance, explaining 37% of the overall variance in zooplankton taxa abundance, and 23% of the variance in the most frequent taxa. Our results also showed that zooplankton abundance was higher during the dry season, associated with high salinities (low-water period), than in the wet season characterized by low salinity (flood period). These results are similar to those noted in Ivorian lagoons of Aby-Tendo-Ehy, Fresco and Ebrié [65–67]. The high zooplanktonic abundance during the low-water period was related to the dominance of copepods and in particular their larvae (nauplii) which represented ~60% of the relative zooplanktonic abundance. During this period, the dominance of brackish water conditioned the proliferation of these halotolerant taxa of copepods from the coastal ocean. The rotifers were then in the minority.

Copepods in the Nokoué Lagoon included both freshwater and marine species, which may have contrasting responses to environmental parameters. However, due to limitations in our dataset, we were unable to differentiate between unidentified Calanoida and Cyclopoida species in terms of their affinity for freshwater or marine environments. Nevertheless, many of these unidentified copepod species were observed throughout the year and were able to tolerate highly contrasting salinity gradients. Based on our analysis, Cyclopoid sp. 1 and Calanoid sp. 1 were likely to correspond to marine species due to their higher abundance in higher salinity conditions (Figure 7). Identified copepod species were divided into freshwater and marine groups based on their known affinity for salinity. However, the low frequencies of occurrence and relative abundances of these identified species (Table 1) limited the ability to draw robust conclusions about the relationship between salinity and copepod diversity. Further research, with a longer monitoring period and a more comprehensive zooplankton dataset, is needed to fully understand the contrasting relationships that freshwater and marine copepod species may have with environmental parameters in the Nokoué Lagoon. It would also be interesting to investigate the percentage of freshwater and marine species present in the copepod community and how this changes in response to environmental conditions, which could provide valuable insights into the ecological dynamics of copepods in the lagoon.

Finally, on an interannual scale, between June and September 2015, 90% of the zooplanktonic abundance was copepods and 10% was rotifers [20]. Between July and September 2020, our study showed a reduced proportion of copepods (~70%) and an increase proportion of rotifers (30%). This interannual variability could be explained by the difference in salinity between the two periods. In June-September 2015, the average salinity was of 6.2 [20], while it was only ~3 in July-September 2020 (Figure 2). Based on the highlighted relationships between salinity and zooplankton species distribution (Figures 6 and 7), it was therefore consistent to observe a lower abundance of copepods between July and September 2020. Although interannual variations in salinity probably explain the changes in copepod and rotifer distributions between 2015 and 2020, a temporal shift in zooplanktonic successions during these 2 years cannot be ruled out either, which cannot be assessed with our bimonthly sampling approach.

5. Conclusions

Based on the analysis of biological data from six bimonthly campaigns, we identified 109 distinct zooplankton taxa in Nokoué Lagoon. Average zooplanktonic abundance was ~60 ind L^{-1} but increased locally to 100–300 ind L^{-1}. This abundance was largely dominated by copepods and rotifers, which represented, on average, 68.1% and 29.1% of organisms, respectively.

Environmental parameters showed significant seasonal variations, especially in terms of the salinity. Indeed, Nokoué Lagoon was filled with fresh water during the flood period in November 2019 and progressively salinized to reach average salinity values of ~22–25 during the low water period in March-May 2020. We showed, through redundancy analyses, that salinity was the key parameter that structured the zooplanktonic ecosystem, both for taxon diversity and abundance. Therefore, on a seasonal scale, a strong shift was observed in the zooplanktonic community. Indeed, during the flood period, abundance was quite high and zooplanktonic diversity was maximal. During this short period, rotifers were dominant and about 30 zooplanktonic species could be observed in each sampling station. In contrast, during the low water period, the diversity became minimal (less than 10 species in each station), and the abundance slightly decreased. However, some zooplankton hotspots were observed in the west and east of the lagoon, likely independent of salinity. These localized areas of high abundance may have been under the influence of other processes, such as the Sô and Ouémé river inputs and/or the presence of *acadjas* brush parks whose effects on zooplankton could not be determined with our dataset. More specific studies as well as the continuation of regular and long-term sampling will help us to understand further such aspects of zooplanktonic structuring and its fine-scale spatio-temporal variations.

The results obtained in this study provided valuable information on the seasonal variations of zooplankton in Nokoué Lagoon, which is one of the most biologically productive in West Africa. Future studies should focus on the simultaneous analysis of the different compartments of the trophic chain (phytoplankton, zooplankton, fish) and their responses to environmental variables. Such an approach will complete our study and lead to a comprehensive view of the functioning of this rich ecosystem that provides 70% of the Beninese fishery resource.

Author Contributions: Contributed to conception and design: A.C., H.H.A. and G.D. Contributed to acquisition of data: F.T.O., T.T.A., F.U.D.-S., V.O.O. and M.B.D. Contributed to zooplankton identification: F.T.O., T.T.A., F.U.D.-S. and F.A. Contributed to analysis and interpretation of data: F.T.O., A.C. and F.A. Contributed to statistical analyses, article writing and figure design: A.C. and F.T.O. Contributed to the proofreading and correction of the article: all authors. All authors have read and agreed to the published version of the manuscript.

Funding: Field campaigns and instrumentation were supported by IRD. This work is a contribution to the « JEAI SAFUME » project funded by IRD. V. OKPEITCHA was funded by OmiDelta project of the Embassy of the Netherlands in Benin, through a scholarship grant of the National Institute of Water (INE). F. OUINSOU and T. AVOCEGAN received funding from IRD.

Institutional Review Board Statement: Not applicable.

Informed Consent Statement: Not applicable.

Data Availability Statement: The datasets generated during and/or analyzed during the current study are available from the corresponding author on reasonable request.

Acknowledgments: Special thanks to the members and crew participating to the bimonthly surveys, and in particular A. ASSOGBA, A. B. TIGO, M. BENOIST, and J. AZANKPO. Collaboration of Team 2/ODA-INE is also acknowledged.

Conflicts of Interest: The authors declare no conflict of interest.

References

1. Lalèyè, P.; Villanueva, M.; Entsua-Mensah, C.M.; Moreau, J. A review of the aquatic living resources in Gulf of Guinea lagoons with particular emphasis on fisheries management. *J. Afrotropical Zool.* **2007**, *10*, 123–136.
2. Gnohossou, P. La Faune Benthique d'une Lagune Ouest Africaine (le Lac Nokoue au Benin), Diversite, Abondance, Variations Temporelles et Spatiales, Place dans la Chaine Trophique. PhD Thesis, Institut National Polytechnique de Toulouse, Toulouse, France, 2006; p. 169.
3. Djihouessi, M.B.; Aina, M.P. A review of hydrodynamics and water quality of Lake Nokoué: Current state of knowledge and prospects for further research. *Reg. Stud. Mar. Sci.* **2018**, *18*, 57–67. [CrossRef]
4. Mama, D.; Aina, M.; Alassane, A.; Boukari, O.; Chouti, W.; Deluchat, V.; Bowen, J.; Afouda, A.; Baudu, M. Caractérisation physico-chimique et évaluation du risque d'eutrophisation du lac Nokoué (Bénin). *Int. J. Biol. Chem. Sci.* **2011**, *5*, 20–76. [CrossRef]
5. Okpeitcha, V.; Chaigneau, A.; Morel, Y.; Stieglitz, T.; Pomalegni, Y.; Sohou, Z.; Mama, D. Seasonal and interannual variability of salinity in a large West-African lagoon (Nokoué Lagoon, Benin). *Estuar. Coast. Shelf Sci.* **2021**, *264*, 107689. [CrossRef]
6. Odountan, O.; de Bisthoven, L.J.; Koudenoukpo, C.; Abou, Y. Spatio-temporal variation of environmental variables and aquatic macroinvertebrate assemblages in Lake Nokoué, a RAMSAR site of Benin. *Afr. J. Aquat. Sci.* **2019**, *44*, 219–231. [CrossRef]
7. Agadjihouede, H.; Akele, D.G.; Gougbedji, A.U.M.; Laleye, P.A. Exploitation de l'huître des mangroves Crassostrea Gasar (Adanson, 1757) dans le lac Nokoué au Bénin. *Eur. Sci. J.* **2017**, *13*, 352. [CrossRef]
8. Lalèyè, P.; Niyonkuru, C.; Moreau, J.; Teugels, G. Spatial and seasonal distribution of the ichthyofauna of Lake Nokoué, Bénin, west Africa. *Afr. J. Aquat. Sci.* **2003**, *28*, 151–161. [CrossRef]
9. Marcarelli, A.M.; Wurtsbaugh, W.A.; Griset, O. Salinity controls phytoplankton response to nutrient enrichment in the Great Salt Lake, Utah, USA. *Can. J. Fish. Aquat. Sci.* **2006**, *63*, 2236–2248. [CrossRef]
10. Zakaria, H.Y.; Radwan, A.A.; Said, M.A. Influence of salinity variations on zooplankton community in El-Mex Bay, Alexandria, Egypt. *Egypt. J. Aquat. Res.* **2017**, *33*, 52–67.

11. Nkwoji, J.A.; Onyema, I.C.; Igbo, J.K. Wet season spatial occurrence of phytoplankton and zooplankton in Lagos Lagoon, Nigeria. *Sci. World J.* **2010**, *5*, 7–14. [CrossRef]
12. Paturej, E.; Gutkowska, A. The effect of salinity levels on the structure of zooplankton communities. *Arch. Biol. Sci.* **2015**, *67*, 483–492. [CrossRef]
13. Beaugrand, G. The North Sea regime shift: Evidence, causes, mechanisms and consequences. *Prog. Oceanogr.* **2004**, *60*, 245–262. [CrossRef]
14. Bonnet, D.; Frid, C. Seven copepod species considered as indicators of water-mass influence and changes: Results from a Northumberland coastal station. *ICES J. Mar. Sci.* **2004**, *61*, 485–491. [CrossRef]
15. Giamali, C.; Kontakiotis, G.; Antonarakou, A.; Koskeridou, E. Ecological Constraints of Plankton Bio-Indicators for Water Column Stratification and Productivity: A Case Study of the Holocene North Aegean Sedimentary Record. *J. Mar. Sci. Eng.* **2021**, *9*, 1249. [CrossRef]
16. Bērziņš, B.; Pejler, B. Rotifer occurrence in relation to pH. *Hydrobiologia* **1987**, *147*, 107–116. [CrossRef]
17. Kuczynski, D. The rotifer fauna of Argentine Patagonia as a potential limnological indicator. *Hydrobiologia* **1987**, *150*, 3–10. [CrossRef]
18. Branco, C.W.; de Assis Esteves, F.; Kozlowsky-Suzuki, B. The zooplankton and other limnological features of a humic coastal lagoon (Lagoa Comprida, Mace, RJ) in Brazil. *Hydrobiologia* **2000**, *437*, 71–81. [CrossRef]
19. Branco, C.W.; Kozlowsky-Suzuki, B.; Esteves, F.A. Environmental changes and zooplankton temporal and spatial variation in a disturbed Brazilian coastal lagoon. *Braz. J. Biol.* **2007**, *67*, 251–262. [CrossRef]
20. Adandedjan, D.; Makponse, E.; Hinvi, L.C.; Laleye, P. Données préliminaires sur la diversité du zooplancton du lac Nokoué (Sud-Bénin). *J. Appl. Biosci.* **2017**, *115*, 11476–11489. [CrossRef]
21. Le Barbé, L.; Alé, G.; Millet, B.; Texier, H.; Borel, Y.; Gualde, R. *Les Ressources en Eaux Superficielles de la République du Bénin*; Monographies hydrologiques; ORSTOM: Paris, France, 1993; Volume 11, 540p.
22. Mama, D.; Chouti, W.; Alassane, A.; Changotade, O.; Alapini, F.; Boukari, M. Etude dynamique des apports en éléments majeurs et nutritifs des eaux de la lagune de Porto-Novo (Sud Bénin). *Int. J. Biol. Chem. Sci.* **2011**, *5*, 1278–1293. [CrossRef]
23. Chaigneau, A.; Okpeitcha, V.O.; Morel, Y.; Stieglitz, T.; Assogba, A.; Benoist, M.; Allamel, P.; Honfo, J.; Awoulbang Sakpak, T.D.; Rétif, F.; et al. From seasonal flood pulse to seiche: Multi-frequency water-level fluctuations in a large shallow tropical lagoon (Nokoué Lagoon, Benin). *Estuar. Coast. Shelf Sci.* **2022**, *267*, 107767. [CrossRef]
24. Texier, H.; Colleuil, B.; Profizi, J.-P.; Dossou, C. Lake Nokoue, lagoonal environment of South-Benin costal margin: Bathymetry, sedimentary facies, salinites, molluscus and vegetation. *Bull. Inst. Géol. Bassin D'aquitaine* **1980**, *28*, 115–142.
25. Texier, H.; Dossou, C.; Colleuil, B. Étude de l'environnement lagunaire du domaine margino-littoral sud-béninois. Étude hydrologique préliminaire du lac Nokoué. *Bull. L'institut Géologie Du Bassin D'aquitaine* **1979**, *25*, 149–166.
26. Wetzel, R.A.; Likens, G.E. *Limnological Analyses*; Springer: New York, NY, USA, 1991; Volume 391, pp. 15–166.
27. Strickland, J.D.H.; Parsons, T.R. *A Practical Handbook of Seawater Analysis*; Fisheries Research Board of Canada: Ottawa, ON, Canada, 1972.
28. Ahlström, E.H. A revision of the Rotatorian genera Brachionus an Platyias with descriptions new species and two new varieties. *Bull. Am. Mus. Nat. Hist.* **1940**, *77*, 148–184.
29. Durand, J.R.; Leveque, C. *Flore et Faune Aquatiques de l'Afrique Sahélo-Soudanienne: Tome 1*; ORSTOM: Paris, France, 1980.
30. Dussart, B.H. Les Copépodes. In *Flore et Faune Aquatiques de l'Afrique Sahélo Soudanienne: Tome 1*; ORSTOM: Paris, France, 1980; pp. 333–356.
31. Conway, D.V.P. Marine zooplankton of southern Britain. Part 1: Radiolaria, Heliozoa, Foraminifera, Ciliophora, Cnidaria, Ctenophora, Platyhelminthes, Nemertea, Rotifera and Mollusca. *Mar. Biol. Assoc. U. K.* **2012**, *25*, 138.
32. Conway, D.V.P. Marine zooplankton of southern Britain. Part 2: Arachnida, Pycnogonida, Cladocera, Facetotecta, Cirripedia and Copepoda. *Mar. Biol. Assoc. U. K.* **2012**, *26*, 163.
33. Conway, D.V.P. Marine zooplankton of southern Britain. Part 3: Ostracoda, Stomatopoda, Nebaliacea, Mysida, Amphipoda, Isopoda, Cumacea, Euphausiacea, Decapoda, Annelida, Tardigrada, Nematoda, Phoronida, Bryozoa, Entoprocta, Brachiopoda, Echinodermata, Chaetognatha, Hemichordata and Chordata. *Mar. Biol. Assoc. U. K.* **2015**, *27*, 271.
34. Conway, D.V.P.; White, R.G.; Hugues-Dit-Ciles, J.; Gallienne, C.P.; Robins, D.B. Guide to the coastal and surface zooplankton of the south-western Indian Ocean. *Mar. Biol. Assoc. U. K.* **2003**, *15*, 354.
35. Carling, K.; Ater, I.; Pellam, M.; Bouchard, A.; Mihuc, T. *A Guide to the Zooplankton of Lake Champlain*; Plattsburgh State University of New York: Plattsburgh, NY, USA, 2004; Volume 1, pp. 38–66.
36. Fontaneto, D.; De Smet, W.H.; Melone, G. Identification key to the genera of marine rotifers worldwide. *Meiofauna Mar.* **2008**, *16*, 75–99.
37. Yamani, F.Y.; Skryabin, V.; Gubanova, A.; Khvorov, S.; Prusova, I. *Marine Zooplankton: Practical Guide for the Northwestern Arabian Gulf*; Kuwait Institute for Scientific Research: Kuwait City, Kuwait, 2011; p. 12.
38. Haney, J.F.; Aliberti, M.A.; Allan, E.; Allard, S.; Bauers, D.J.; Beagen, W.S.; Bradtr, R.; Carlson, B.; Carlson, S.C.; Doan, U.M.; et al. *An-Image-Based Key to the Zooplankton of North America*; Version 5.0; University of New Hampshire Center for Freshwater Biology: Durham, NH, USA, 2013.
39. LaMay, M.; Hayes-Pontius, E.; Ater, I.M.; Mihuc, T.B. A revised key to the zooplankton of Lake Champlain. *Sci. Discipulorum* **2013**, *6*, 141.

40. Ezz, S.M.A.; Aziz, A.N.E.; Abou Zaid, M.M.; El Raey, M.; Abo-Taleb, H.A. Environmental assessment of El-Mex Bay, Southeastern Mediterranean by using Rotifera as a plankton bio-indicator. *Egypt. J. Aquat. Res.* **2014**, *40*, 43–57.
41. Swadling, K.; Slotwinski, A.; Davies, C.; Beard, J.; McKinnon, A.D.; Coman, F.; Murphy, N.; Tonks, M.; Rochester, W.; Conway, D.V.P.; et al. Australian Marine Zooplankton: A Taxonomic Guide and Atlas. In *Image Key-Zooplankton*; The University of Tasmania: Hobart, Australia, 2013. Available online: https://www.imas.utas.edu.au/zooplankton/image-key (accessed on 1 February 2023).
42. Glime, J.M. Arthropods: Crustacea—Copepoda and Cladocera. Chapter 10–1. *Bryophyt. Ecol.* **2017**, *2*, 10–120.
43. Glime, J.M. Invertebrates: Rotifer Taxa—Monogononta Chapter 4–7 a-b-c. *Bryophyt. Ecol.* **2017**, *2*, 7–37.
44. Santhanam, P.; Perumal, P.; Begum, A. A Method of Collection, Preservation and Identification of Marine Zooplankton. In *Basic and Applied Zooplankton Biology*; Springer: Berlin/Heidelberg, Germany, 2019; pp. 1–44.
45. Wilke, T.; Ahlrichs, W.H.; Bininda-Emonds, O.R.P. A weighted taxonomic matrix key for species of the rotifer genus *Synchaeta (Rotifera, Monogononta, Synchaetidae)*. *ZooKeys* **2019**, *871*, 40. [CrossRef] [PubMed]
46. Available online: http://rotifera.hausdernatur.at/ (accessed on 1 February 2023).
47. Available online: https://www.shetlandlochs.com/species/eukaryota/animalia/rotifera/eurotatoria/ploima/asplanchnidae/asplanchna (accessed on 1 February 2023).
48. Shannon, C.E.; Weaver, W.W. *The Mathematical Theory of Communications*; University of Illinois Press: Urbana, IL, USA, 1963; 117p.
49. Peet, R.K. The measurement of species diversity. *Annu. Rev. Ecol. Syst.* **1974**, *5*, 285–307. [CrossRef]
50. Pielou, E.C. The measurement of diversity in different types of biological collections. *J. Theor. Biol.* **1966**, *13*, 131–144. [CrossRef]
51. Magurran, A.E. *Ecological Diversity and Its Measurement*; Princeton University Press: Princeton, NJ, USA, 1988.
52. Grall, J.; Coïc, N. Synthèse des méthodes d'évaluation de la qualité du benthos en milieu côtier. *Rapp. IFREMER* **2005**, *91*, 2711826.
53. Bretherton, F.; Davis, R.; Fandry, C. A technique for objective analysis and design of oceanographic experiments applied to MODE-73. *Deep-Sea Res. Oceanogr. Abstr.* **1976**, *23*, 559–582. [CrossRef]
54. McIntosh, P. Oceanographic data interpolation: Objective analysis and splines. *J. Geophys. Res.* **1990**, *95*, 13529–13541. [CrossRef]
55. Wong, A.P.S.; Johnson, G.C.; Owens, W.B. Delayed-mode calibration of autonomous CTD profiling float salinity data by y–S climatology. *J. Atmos. Ocean. Technol.* **2003**, *20*, 308–318. [CrossRef]
56. Legendre, P.; Legendre, L. *Numerical Ecology*; Elsevier: Amsterdam, The Netherland, 1998; 853p.
57. Legendre, P.; Gallagher, E.D. Ecologically meaningful transformations for ordination of species data. *Oecologia* **2001**, *129*, 271–280. [CrossRef]
58. Ter Braak, C.J.; Prentice, I.C. A theory of gradient analysis. *Adv. Ecol. Res.* **1988**, *18*, 271–317.
59. Hastie, T.J. Generalized Additive Models. In *Statistical Models in S.*; Routledge: New-York, USA, 2017; pp. 249–307.
60. Ter Braak, C.J.; Smilauer, P. *CANOCO Reference Manual and Canodraw for Windows User's Guide: Softare for Canonical Community Ordination (Version 4.5)*; Microcomputer Power: Ithaca, NY, USA, 2002.
61. Wood, S. *Generalized Additive Models: An Introduction with R.*; Chapman & Hall/CRC: Boca Raton, FL, USA, 2006; 422p.
62. Wood, S. *mgcv: Mixed GAM Computation Vehicle with GCV/AIC/REML Smoothness Estimation*; University of Bath: Bath, UK, 2012.
63. Badsi, H.; Ali, H.O.; Loudiki, M.; Hafa, M.E.; Chakli, R.; Aamiri, A. Ecological factors affecting the distribution of zooplankton community in the Massa Lagoon (Southern Morocco). *Afr. J. Environ. Sci. Technol.* **2010**, *4*, 751–762.
64. Okogwu, I.O. Seasonal variations of species composition and abundance of zooplankton in Ehoma Lake, a floodplain lake in Nigeria. *Rev. Biol. Trop.* **2010**, *58*, 171–182. [CrossRef] [PubMed]
65. Appiah, Y.S.; Étilé, R.N.; Kouamé, K.A.; Paul, E. Zooplankton diversity and its relationships with environmental variables in a West African tropical coastal lagoon (Ebrié lagoon, Côte d'Ivoire, West Africa). *J. Biodivers. Environ. Sci.* **2018**, *13*, 1–16.
66. Étilé, R.N.; Aka, M.N.; Blahoua, G.K.; Kouamélan, P.E. Zooplankton diversity and distribution in a Fresco Lagoon (West Africa, Côte d'Ivoire). *Int. Res. J. Environ. Sci.* **2018**, *7*, 9–20.
67. Monney, I.A.; Etile, R.N.; Ouattara, I.N.; Kone, T. Seasonal distribution of zooplankton in the Aby-Tendo-Ehy lagoons system (Côte d'Ivoire, West Africa). *Int. J. Biol. Chem. Sci.* **2016**, *9*, 23–62. [CrossRef]
68. Morel, Y.; Chaigneau, A.; Okpeitcha, V.O.; Stieglitz, T.; Assogba, A.; Duhaut, T.; Rétif, F.; Peugeot, C.; Sohou, Z. Terrestrial or oceanic forcing? Water level variations in coastal lagoons constrained by river inflow and ocean tides. *Adv. Water Resour.* **2022**, *169*, 104309. [CrossRef]
69. Houssou, A.; Montchowui, E.; Bonou, C. Composition and structure of zooplankton community in ouémé river basin, republic of Benin. *J. Entomol. Zool. Stud.* **2017**, *10*, 336–344.
70. Adjahouinou, D.C.; Liady, N.D.; Fiogbe, E.D. Diversité phytoplanctonique et niveau de pollution des eaux du collecteur de Dantokpa (Cotonou-Bénin). *Int. J. Biol. Chem. Sci.* **2012**, *6*, 1938–1949. [CrossRef]
71. Sukumaran, P.K.; Das, A.K. Distribution and abundance of rrotifera in relation to the water quality of selected tropical reservoirs. *Indian J. Fish.* **2004**, *51*, 295–301.
72. Özbay, H.; Altindağ, A. Zooplankton abundance in the River Kars, Northeast Turkey: Impact of environmental variables. *Afr. J. Biotechnol.* **2009**, *8*, 5814–5818.
73. Onana, F.M.; Zébazé Togouet, S.H.; Nyamsi, T.; Domche, T.; Ngassam, P. Distribution spatio-temporelle du zooplancton en relation avec les facteurs abiotiques dans un hydrosystème urbain: Le ruisseau de Kondi, Cameroun. *J. Appl. Biosci.* **2014**, *82*, 7326–7338. [CrossRef]
74. Zannatul, F.; Muktadir, A.K.M. A Review: Potentiality of Zooplankton as Bioindicator. *Am. J. Appl. Sci.* **2009**, *10*, 1815–1819.

75. Koudenoukpo, C.Z.; Chikou, A.; Zebaze, S.H.T.; Mvondo, N.; Hazoume, R.U.; Houndonougbo, P.K.; Laleye, P.A. Zooplanctons et Macroinvertébrés aquatiques: Vers un assemblage de bioindicateurs pour un meilleur monitoring des écosystèmes aquatiques en région tropicale. *Int. J. Innov. Appl. Stud.* **2017**, *20*, 276.
76. Segers, H. Global diversity of rotifère (Phylum Rotifera) in freshwater. *Hydrobiologia* **2008**, *595*, 49–59. [CrossRef]
77. Sarma, S.S.S.; Nandini, S.; Morales-Ventura, J.; Delgado-Martínez, I.; González-Valverde, L. Effects of NaCl salinity on the population dynamics of freshwater zooplankton (rotifers and cladocerans). *Aquat. Ecol.* **2006**, *40*, 349–360. [CrossRef]
78. Epp, R.W.; Winston, P.W. Osmotic regulation in the brackish-water rotifer *Brachionus plicatilis* (MULLER). *J. Exp. Biol.* **1977**, *68*, 151–156. [CrossRef]
79. Lowe, C.D.; Kemp, S.J.; Díaz-Avalos, C.; Montagnes, D.J.S. How does salinity tolerance influence the distributions of *Brachionus plicatilis* sibling species? *Mar. Biol.* **2007**, *150*, 377–386. [CrossRef]
80. Walker, K.F. A synopsis of ecological information on the saline lake rotifer *Brachionus plicatilis* Muller 1786. *Hydrobiologia* **1981**, *81*, 159–167. [CrossRef]
81. Miracle, R.M.; Serra, M.; Oltra, R.; Vicente, E. Differencial distribution of *Brachionus* species in the coastal lagoons. *Verh. Internat. Verein. Limnol.* **1988**, *25*, 2006–2015.
82. Arcifa, M.S.; Castilho, M.S.M.; Carmouze, J.-P. Composition et évolution du zooplancton dans une lagune tropicale (Brésil) au cours d'une période marquée par une mortalité de poissons. *Hydrobiol. Trop.* **1994**, *27*, 251–263.
83. Remane, A. Die Brackwasserfauna. Verhandlungen der Deutschen Zoologischen Gesellschaft. *Greifswald* **1934**, *36*, 34–74.
84. Khlebovich, V.V.; Abramova, E.N. Some problems of crustacean taxonomy related to the phenomenon of horohalinicum. *Hydrobiologia* **2000**, *417*, 109–113. [CrossRef]
85. Telesh, I.V.; Schubert, H.; Skarlato, S.O. Life in the salinity gradient: Discovering mechanisms behind a new biodiversity pattern. *Estuar. Coast. Shelf Sci.* **2013**, *135*, 317–327. [CrossRef]
86. Boesch, D.F.; Diaz, R.J.; Virnstein, R.W. Effets de la tempête tropicale Agnes sur les communautés macrobenthiques à fond meuble des estuaires James et York et de la partie inférieure de la baie de Chesapeake. *Chesap. Sci.* **1976**, *17*, 246–259. [CrossRef]
87. Attrill, M.J. Un modèle linéaire testable pour les tendances de la diversité dans les estuaires. *J. D'écologie Anim.* **2002**, *71*, 262–269. [CrossRef]
88. Telesh, I.V.; Schubert, H.; Skarlato, S.O. Revisiting Remane's concept: Evidence for high plankton diversity and a protistan species maximum in the horohalinicum of the Baltic Sea. *Mar. Ecol. Prog. Ser.* **2011**, *421*, 1–11. [CrossRef]
89. Whitfield, A.K.; Elliott, M.; Basset, A.; Blaber, S.J.M.; West, R.J. Paradigms in estuarine ecology—A review of the Remane diagram with a suggested revised model for estuaries. *Estuar. Coast. Shelf Sci.* **2012**, *97*, 78–90. [CrossRef]
90. Laprise, R.; Dodson, J.J. Environmental variability as a factor controlling spatial patterns in distribution and species diversity of zooplankton in the St. Lawrence Estuary. *Mar. Ecol.-Prog. Ser.* **1994**, *107*, 67. [CrossRef]
91. Crump, B.C.; Armbrust, E.V.; Baross, J.A. Phylogenetic analysis of particle-attached and free-living bacterial communities in the Columbia River, its estuary, and the adjacent coastal ocean. *Appl. Environ. Microbiol.* **1999**, *65*, 3192–3204. [CrossRef]
92. Dolan, J.R.; Gallegos, C.L. Estuarine diversity of tintinnids (planktonic ciliates). *J. Plankton Res.* **2001**, *23*, 1009–1027. [CrossRef]
93. Hewson, I.; Fuhrman, J.A. Richness and diversity of bacterioplankton species along an estuarine gradient in Moreton Bay, Australia. *Appl. Environ. Microbiol.* **2004**, *70*, 3425–3433. [CrossRef]
94. Telesh, I.V. Plankton of the Baltic estuarine ecosystems with emphasis on Neva Estuary: A review of present knowledge and research perspectives. *Mar. Pollut. Bull.* **2004**, *49*, 206–219. [CrossRef] [PubMed]
95. Duggan, S.; McKinnon, A.D.; Carleton, J.H. Zooplankton in an Australian tropical estuary. *Estuaries Coasts* **2008**, *31*, 455–467. [CrossRef]
96. Welcomme, R.L. An evaluation of acadja method of fishing as practiced in the coastal lagoons of Dahomey (West africa). *J. Fish Bid.* **1972**, *4*, 39–55. [CrossRef]
97. Chaffra, A.S.; Agbon, A.C.; Tchibozo, E.A.M. Cartographie par télédétection des Acadjas, une technique de pêche illicite sur le lac Nokoué au Bénin. *Sci. Tech.* **2020**, *5*, 11–29.
98. Welcomme, R.L.; Kapetsky, I.C. Acadjas: The Brush Park Fisheries of Benin, West Africa. *ICLARM Newsl.* **1981**, *4*, 3–4.
99. Guiral, D.; Gourbault, N.; Helleouet, M.-N. Étude sédimentologique et méiobenthos d'un écosystème lagunaire modifié par un récif artificiel à vocation aquacole: L'acadja. *Oceanol. Acta* **1995**, *18*, 543–555.
100. Visser, A.W.; Stips, A. Turbulence and zooplankton production: Insights from PROVESS. *J. Sea Res.* **2002**, *47*, 317–329. [CrossRef]
101. Guiral, D.; Arfi, R.; Da, K.P.; Konan-Brou, A.A. Communautés, biomasses et productions algales au sein d'un récif artificiel (acaja) en milieu lagunaire tropical. *Rev. Hydrobiol. Trop.* **1993**, *26*, 219–228.

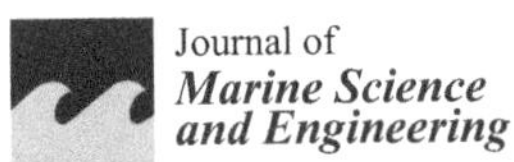
Journal of
Marine Science and Engineering

MDPI

Article

The Distribution of Ferritins in Marine Copepods

Vittoria Roncalli [1,*], Marco Uttieri [1,2] and Ylenia Carotenuto [1]

[1] Integrative Marine Ecology Department, Stazione Zoologica Anton Dohrn, Villa Comunale, 80121 Napoli, Italy; marco.uttieri@szn.it (M.U.); ylenia.carotenuto@szn.it (Y.C.)
[2] NBFC, National Biodiversity Future Center, Piazza Marina 61, 90133 Palermo, Italy
* Correspondence: vittoria.roncalli@szn.it

Abstract: Iron is an essential element for the functioning of cellular processes. Ferritins, the major intracellular iron storage proteins, convert the free Fe^{2+} into the nontoxic Fe^{3+} which can be stored and transported where needed. To date, little is known about the iron metabolism in copepods; however, in these crustaceans, ferritins have been used as biomarkers of stress and diapause. A limiting factor of these studies has been the use of a single ferritin transcript as a biomarker. In this paper, we in silico mined the publicly available copepod transcriptomes to characterize the multiplicity of the ferritin transcripts in different orders and families. We also examined the expression of ferritin in three ecologically important copepods—*Calanus finmarchicus*, *C. helgolandicus* and *Temora stylifera*—during development and under stress conditions. A full-length transcript encoding ferritin heavy chain has been identified in all 27 mined transcriptomes, with 50% of the species possessing multiple transcripts. Ferritin expression increased in *C. finmarchicus* during the early–late development transition, and in *T. stylifera* females exposed to oxylipins at sea. Overall, our results suggest that copepod ferritins can be involved in iron storage, larval development and stress response, thus representing potential biomarker genes for ocean health status monitoring.

Keywords: zooplankton; transcriptome; gene discovery; iron metabolism; stress response; diapause

Citation: Roncalli, V.; Uttieri, M.; Carotenuto, Y. The Distribution of Ferritins in Marine Copepods *J. Mar. Sci. Eng.* **2023**, *11*, 1187. https://doi.org/10.3390/jmse11061187

Academic Editor: Albert Calbet

Received: 3 May 2023
Revised: 30 May 2023
Accepted: 1 June 2023
Published: 7 June 2023

1. Introduction

In all animals, iron (Fe) is an essential element required for the functioning of many cellular processes and to meet nutritional requirements. It participates in various metabolic processes, including DNA synthesis, electron and oxygen transport and cellular oxidation mechanisms [1]. Iron must be absorbed from the diet into gut cells and then transported and stored for future use [2]. However, excess free Fe in living cells can be potentially toxic. In the presence of oxygen, Fe can be rapidly reduced to the ferrous ion (Fe^{2+}) form and this may catalyze the generation of reactive oxygen species (ROS). High levels of ROS cause oxidative stress, which is harmful to cellular compounds such as lipids, DNA and proteins [3,4]. Thus, the maintenance of a balanced Fe metabolism is essential for the organism's homeostasis [2].

In order to avoid cell harm, organisms possess iron storage proteins (transferrins and ferritins) that keep Fe in their cavity and transport it where needed [2,3]. Ferritins are globular proteins composed of multiple subunits, with 24 equivalent subunits assembled into a "cage-like" oligomer [2]. In the ferroxidase center, they convert the Fe^{2+} into a non-toxic, soluble and biologically non-reactive ferric ion (Fe^{3+}). In vertebrates, ferritins have been classified in heavy (H) chain and light (L) chain subunits; in arthropods, ferritins are homologs to the vertebrate proteins (heavy-chain homolog, HCH, and light-chain homolog, LCH) [2,5]. The major difference is that H-chain ferritins possess the ferroxidase center and are capable of ferroxidase activity, while L-chain ferritins, lacking the ferroxidase center, are characterized by amino acid residues that induce nucleation of iron and may be responsible for the electron transfer across the protein cage [6]. A third type of ferritin, mitochondrial ferritin—a homopolymer with ferroxidase activity targeted to mitochondria—has been identified in mammals and *Drosophila melanogaster* (ferritin 3, heavy-chain homolog, Fer3HCH), (reviewed in [5]). Although intensively studied in mammals and yeast, much less is known about the structure and function of ferritins in arthropods. Three ferritins have been identified in the fruit fly *D. melanogaster*: the cytoplasmic ferritin 1 (HCH), the ferritin 2 (LCH) and the mitochondrial ferritin [2]. Their role in dietary iron efflux or delivery was confirmed by knockdown experiments of either HCH or LCH ferritin subunits by RNAi, which resulted in the downregulation of the protein level of both subunits and in Fe accumulation in the midgut (reviewed in [5]). In the cladoceran *Daphnia pulex*, genome annotation and phylogenetic analysis revealed that ferritins clearly expanded compared with insects with a total of seven distinct ferritin *loci* [7].

Scanty information is available on the iron metabolism in zooplankton organisms, particularly in copepods; however, in these crustaceans, ferritins have been intensively used as biomarkers of stress. In the neritic calanoid species *Acartia tonsa*, an increase in transcriptional expression of ferritin has been reported in individuals after exposure to nano-contaminants such as Ni and CdSe/ZnS quantum dots [8], after acclimation to high pCO_2 conditions (1200 ppm) [9], and in response to infestation by the epibiotic euglenid *Colacium vesiculosum* [10]. Higher expression of a ferritin transcript was also reported in *A. tonsa* quiescent eggs compared with the subitaneous stage [11]. In the calanoid *Calanus finmarchicus*, ferritin has been suggested as a biomarker of stress associated with the diapause phase; a significant upregulation of ferritin was found in copepodites (CV) and females collected from deep water compared with individuals from the surface [12,13]. However, in all these studies, only a single ferritin gene has been used as a biomarker. In a recent study, genome analysis revealed the presence of a total of four ferritin genes (LsFer1–4) in the salmon louse *Lepeophtheirus salmonis*, with three encoding heavy chains (LsFer1,3,4) and one (LsFer2) encoding a light chain [14]. Based on these results, we might suggest that in copepods, the use of a single biomarker for ferritin might be limiting.

In consideration of the critical role of ferritin in Fe homeostasis, the aim of this study is to expand the understanding of the ferritin diversity and function in copepods. Mining the publicly available high-quality transcriptomes for several calanoid families [15–17], we examined the presence of transcripts encoding ferritins and compared them to homologous ferritins in *D. melanogaster* and in salmon louse *L. salmonis*. To provide a better understanding of the functioning of these genes, the expression of ferritins was examined using existing RNASeq data *in C. finmarchicus*, *Calanus helgolandicus* and *Temora stylifera*. Expression was examined across development (*C. finmarchicus*) [18] and after exposure to toxic algae (*C. finmarchicus, C. helgolandicus, T. stylifera*) [19–22]. This study expands the knowledge of the diversity of the ferritin family in copepods and suggests species-specific and stage-specific functional roles in these organisms. Our results shed light on the need to characterize the gene family of interest in particular when the genes are widely used as biomarkers in eco-physiological studies.

2. Materials and Methods

2.1. In Silico Workflow

Searches for putative transcripts encoding ferritins (HCH and LCH) were performed using a well-established vetting workflow that includes a mining step, a reciprocal blast and an examination of the protein structural domain [23,24]. Query sequences from the fruit fly *Drosophila melanogaster* (NP_524873, AAF07876, NP_572854) were used to mine the transcriptome shotgun assembly (TSA) database on the National Center for Biotechnology Information (NCBI), limiting the searches (tblastn algorithm) to Copepoda (taxid:6830) (search February 2023). For *Rhincalanus gigas*, raw reads were downloaded from NCBI [25] and assembled as described in [26]. Additional mining was performed using queries from the copepod *Lepeophtheirus salmonis* (order Siphonostomatoida) (BT121711, BT121232, MK887318, BT077723, BT121164) [14]. The results of both searches were compared and integrated.

All resulting transcripts were fully translated using ExPASY [27] and blasted against the NCBI non-redundant (nr) protein database (blastp algorithm) limited to Arthropoda (taxid:6656). The presence of the expected protein structural domain for ferritin (Pfam:PF00210) was examined using SMART software [28]. In the species with multiple transcripts, amino acid sequences were aligned and amino acid identity was calculated between pairs. Sequences with ≥95% amino acid identity were considered the same protein, and among those the longest sequence was kept. A final alignment was performed for the copepod transcripts identified in this study encoding full-length proteins and having the expected ferritin domain, with HCHs and LCHs from *D. melanogaster*, *L. salmonis* and *Homo sapiens*, in order to verify the presence of amino acids characterizing the ferroxidase center [14]. Amino acid sequence alignment was performed using MAFFT software [29].

2.2. Cladogram of Copepod Ferritin Genes

A phylogenetic analysis was performed to confirm the annotation of transcripts identified in this study and to establish their relationship to each other, to other copepod species and arthropods. An unrooted phylogenetic tree was generated with amino acid sequences from this study, and sequences previously identified from the copepods *Acartia pacifica* and *Calanus sinicus* (order: Calanoida) [8], *Caligus rogercresseyi*, *C. clemensi*, *L. salmonis* (order: Siphonostomatoida) [14] and *Tigriopus californicus* (order: Harpacticoida) [30]. We also considered sequences from the insects *D. melanogaster* and *Anopheles aegypti*, homologs from *Homo sapiens* [14] and the water flea *Daphnia pulex* [7]. Among the seven ferritin sequences identified in *D. pulex*, two partial sequences (DQ983427, DQ983426) were excluded from the analysis. All sequences were first aligned using ClustalW software (Galaxy version 2.1), and then a maximum-likelihood phylogenetic tree was built using the evolution model Kimura computing bootstrap for 1000 samples (RapidNJ, Galaxy version 2.3.2).

2.3. Expression of Ferritin in Calanus finmarchicus, Calanus helgolandicus and Temora stylifera

Relative expression of transcripts encoding for ferritins was examined in the copepods *Calanus finmarchicus*, *Calanus helgolandicus* and *Temora stylifera* (order: Calanoida), using existing RNASeq data. In *C. finmarchicus*, the expression of ferritins was examined across development [18] and in females exposed to a toxic dinoflagellate [19]. For the developmental dataset, adult *C. finmarchicus* and stage CV copepodites were collected from the Gulf of Maine in 2012. Wild-caught females were laboratory maintained to obtain the target developmental stages: embryos, early nauplii, early copepodites (CI) and late copepodites (CIV). From each stage, total RNA was extracted from three biological replicates (two for CI and CIV), and cDNA libraries were multiplexed and sequenced on an Illumina HiSeq 2000 platform (PE 100 bp) [18]. The second dataset includes *C. finmarchicus* females exposed to the saxitoxin-producing dinoflagellate *Alexandrium fundyense* [19]. Briefly, adult females collected from the Gulf of Maine (Mount Desert Rock, 2012) were laboratory incubated over a week with a low dose (LD: 50 cell mL^{-1}) and high dose (HD: 200 cell mL^{-1}) of the toxic dinoflagellate and with the non-toxic cryptophyte *Rhodomonas baltica* (8000 cells mL^{-1}) as the control diet. Copepods were kept under 10 °C, on a 14:10 light:dark cycle and at two and five days, three biological replicates (15 females each) were harvested and processed for RNASeq. The two time points were chosen to test the hypothesis that the toxic algae would induce a detoxification response after two days that would persist over time (five days). Total RNA was extracted from each sample, followed by the generation of multiplexed cDNA libraries that were sequenced on an Illumina HiSeq 2000 platform (PE 100 bp) [19].

In *C. helgolandicus*, ferritin expression was derived from RNASeq data of laboratory-incubated females feeding on the oxylipin-producing toxic diatom *Skeletonema marinoi* (SKE) and the dinoflagellate control diet *Prorocentrum minimum* (PRO) for five days [20]. Briefly, females were collected in the Gulf of Naples (Central Tyrrhenian Sea, Western Mediterranean Sea, 2012) and laboratory incubated for five days with either *S. marinoi* (45,000 cells mL^{-1}) or *P. minimum* (5000 cells mL^{-1}) (three replicates per diet, 10 females each). Copepods were kept under 18 °C, on a 12:12 light:dark cycle, fed daily either with SKE or PRO and harvested after five days for RNASeq. Total RNA was extracted from each sample followed by the generation of multiplexed cDNA libraries that were sequenced on an Illumina platform HiSeq (PE 50 bp) [20]. Lastly, the *T. stylifera* dataset included females collected in the Gulf of Naples during two consecutive weeks in May 2017 (test: 23rd and control: 30th) when low–high reproductive fitness was associated to high–low oxylipin content in the natural phytoplankton assemblage, respectively [21]. Phytoplankton-derived oxylipins were measured from surface water samples collected on the two dates, as described in [22]. Wild-caught *T. stylifera* females were immediately flash-frozen in liquid nitrogen and stored for RNA extraction. Three replicates (10 females each) were harvested on both weeks, extracted for total RNA, processed for cDNA library preparation and sequenced on a HiSeq 2500 platform (PE 100 bp) [21]. For each copepod species, expression levels were quantified by mapping the RNASeq libraries against their species-specific reference transcriptome using Bowtie and reads were normalized by length using the RPKM methods reads per kilobase per million mapped reads (RPKM). RPKM in each species were compared among transcripts and among stages or treatments using two-way ANOVA ($p < 0.0001$) followed by Tukey's post hoc test and multiple unpaired *t*-tests ($p < 0.05$). Statistical analyses were performed using GraphPad Prism, version 9 (GraphPad Software, San Diego, CA, USA).

3. Results

3.1. Identification of Ferritins in Copepods

Ferritin-encoding transcripts were identified in 27 copepod species, including 22 from the order Calanoida, three from the order Harpacticoida and two from the order Cyclopoida (Table 1). Within the Calanoida order, transcripts were identified among different families, the majority within the Calanidae (e.g., *Calanus finmarchicus, C. helgolandicus, C. propinquus, Calanoides acutus, Neocalanus flemingeri, N. cristatus* and *N. plumchrus*), followed by the Temoridae (*Temora stylifera, T. longicornis, Epischura baikalensis*), the Acartiidae (*Acartia clausi, A. tonsa*), the Pontellidae (*Labidocera madurae*), the Pseudodiaptomidae (*Pseudodiaptomus annandalei*), the Centropagidae (*Centropages hamatus*) and the Rhincalanidae (*Rhincalanus gigas*) (Table 1). For all transcripts, reciprocal blast confirmed their annotation as a heavy chain (HCH) or light chain (LHC), with the majority being highly similar (top hit reciprocal blast) to homologs from *L. salmonis* (LS Fe1; BT121711), *Calanus sinicus* (APC62655) and *Eurytemora affinis* (XM_023489461). For the majority of the transcriptomes mined in this study, the resulting transcripts encoded full-length proteins with the typical conserved ferritin domain (Pfam00210; *p*-values < 0.05) (Table S1). Few partial transcripts, with no significant domain, were identified in *N. flemingeri* (2), *C. helgolandicus* (1), *C. hamatus* (1) and *C. acutus* (1), and these were not included in the downstream analyses.

Table 1. Summary of ferritins in copepods. For each copepod, genus, species, order, family and number of transcripts encoding ferritin were listed. The table includes only ferritins that were identified as full length that passed the reciprocal blast and protein domain analysis (Table S1).

Species	Order	Family	Ferritin Transcripts
Acartia clausi	Calanoida	Acartiidae	1
Acartia tonsa	Calanoida	Acartiidae	2
Calanoides acutus	Calanoida	Calanidae	1
Calanus finmarchicus	Calanoida	Calanidae	4
Calanus glacialis	Calanoida	Calanidae	1
Calanus helgolandicus	Calanoida	Calanidae	4
Calanus hyperboreus	Calanoida	Calanidae	2
Calanus marshallae	Calanoida	Calanidae	3
Calanus pacificus	Calanoida	Calanidae	4
Calanus propinquus	Calanoida	Calanidae	2

Table 1. *Cont.*

Species	Order	Family	Ferritin Transcripts
Centropages hamatus	Calanoida	Centropagidae	3
Epischura baikalensis	Calanoida	Temoridae	1
Eucyclops serrulatus	Cyclopoida	Cyclopidae	3
Hemidiaptomus amblyodon	Calanoida	Diaptomidae	1
Labidocera madurae	Calanoida	Pontellidae	1
Neocalanus cristatus	Calanoida	Calanidae	1
Neocalanus flemingeri	Calanoida	Calanidae	6
Neocalanus plumchrus	Calanoida	Calanidae	1
Paracyclopina nana	Cyclopoida	Cyclopettidae	1
Platychelipus ittoralis	Harpacticoida	Laophontidae	4
Pleuromamma xiphias	Calanoida	Metridinidae	1
Pseudodiaptomus annandalei	Calanoida	Pseudodiaptomidae	2
Rhincalanus gigas	Calanoida	Rhincalanidae	3
Temora longicornis	Calanoida	Temoridae	6
Temora stylifera	Calanoida	Temoridae	4
Tisbe furcata	Harpacticoida	Tisbidae	1
Tisbe holothuriae	Harpacticoida	Tisbidae	1

The number of transcripts encoding ferritins changed across copepods, ranging from one to a maximum of six. A single ferritin transcript was identified in 12 copepod species, whereas the highest diversification was found in *N. flemigeri* and *T. longicornis*, both having six different transcripts (Table 1). *C. finmarchicus*, *Calanus pacificus*, *C. helgolandicus* and *T. stylifera* showed four different ferritins. Lastly, three transcripts were identified in *Calanus marshallae*, *C. hamatus* and *R. gigas*.

Alignment of the copepod sequences identified in this study with ferritins from *D. melanogaster*, *H. sapiens* and *L. salmonis* showed that 57 out of 64 transcripts identified in this study conserved all the amino acids of the ferroxidase site in HCH from the reference sequences (Figure S1). The exception to this is the single ferritin from *C. finmarchicus* (GAXK01169093), *C. helgolandicus* (GJFL01003552) and *N. flemingeri* (GFUD01021847), and two ferritins from *T. longicornis* (GINW01248260, GINW01100697) and *C. pacificus* (GJQY01004559, GJQY01232255). While in the *N. flemingeri* and *C. pacificus* ferritins only a few substitutions were observed, in the other sequences five (out of seven) of the amino acids making the ferroxidase center were not highly conserved; both *C. finmarchicus* and *C. helgolandicus* showed in all the sites the same substitutions (Figure S1). The amino acid variability found in the ferroxidase center for *C. finmarchicus*, *C. helgolandicus* and *T. longicornicus* sequences might suggest that these ferritins belong to the LCH category, although further analyses are needed.

The unrooted tree generated from a total of 100 transcripts showed that ferritins clustered into several phylogenetic groups (Figure 1). LCH clustered in a separate group (Figure 1 light blue, bootstrap 98) including LCHs from *D. melanogaster*, *Aedes aegypti*, the copepods *C. clemensi* and *L. salmonis* (previously known), two *T. longicornis* transcripts, one from *C. finmarchicus* and one from *C. helgolandicus* (this study). The sequences in this group were the ones showing changes in the ferroxidase center in Figure S1. HCH ferritins are separated into several groups. The largest group (Figure 1 yellow, bootstrap 85) included known HCHs from the insects *D. melanogaster* and *A. aegypti*, the cladoceran *D. pulex*, the copepods *L. salmonis*, *C. clemensi*, *C. rogergressery*, *E. affinis*, *T. californicus* and 17 sequences identified in this study; these included ferritins from several species such as *C. marshallae*, *N. flemingeri*, *L. madurae*, *A. clausi* and *A. tonsa*, *P. annanadalei*, *P. nana*,

T. longicornis and *T. stylifera* (two out of four). The second largest group (Figure 1 green, bootstrap 81) included only ferritins identified in this study (23) from *C. finmarchicus* (two out of four), *C. helgolandicus* (one), *C. marshallae*, (two out of three), *N. flemingeri* (three out of six), *T. furcata* and *Tisbe holoturiae*, *P. annandalei* and *Pleuromamma xiphias*. The remaining HCHs clustered in smaller groups. Among those, there was a small group including all *D. pulex* sequences (four out of five) and a single ferritin from *C. propinquus* (Figure 1, orange, bootstrap 58), and another group including six *T. longicornis* ferritins and two *H. sapiens* homologs (Figure 1, purple, bootstrap 51). None of the ferritin identified in this study clustered with the *D. melanogaster* mitochondrial ferritin which was separated from all other sequences.

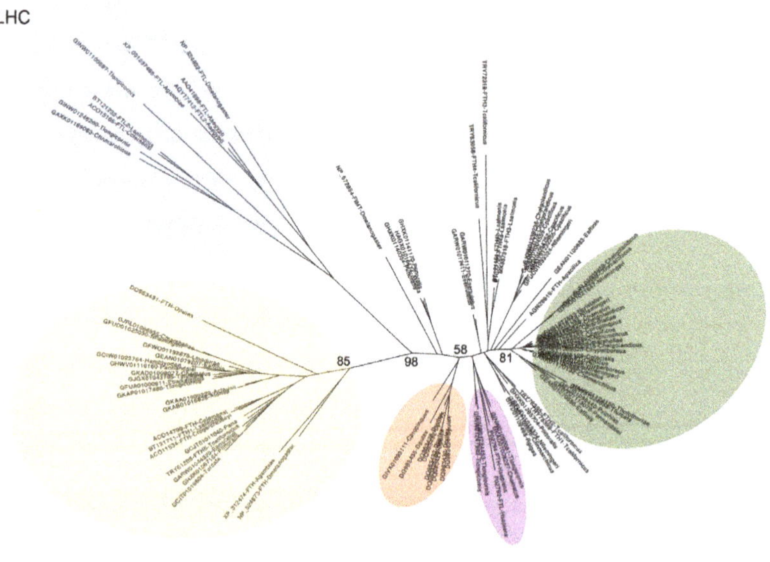

Figure 1. Cladogram of ferritin-related genes identified in this study. The analysis includes sequences from this study (Table 1) from the insect *D. melanogaster*, the crustacean *D. pulex* and other copepods previously identified (see text). For the analysis, amino acid sequences were aligned using ClustalW and then a maximum-likelihood phylogenetic tree was built using the evolution model Kimura computing bootstrap for 1000 samples (RapidNJ, Galaxy version 2.3.2). Colors refer to the groups described in the text. Bootstrap is only indicated for the groups described. Scale bars: 0.3 estimated substitutions per site. LCH: light chain subunit.

3.2. *Expression of ferritin in C. finmarchicus, Calanus helgolandicus and Temora stylifera*

The expression of ferritin-encoding transcripts significantly changed in *C. finmarchicus* across the different developmental stages; a significant difference was also found among the different transcripts in each stage (two-way ANOVA, $p < 0.001$) (ta 2). Two ferritins, the HCH (GAXK01142559) and the LCH (GAXK01169093), showed a similar and low expression in all developmental stages, whereas HCH ferritin (GAXK01168686) was always highly expressed in copepodites (C1-C5) compared to embryos, copepodites and adults (568 RPKM vs. 82 RPKM, on average; Tukey's multiple comparison test, $p < 0.0001$); similarly, the other HCH transcript (GAXK01170132) showed significantly higher expression in the later developmental stages (500 RPKM, on average, in C4 and AF), with a significant peak in expression in the C5 stage (Tukey's multiple comparison test, $p < 0.0001$) (Figure 2).

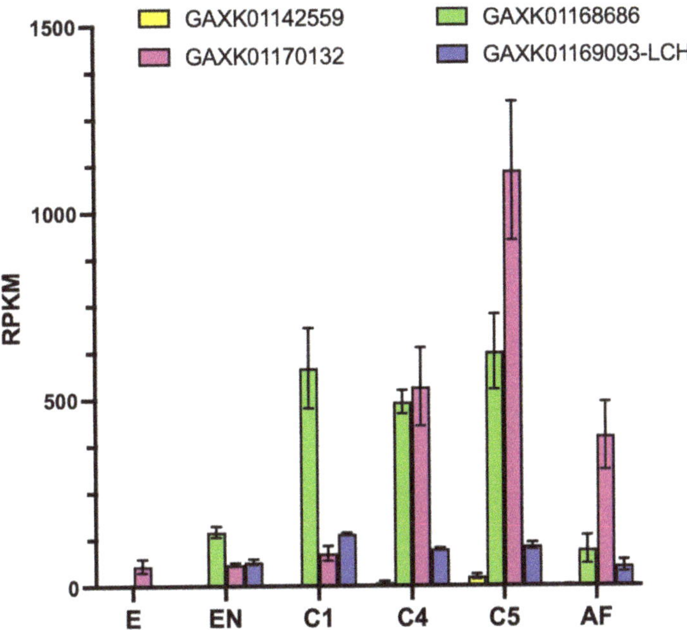

Figure 2. Relative expression of ferritin transcripts in *C. finmarchicus* across development. Relative expression normalized by length (RPKM) across six developmental stages: embryos, early nauplii (NII-NIII), copepodites I, IV and V (C1, C4 and C5), and adult females (AF). Bars are mean ± standard deviation (n = 3 replicates, n = 2 C1 and C4). Colors indicate different transcripts and their NCBI accession number.

No significant differences were found in the expression of all ferritins in *C. finmarchicus* females feeding for two days on the toxic *Alexandrium fundyense* at different concentrations (low and high dose) compared to control conditions (Figure 3a,b). However, in spite of treatments and time points, the expression of the different transcripts was significantly different (two-way ANOVA, $p < 0.0001$). At two days, similarly to the previous dataset, two transcripts showed a null or very low expression in all treatments, whereas the transcripts GAXK01168686 and GAXK01170132 were significantly more abundant (average 300 RPKM) compared to the other two (Tukey's multiple comparison test, $p < 0.001$) (Figure 3a). The same results were observed after five days of exposure, with the same two transcripts (GAXK01168686 and GAXK01170132) being significantly highly expressed in all conditions compared with the other two (Tukey's multiple comparison test, $p < 0.001$) (Figure 3b).

A similar result was also found in *C. helgolandicus*, where expression of the four ferritins did not change significantly between females feeding the toxic *Skeletonema marinoi* diet and the control treatment *Prorocentrum minimum*. However, similar to what was observed in *C. finmarchicus*, in spite of the treatments, one HCH ferritin (GJFL01006140) and the LCH ferritin (GJFL01003552) showed a significantly high expression (up to 20-fold and 60-fold difference) compared with the other two transcripts (Figure 4a). In *T. stylifera* females collected from the field during different weeks, the same transcript-dependent expression was observed: one transcript (GJGX01146428) showed ca. 100-fold higher expression compared to the other three; the same transcript was also significantly upregulated in females exposed to a high content of harmful oxylipins (test) (multiple unpaired t-test, $p < 0.05$) (Figure 4b).

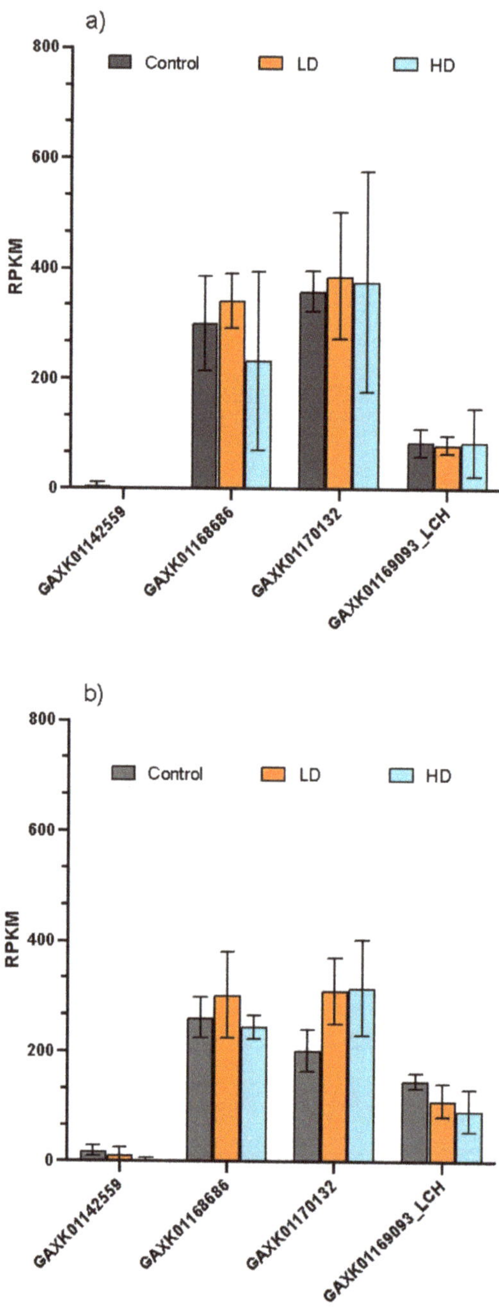

Figure 3. Relative expression of ferritin transcripts in *C. finmarchicus* feeding on a toxic diet. Relative expression normalized by length (RPKM) in *C. finmarchicus* females fed the non-toxic cryptophyte *Rhodomonas baltica* (control) and the toxic dinoflagellate *Alexandrium fundyense* at low (orange) and high (cyan) doses, for 2 (**a**) and 5 (**b**) days. Different names indicate different transcripts. Bars are mean ± standard deviation (n = 3 replicates).

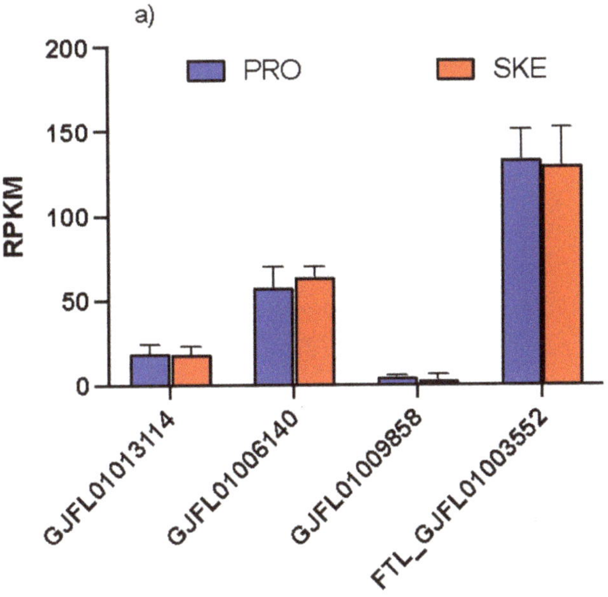

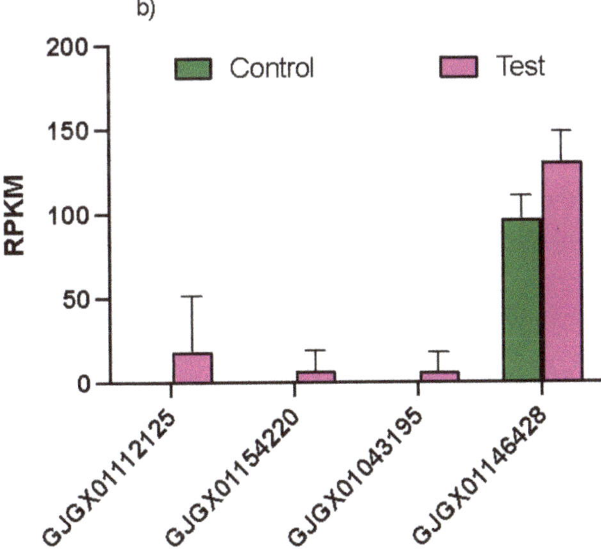

Figure 4. Relative expression of ferritin transcripts in *C. helgolandicus* and *T. stylifera*. (**a**) Relative expression normalized by length (RPKM) in *C. helgolandicus* females fed the dinoflagellate *Prorocentrum minimum* (PRO) and the toxic diatom *Skeletonema marinoi* (SKE) for 5 days. (**b**) Relative expression normalized by length (RPKM) in *T. stylifera* collected in the Gulf of Naples when low (control) and high (test) content of harmful oxylipins were measured in the phytoplankton. Different names indicate different transcripts. Bars are mean ± standard deviation (n = 3 replicates).

4. Discussion

The evolution of life on Earth has been dependent on the availability of iron, leading to the evolution of ferritins, multimeric iron storage proteins that are ubiquitous in the living kingdom [1]. Most cells express ferritin genes, but protein concentration can vary among cell types. Ferritin sequence and structures are highly conserved among plants and animals, suggesting evolutionary convergence in eukaryotes, while bacterial ferritins diverge in sequence but not in structure (as reviewed in [31]).

Ferritins are also present in copepods, the most abundant metazoans on the planet [32]. Gene expression changes in ferritin have been reported in studies in the literature (Table 2) when copepods are exposed to abiotic stress, such as metals [8,33], heat shocks [10], high pCO_2 levels [9] or water-accommodated fractions [34]. Changes in gene expression have also been recorded in response to biological drivers, namely, infestation by an epibiotic parasite [10], during different phases of dormancy, such as induction and recovery from quiescence [11], or during the diapause "stationary" phase [12,13]. In addition, it has been reported an impairment of feeding and reproduction upon the knockdown of ferritin, as well as a reduction in the protein level during starvation [14]. Interestingly, no changes in ferritin expression were associated with different crowding levels [35]. Overall, the reported differences in the regulation of ferritins opened questions related to stressor-specific responses associated with different transcripts or species-specific responses.

Table 2. Summary of ferritin gene expression changes in copepods reported in the literature and our study. For each study, the species tested, the stressor and the regulation (+ = upregulation, − = downregulation and ns = non-significant change) of the ferritin gene are shown.

Species	Stressor	Regulation	Reference Paper
P. annandalei	Ni nanoparticles	+	33
A. tonsa	Ni nanoparticles	+	8
	CdSe/ZnS quantum dots	+	8
	Heat shock (salinity dependent)	+	10
	Infestation *C. vesiculosum*	+	10
	Entry/emergence quiescence eggs	+	11
	Crowding	ns	11
	High pCO^2 (only costal)	+	9
	WAF	ns	34
Calanus spp.	Diapause (CV females)	+	12 13 42
L. salmonis	Starvation	−	14
C. finmarchicus	Diapause preparation (CV) vs. reproductive program	ns	45
	Our study		
C. finmarchicus	Copepodites	+	
	A. fundyense	ns	
C. helgolandicus	Oxylipin (lab)	ns	
T. stylifera	Oxylipin (field)	+	
N. flemingeri	Diapause emergence	ns	

Indeed, the role of ferritin in the metabolism and physiology of copepods is presently understudied. Based on the relatively sparse available evidence (as summarized in Table 2), we speculate about different mechanisms potentially at play, which may differ not only among stressors but also among species. In this context, we interpret our results also by direct comparison with insect ferritins, considering their closeness in terms of absolute success [36].

Insect ferritins can be divided into three groups depending on their structure, location and function [37]. HCH and LCH differentiate based on the presence in HCH of the ferroxidase center, which confers the role of reducing the ferrous ion (Fe^{2+}) to its non-reactive ferric ion (Fe^{3+}). Insects also present a third type of protein, mitochondrial ferritin (HCH). Endoplasmic reticulum ferritins (HCH) act as iron transporters, mitochondrial ferritins act as antioxidants, while hemolymph ones (LCH) have a dual function [37]. In copepods, a full-length transcript encoding ferritin HCH has been identified in all 27 mined transcriptomes, with 50% of the species possessing multiple transcripts. Consistently with what was reported in the genome of *Lepeophtheirus salmonis* [14], mitochondrial ferritins were not found in any of the mined copepod transcriptomes. As opposed to insects [37], the location of ferritins has not been explored yet in copepods. From our results, mining transcripomes generated from whole or even pool of individuals, it is not possible to infer the specific location of the identified ferritins. Nonetheless, by analogy with insects, we can speculate that copepod ferritins are associated with the endoplasmic reticulum (HCH) and the hemolymph (LCH).

Arthropod immunity is based solely on an innate system [38]. In this framework, ferritin plays a key role, acting as a stress-induced protein and permitting the accumulation of iron which is no longer available to pathogens [38]. This process is compliant with the upregulation of ferritin in *Acartia tonsa* infested by the epibiotic euglenid *Colacium vesiculosum* [10]. Blood-sucking arthropods need further protection to reduce the risks associated with the massive ingestion of iron and heme [39]. In the yellow fever mosquito *Aedes aegypti*, ferritin expression is upregulated upon iron uptake, likely as cytotoxic protection [39,40]. Consistently, in the parasitic copepod *L. salmonis* ferritin levels in the midgut decrease during starvation [14], evidencing the direct link between gene expression and blood-feeding.

Involvement of ferritin has also been suggested during dormancy in the fly *D. melanogaster*, in the crustacean branchiopod *Artemia* sp., and in the copepods *C. finmarchicus* and *A. tonsa*. High expression of ferritins in *Artemia* sp. and *A. tonsa* embryos was hypothesized as preventing embryogenesis to continue and inhibiting development during the resting state [11,41]. Indeed, an increase in ferritin corresponds to higher chelation of iron stores and a reduction of iron available for processes such as embryogenesis [42]. Consistent with its role as an enhancer of stress resistance, high expression of ferritin has been found in both *D. melanogaster* and *C. finmarchicus* (C5, females) when at diapause [12,13,42,43]. Diapause, a type of dormancy, is characterized by a delay in development, a decrease in metabolism and an increase in stress resistance [44]. In *Calanus* spp., it has been reported that ferritin expression has its highest value during the early stages of diapause and it decreases at the beginning of the maintenance phase [12,13,42]. However, in another study, ferritin was not found among the genes differentiating C5s preparing for diapause from the ones on the reproductive program [45]. Furthermore, it seems that the ferritin role could be species-specific, as suggested by the fact that in *Neocalanus flemingeri* diapausing females ferritins were not among the genes characterizing the diapause phenotype [46].

In six out of the eleven papers published in the literature on copepod ferritins, the target species is the calanoid *A. tonsa* [8–11,34,35]. A renowned global-scale invasive species (e.g., [47,48]), *A. tonsa* is an established model organism in ecotoxicology studies (e.g., [49]) and it has proven valuable as feed for fish larvae in mariculture (e.g., [50]). The remaining papers surveyed are centered on two pelagic (*Calanus* spp.: [12,13,42]) and one parasitic (*L. salmonis*: [14]) species, relevant for their ecological (*Calanus* spp.) and economic (*L. salmonis*) impacts. Ferritin expression studies in these species allow the development of sensitive biomarkers for marine pollution monitoring, but also for the implementation of ecological indicators. The present study surveys the ferritin distribution in 27 copepods from the Calanoida, Harpacticoida and Cyclopoida orders, and it explores the variability of ferritin expression in three key ecological species: *C. finmarchicus, C. helgolandicus* and *T. stylifera*. Our study highlights the fact that the use of a single transcript as a biomarker can be highly limiting. In more than 50% of the examined copepods, in fact, we found more than one transcript encoding ferritin and these included *C. finmarchicus, C. helgolandicus* and *T. stylifera*. Moreover, our expression studies highlighted the fact that in spite of the stressor or developmental stage, two transcripts (out of four) were always more abundant compared with the others in all three species. Most protein-coding genes have dominant transcripts that are expressed at a considerably higher level than any minor transcripts across different conditions [51,52]. It is likely that these dominant transcripts are the main contributors to the proteome of the individual; thus, they might be used as good candidates for physiological and ecological studies on copepods.

Based on our results, we suggest that the role of ferritin in stress is still unclear. Changes in the expression of ferritins were only found in *T. stylifera* in response to different levels of the cytotoxic oxylipins in the natural phytoplankton assemblage; in contrast to the two *Calanus* species exposed to toxic algae, expression of ferritin did not change compared with a control diet. However, in *C. helgolandicus,* exposure to the harmful *S. marinoi* was associated with the upregulation of detoxification (GSTs), protein repair (HSP60 and HSP70), and immune system (prophenoloxidase activating enzyme, PPAE) genes [20]. Thus, we can speculate that the lack of a significant regulation of the ferritins could be associated with the fact that the concerted over-expression of detoxification genes could be sufficient to protect the copepod from the direct toxic effect of the diatom. In contrast, in *T. stylifera,* a high concentration of oxylipins induced the downregulation of stress response and oxidation–reduction genes [21]. Therefore, the upregulation of ferritin transcripts in *T. stylifera* could act as a compensatory defensive mechanism protecting the copepods from the oxidative damage and apoptosis associated with the ingestion of oxylipins [48]. Concerning *C. finmarchicus*, the differences in the expression found across development and not in response to the toxic algae could suggest a role of ferritin as a developmental marker. Two dominant ferritins were always poorly expressed in embryos and nauplii, compared to copepodites, confirming that iron mobilization due to low ferritin expression might stimulate embryogenesis and early larval development in copepods [43]. These ferritin-encoding dominant transcripts, thus, could be used as potential biomarkers of early–late development transition in *C. finmarchicus*.

5. Conclusions

Ferritins are the major intracellular iron storage proteins, highly conserved and present in most prokaryotes and eukaryotes including fungi, algae, plants, and animals [1]. Their expression in copepods is regulated by numerous endogenous and exogenous factors and may also present species-specific modulations. Our study expands the knowledge on the ferritin diversity in copepods; in all species, we found a high chain ferritin (HCH) and we confirm the lack of mitochondrial ferritin in crustaceans. We suggest that copepod ferritins can be involved in multiple processes such as iron storage, larval development and stress, highlighting that ferritin regulation can be species-specific, stressor-specific and stage-specific.

Supplementary Materials: The following supporting information can be downloaded at: https://www.mdpi.com/article/10.3390/jmse11061187/s1. Table S1: Summary of reciprocal blast and protein domain analysis for the investigated ferritins. For each species, NCBI accession number (BioProject and the single transcript), E-value of the reciprocal blast, annotation result, top hit (species) and its accession number. Additionally, information includes the presence of the Pfam domain (PF00210) and its E value. The list only includes full-length sequences. Figure S1: Alignment of ferritin sequences identified for copepods in this study with known sequences from *D. melanogaster* (NP_524873, NP_524802, NP_572854), *L. salmonis* (LsFer1–4) (BT121711, BT121232, MK887318, BT077723, BT121164) and homologs from *H. Sapiens* (P02794, P02792). Copepod sequences are in alphabetical order as in Table 1. Light chain subunits (LCH) are indicated in green. Amino acids that make up the ferroxidase center of the heavy chain in *H. Sapiens* are bolded in red. Highlighted in yellow are the amino acids which were not conserved.

Author Contributions: Conceptualization, V.R.; writing—original draft preparation, V.R., M.U. and Y.C.; writing—review and editing, V.R., M.U. and Y.C. All authors have read and agreed to the published version of the manuscript.

Funding: M.U. acknowledges the support of NBFC to Stazione Zoologica Anton Dohrn, funded by the Italian Ministry of University and Research, PNRR, Missione 4 Componente 2, "Dalla ricerca all'impresa", Investimento 1.4, Project CN00000033.

Institutional Review Board Statement: Not applicable.

Informed Consent Statement: Not applicable.

Data Availability Statement: The National Center for Biotechnology Information (NCBI) Bioproject numbers for the datasets examined in the present study are indicated in Table S1. Supplementary File S1 includes FASTA files for the transcript-encoding protein identified in this study.

Acknowledgments: We would like to thank Hartline and Lenz from the University of Hawaii at Manoa for the intellectual discussion.

Conflicts of Interest: The authors declare no conflict of interest.

References

1. Theil, E.C. Ferritin: Structure, gene regulation, and cellular function in animals, plants, and microorganisms. *Annu. Rev. Biochem.* **1987**, *56*, 289–315. [CrossRef] [PubMed]
2. Nichol, H.; Law, J.H.; Winzerling, J.J. Iron metabolism in insects. *Annu. Rev. Entomol.* **2002**, *47*, 535–559. [CrossRef] [PubMed]
3. Arosio, P.; Ingrassia, R.; Cavadini, P. Ferritins: A family of molecules for iron storage, antioxidation and more. *Biochim. Et Biophys. Acta (BBA)-Gen. Subj.* **2009**, *1790*, 589–599. [CrossRef] [PubMed]
4. Emerit, J.; Beaumont, C.; Trivin, F. Iron metabolism, free radicals, and oxidative injury. *Biomed. Pharmacother.* **2001**, *55*, 333–339. [CrossRef]
5. Tang, X.; Zhou, B. Ferritin is the key to dietary iron absorption and tissue iron detoxification in *Drosophila melanogaster*. *FASEB J.* **2013**, *27*, 288–298. [CrossRef]
6. Hamburger, A.E.; West, A.P., Jr.; Hamburger, Z.A.; Hamburger, P.; Bjorkman, P.J. Crystal structure of a secreted insect ferritin reveals a symmetrical arrangement of heavy and light chains. *J. Mol. Biol.* **2005**, *349*, 558–569. [CrossRef]
7. Colbourne, J.K.; Eads, B.D.; Shaw, J.; Bohuski, E.; Bauer, D.J.; Andrews, J. Sampling Daphnia's expressed genes: Preservation, expansion and invention of crustacean genes with reference to insect genomes. *BMC Genom.* **2007**, *8*, 217. [CrossRef]
8. Zhou, C.; Hou, J.; Lin, D. A ferritin gene in the marine copepod Acartia tonsa as a highly sensitive biomonitor for nanocontamination. *Aquat. Toxicol.* **2022**, *253*, 106353. [CrossRef]
9. Aguilera, V.M.; Vargas, C.A.; Lardies, M.A.; Poupin, M.J. Adaptive variability to low-pH river discharges in *Acartia tonsa* and stress responses to high PCO2 conditions. *Mar. Ecol.* **2016**, *37*, 215–226. [CrossRef]
10. Petkeviciute, E.; Kania, P.W.; Skovgaard, A. Genetic responses of the marine copepod *Acartia tonsa* (Dana) to heat shock and epibiont infestation. *Aquac. Rep.* **2015**, *2*, 10–16. [CrossRef]
11. Nilsson, B.; Jepsen, P.M.; Rewitz, K.; Hansen, B.W. Expression of hsp70 and ferritin in embryos of the copepod *Acartia tonsa* (Dana) during transition between subitaneous and quiescent state. *J. Plankton Res.* **2014**, *36*, 513–522. [CrossRef]
12. Aruda, A.M.; Baumgartner, M.F.; Reitzel, A.M.; Tarrant, A.M. Heat shock protein expression during stress and diapause in the marine copepod Calanus finmarchicus. *J. Insect Physiol.* **2011**, *57*, 665–675. [CrossRef] [PubMed]
13. Tarrant, A.M.; Baumgartner, M.F.; Verslycke, T.; Johnson, C.L. Differential gene expression in diapausing and active Calanus finmarchicus (Copepoda). *Mar. Ecol. Prog. Ser.* **2008**, *355*, 193–207. [CrossRef]
14. Heggland, E.I.; Tröße, C.; Eichner, C.; Nilsen, F. Heavy and light chain homologs of ferritin are essential for blood-feeding and egg production of the ectoparasitic copepod Lepeophtheirus salmonis. *Mol. Biochem. Parasitol.* **2019**, *232*, 111197. [CrossRef] [PubMed]
15. Lauritano, C.; Carotenuto, Y.; Roncalli, V. Glutathione S-transferases in marine copepods. *J. Mar. Sci. Eng.* **2021**, *9*, 1025. [CrossRef]
16. Roncalli, V.; Uttieri, M.; Capua, I.D.; Lauritano, C.; Carotenuto, Y. Chemosensory-Related Genes in Marine Copepods. *Mar. Drugs* **2022**, *20*, 681. [CrossRef]
17. Roncalli, V.; Lauritano, C.; Carotenuto, Y. First report of OvoA gene in marine arthropods: A new candidate stress biomarker in copepods. *Mar. Drugs* **2021**, *19*, 647. [CrossRef]
18. Cieslak, M.C.; Castelfranco, A.M.; Roncalli, V.; Lenz, P.H.; Hartline, D.K. t-Distributed Stochastic Neighbor Embedding (t-SNE): A tool for eco-physiological transcriptomic analysis. *Mar. Genom.* **2020**, *51*, 100723. [CrossRef]

19. Roncalli, V.; Cieslak, M.C.; Lenz, P.H. Transcriptomic responses of the calanoid copepod Calanus finmarchicus to the saxitoxin producing dinoflagellate *Alexandrium fundyense*. *Sci. Rep.* **2016**, *6*, 25708. [CrossRef]
20. Asai, S.; Sanges, R.; Lauritano, C.; Lindeque, P.K.; Esposito, F.; Ianora, A.; Carotenuto, Y. De novo transcriptome assembly and gene expression profiling of the copepod *Calanus helgolandicus* feeding on the PUA-producing diatom *Skeletonema marinoi*. *Mar. Drugs* **2020**, *18*, 392. [CrossRef]
21. Russo, E.; Lauritano, C.; d'Ippolito, G.; Fontana, A.; Sarno, D.; von Elert, E.; Ianora, A.; Carotenuto, Y. RNA-Seq and differential gene expression analysis in *Temora stylifera* copepod females with contrasting non-feeding nauplii survival rates: An environmental transcriptomics study. *Bmc Genom.* **2020**, *21*, 693. [CrossRef]
22. Russo, E.; d'Ippolito, G.; Fontana, A.; Sarno, D.; D'Alelio, D.; Busseni, G.; Ianora, A.; von Elert, E.; Carotenuto, Y. Density-dependent oxylipin production in natural diatom communities: Possible implications for plankton dynamics. *ISME J.* **2020**, *14*, 164–177. [CrossRef] [PubMed]
23. Christie, A.E.; Fontanilla, T.M.; Nesbit, K.T.; Lenz, P.H. Prediction of the protein components of a putative *Calanus finmarchicus* (Crustacea, Copepoda) circadian signaling systems using a de novo assembled transcriptome. *Comp. Biochem. Physiol. D-Genom. Proteom.* **2013**, *8*, 165–193. [CrossRef] [PubMed]
24. Roncalli, V.; Cieslak, M.C.; Passamaneck, Y.; Christie, A.E.; Lenz, P.H. Glutathione S-transferase (GST) gene diversity in the crustacean Calanus finmarchicus–contributors to cellular detoxification. *PLoS ONE* **2015**, *10*, e0123322. [CrossRef] [PubMed]
25. Lauritano, C.; Roncalli, V.; Ambrosino, L.; Cieslak, M.C.; Ianora, A. First de novo transcriptome of the copepod *Rhincalanus gigas* from Antarctic waters. *Biology* **2020**, *9*, 410. [CrossRef] [PubMed]
26. Hartline, D.K.; Cieslak, M.C.; Castelfranco, A.M.; Lieberman, B.; Roncalli, V.; Lenz, P.H. De novo transcriptomes of six calanoid copepods (Crustacea): A resource for the discovery of novel genes. *Sci. Data* **2023**, *10*, 242. [CrossRef]
27. Gasteiger, E.; Gattiker, A.; Hoogland, C.; Ivanyi, I.; Appel, R.D.; Bairoch, A. ExPASy: The proteomics server for in-depth protein knowledge and analysis. *Nucleic Acids Res.* **2003**, *31*, 3784–3788. [CrossRef]
28. Letunic, I.; Khedkar, S.; Bork, P. SMART: Recent updates, new developments and status in 2020. *Nucleic Acids Res.* **2021**, *49*, D458–D460. [CrossRef]
29. Katoh, K.; Misawa, K.; Kuma, K.i.; Miyata, T. MAFFT: A novel method for rapid multiple sequence alignment based on fast Fourier transform. *Nucleic Acids Res.* **2002**, *30*, 3059–3066. [CrossRef]
30. Barreto, F.S.; Watson, E.T.; Lima, T.G.; Willett, C.S.; Edmands, S.; Li, W.; Burton, R.S. Genomic signatures of mitonuclear coevolution across populations of Tigriopus californicus. *Nat. Ecol. Evol.* **2018**, *2*, 1250–1257. [CrossRef]
31. Andrews, S.C. The Ferritin-like superfamily: Evolution of the biological iron storeman from a rubrerythrin-like ancestor. *Biochim. Et Biophys. Acta (BBA)-Gen. Subj.* **2010**, *1800*, 691–705. [CrossRef] [PubMed]
32. Uttieri, M. Trends in copepod studies. In *Trends in Copepod Studies—Distribution, Biology and Ecology*; Nova Science Publishers Inc.: New York, NY, USA, 2018; pp. 1–11.
33. Jiang, J.-L.; Wang, G.-Z.; Mao, M.-G.; Wang, K.-J.; Li, S.-J.; Zeng, C.-S. Differential gene expression profile of the calanoid copepod, *Pseudodiaptomus annandalei*, in response to nickel exposure. *Comp. Biochem. Physiol. Part C Toxicol. Pharmacol.* **2013**, *157*, 203–211. [CrossRef] [PubMed]
34. Hafez, T.; Bilbao, D.; Etxebarria, N.; Duran, R.; Ortiz-Zarragoitia, M. Application of a biological multilevel response approach in the copepod Acartia tonsa for toxicity testing of three oil Water Accommodated Fractions. *Mar. Environ. Res.* **2021**, *169*, 105378. [CrossRef] [PubMed]
35. Nilsson, B.; Jakobsen, H.H.; Stief, P.; Drillet, G.; Hansen, B.W. Copepod swimming behavior, respiration, and expression of stress-related genes in response to high stocking densities. *Aquac. Rep.* **2017**, *6*, 35–42. [CrossRef]
36. Schminke, H.K. Entomology for the copepodologist. *J. Plankton Res.* **2007**, *29*, i149–i162. [CrossRef]
37. Pham, D.Q.; Winzerling, J.J. Insect ferritins: Typical or atypical? *Biochim. Et Biophys. Acta (BBA)-Gen. Subj.* **2010**, *1800*, 824–833. [CrossRef]
38. Whiten, S.R.; Eggleston, H.; Adelman, Z.N. Ironing out the details: Exploring the role of iron and heme in blood-sucking arthropods. *Front. Physiol.* **2018**, *8*, 1134. [CrossRef]
39. Dunkov, B.C.; Georgieva, T.; Yoshiga, T.; Hall, M.; Law, J.H. Aedes aegypti ferritin heavy chain homologue: Feeding of iron or blood influences message levels, lengths and subunit abundance. *J. Insect Sci.* **2002**, *2*, 7. [CrossRef]
40. Geiser, D.L.; Chavez, C.A.; Flores-Munguia, R.; Winzerling, J.J.; Pham, D.Q.D. Aedes aegypti ferritin: A cytotoxic protector against iron and oxidative challenge? *Eur. J. Biochem.* **2003**, *270*, 3667–3674. [CrossRef]
41. Chen, T.; Amons, R.; Clegg, J.S.; Warner, A.H.; MacRae, T.H. Molecular characterization of artemin and ferritin from Artemia franciscana. *Eur. J. Biochem.* **2003**, *270*, 137–145. [CrossRef]
42. Skottene, E.; Tarrant, A.M.; Olsen, A.J.; Altin, D.; Østensen, M.-A.; Hansen, B.H.; Choquet, M.; Jenssen, B.M.; Olsen, R.E. The β-oxidation pathway is downregulated during diapause termination in Calanus copepods. *Sci. Rep.* **2019**, *9*, 16686. [CrossRef] [PubMed]
43. Kučerová, L.; Kubrak, O.I.; Bengtsson, J.M.; Strnad, H.; Nylin, S.; Theopold, U.; Nässel, D.R. Slowed aging during reproductive dormancy is reflected in genome-wide transcriptome changes in Drosophila melanogaster. *Bmc Genom.* **2016**, *17*, 50. [CrossRef] [PubMed]
44. Lenz, P.H.; Roncalli, V. Diapause within the Context of Life-History Strategies in Calanid Copepods (Calanoida: Crustacea). *Biol. Bull* **2019**, *237*, 170–179. [CrossRef] [PubMed]

45. Lenz, P.H.; Roncalli, V.; Cieslak, M.C.; Tarrant, A.M.; Castelfranco, A.M.; Hartline, D.K. Diapause vs. reproductive programs: Transcriptional phenotypes in a keystone copepod. *Commun. Biol.* **2021**, *4*, 426. [CrossRef]
46. Roncalli, V.; Cieslak, M.C.; Castelfranco, A.M.; Hopcroft, R.R.; Hartline, D.K.; Lenz, P.H. Post-diapause transcriptomic restarts: Insight from a high-latitude copepod. *BMC Genom.* **2021**, *22*, 409.
47. Barroeta, Z.; Villate, F.; Uriarte, I.; Iriarte, A. Impact of Colonizer Copepods on Zooplankton Structure and Diversity in Contrasting Estuaries. *Estuaries Coasts* **2022**, *45*, 2592–2609. [CrossRef]
48. Camatti, E.; Pansera, M.; Bergamasco, A. The copepod Acartia tonsa dana in a microtidal Mediterranean lagoon: History of a successful invasion. *Water* **2019**, *11*, 1200. [CrossRef]
49. Carotenuto, Y.; Vitiello, V.; Gallo, A.; Libralato, G.; Trifuoggi, M.; Toscanesi, M.; Lofrano, G.; Esposito, F.; Buttino, I. Assessment of the relative sensitivity of the copepods Acartia tonsa and Acartia clausi exposed to sediment-derived elutriates from the Bagnoli-Coroglio industrial area. *Mar. Environ. Res.* **2020**, *155*, 104878. [CrossRef]
50. Støttrup, J.G.; Richardson, K.; Kirkegaard, E.; Pihl, N.J. The cultivation of Acartia tonsa Dana for use as a live food source for marine fish larvae. *Aquaculture* **1986**, *52*, 87–96. [CrossRef]
51. Gonzàlez-Porta, M.; Frankish, A.; Rung, J.; Harrow, J.; Brazma, A. Transcriptome analysis of human tissues and cell lines reveals one dominant transcript per gene. *Genome Biol.* **2013**, *14*, R70. [CrossRef]
52. Taneri, B.; Snyder, B.; Gaasterland, T. Distribution of alternatively spliced transcript isoforms within human and mouse transcriptomes. *J. Omics Res.* **2011**, *1*, 1–5.

Journal of *Marine Science and Engineering*

MDPI

Review

The Distribution of *Pseudodiaptomus marinus* in European and Neighbouring Waters—A Rolling Review

Marco Uttieri [1,2,*], Olga Anadoli [3], Elisa Banchi [2,4], Marco Battuello [5,6], Şengül Beşiktepe [7], Ylenia Carotenuto [1], Sónia Cotrim Marques [8], Alessandra de Olazabal [4], Iole Di Capua [1,2], Kirsten Engell-Sørensen [9], Alenka Goruppi [2,4], Tamar Guy-Haim [10], Marijana Hure [11], Polyxeni Kourkoutmani [12], Davor Lučić [11], Maria Grazia Mazzocchi [1], Evangelia Michaloudi [12], Arseniy R. Morov [10], Tuba Terbıyık Kurt [13], Valentina Tirelli [2,4], Jessica Vannini [1], Ximena Velasquez [10], Olja Vidjak [14] and Marianne Wootton [15]

1 Stazione Zoologica Anton Dohrn, Villa Comunale, 80121 Naples, Italy; ylenia.carotenuto@szn.it (Y.C.); iole.dicapua@szn.it (I.D.C.); grazia.mazzocchi@szn.it (M.G.M.); jessica.vannini@szn.it (J.V.)
2 NBFC—National Biodiversity Future Center, Piazza Marina 61, 90133 Palermo, Italy; ebanchi@ogs.it (E.B.); agoruppi@ogs.it (A.G.); vtirelli@ogs.it (V.T.)
3 Hellenic Centre for Marine Research, 71003 Heraklion, Greece; olganadoli1997@gmail.com
4 National Institute of Oceanography and Applied Geophysics—OGS, Via A. Piccard 54, 34151 Trieste, Italy; adeolazabal@ogs.it
5 Department of Life Sciences and Systems Biology, University of Torino, Via Accademia Albertina 13, 10123 Torino, Italy; marco.battuello@unito.it
6 Pelagosphera, Marine Environmental Services Cooperative, Via Umberto Cosmo 17/bis, 10131 Torino, Italy
7 The Institute of Marine Sciences and Technology, Dokuz Eylul University, 35340 İzmir, Turkey; sengul.besiktepe@deu.edu.tr
8 MARE/ARNET, School of Tourism and Maritime Technology, Polytechnic of Leiria, 2520-614 Peniche, Portugal; sonia.cotrim@ipleiria.pt
9 Fishlab, Hasselager Allé 8, DK-8260 Højbjerg, Denmark; kes@fishlab.dk
10 National Institute of Oceanography, Israel Oceanographic and Limnological Research, Haifa 3100000, Israel; tamar.guy-haim@ocean.org.il (T.G.-H.); morovar@ocean.org.il (A.R.M.); ximvel89@gmail.com (X.V.)
11 Institute for Marine and Coastal Research, University of Dubrovnik, 20000 Dubrovnik, Croatia; marijana.hure@unidu.hr (M.H.); davor.lucic@unidu.hr (D.L.)
12 School of Biology, Aristotle University of Thessaloniki, 54124 Thessaloniki, Greece; kourkoutm@bio.auth.gr (P.K.); tholi@bio.auth.gr (E.M.)
13 Department of Marine Biology, Faculty of Fisheries, Çukurova University, 01330 Adana, Turkey; tubaterbiyik@gmail.com
14 Institute of Oceanography and Fisheries, Šetalište Ivana Meštrovića 63, 21000 Split, Croatia; vidjak@izor.hr
15 CPR Survey, The Marine Biological Association, Plymouth 01752, UK; mawo@mba.ac.uk
* Correspondence: marco.uttieri@szn.it

Citation: Uttieri, M.; Anadoli, O.; Banchi, E.; Battuello, M.; Beşiktepe, Ş.; Carotenuto, Y.; Marques, S.C.; de Olazabal, A.; Di Capua, I.; Engell-Sørensen, K.; et al. The Distribution of *Pseudodiaptomus marinus* in European and Neighbouring Waters—A Rolling Review *J. Mar. Sci. Eng.* **2023**, *11*, 1238. https://doi.org/10.3390/jmse11061238

Academic Editor: Albert Calbet

Received: 19 May 2023
Revised: 10 June 2023
Accepted: 12 June 2023
Published: 16 June 2023

Abstract: Among non-native copepods, the calanoid *Pseudodiaptomus marinus* Sato, 1913 is the species probably spreading at the fastest pace in European and neighbouring waters since its first record in the Adriatic Sea in 2007. In this contribution, we provide an update on the distribution of *P. marinus* in the Mediterranean and Black Seas, along the Atlantic coasts of Europe, in the English Channel and in the southern North Sea. Starting from a previous distribution overview, we include here original and recently (2019–2023) published data to show the novel introduction of this species in different geographical areas, and its secondary spreading in already colonised regions. The picture drawn in this work confirms the strong ability of *P. marinus* to settle in environments characterised by extremely diverse abiotic conditions, and to take advantage of different vectors of introduction. The data presented allow speculations on realistic future introductions of *P. marinus* and on the potential extension of its distribution range.

Keywords: *Pseudodiaptomus marinus*; copepod; non-indigenous species; European waters

1. Introduction

One of the effects of globalisation and anthropic pressure on aquatic natural systems is the huge amount of species (~10,000) transported daily worldwide [1], with a concomitant increase in the rate of introduction of non-indigenous species (NIS) (also known as alien, allochthonous, introduced, non-native species) in new areas [2]. Zooplanktonic organisms are among the most efficient at colonising regions outside their native range [3]. In most cases (90%), these include holoplanktonic species, principally in freshwater (62%) rather than haline (estuarine, brackish and marine; 38%) systems [3].

Among marine zooplankters, copepods are known to be efficient invaders both within and between continents (as discussed in [4]). These millimetre-sized crustaceans can be efficiently introduced through trans-oceanic ships, as they dominate the ballast waters zooplankton community (e.g., [5–7]). Additionally, documented evidence also reports the introduction of alien copepods through natural and/or human-made canals (e.g., [8]), as well as through aquaculture/mariculture (e.g., [9,10]). In the pelagic marine environment, examples of globally successful invaders include the cyclopoids *Oithona davisae* Ferrari & Orsi, 1984 [8,9,11–13] and *Limnoithona tetraspina* Zhang & Li, 1976 [14,15] and the calanoids *Acartia* (*Acanthacartia*) *tonsa* Dana, 1849 [16,17], *Pseudodiaptomus forbesi* Poppe & Richard, 1890 and *Pseudodiaptomus inopinus* Burckhardt, 1913 [14,18,19].

Over the last few years, another representative of the genus *Pseudodiaptomus* has made its appearance in European and neighbouring waters (ENW), namely, *Pseudodiaptomus marinus* Sato, 1913. Following its first record in the Adriatic Sea (northernmost part of the Mediterranean Sea) in 2007 [20], this species has rapidly spread not only across the entire Mediterranean Sea but also in the Black Sea, along the Atlantic coasts of Europe, in the English Channel and in the southern North Sea (as reviewed in [21,22]), with a >450% increase over a four-year time window (2015–2019) [22]. In some European countries, such as Croatia and Italy, *P. marinus* has been included in the NIS list monitored under the Marine Strategy Framework Directive (MSFD, 2008/56/EC) (Descriptor 1: Biodiversity and Descriptor 2: Non-indigenous species). The supposed primary vector of introduction is ballast waters, while secondary spread through coastal circulation and local ship traffic may favour its further dispersal [22]. The ability of this species to establish in diverse environments is likely supported by specific traits, including a wide salinity tolerance [23], behavioural plasticity (including a day–night alternation of epibenthic and pelagic phases) [24] and genetic diversity [25].

This contribution is intended as a periodic revision of *P. marinus* occurrence in the study area, following previous reviews ([21,22,26]; as of fall 2019). Such rolling reviews can provide an almost real-time view of the spread of this NIS and of its successful establishment in introduced areas. This work stems from the activities of the ICES WGEUROBUS (Towards a EURopean OBservatory of the non-indigenous calanoid copepod *Pseudodiaptomus marinus*) [22]. The emerging scenario manifests the ongoing arrival of *P. marinus* in new regions and its further spreading from already colonised areas via secondary introduction. The results presented provide elements to make hypotheses on the occurrence of *P. marinus* at a global scale and are contextualised in the general framework of global-scale marine invasions.

2. Materials and Methods

Starting from the scenario depicted in fall 2019 [22], a survey of the published literature reporting new records of *P. marinus* in ENW was carried out. For details on materials and methods employed, the interested reader is invited to refer to the cited works. In addition, new records are presented here as original data. In this case, as the materials and methods may differ from site to site, specific details are provided in the description of each record.

The study area has been regionalised according to the description given by the European Environment Agency [27], in line with scientific usage [28,29]. For each region, a west–east direction is followed in the listing of the records.

Figure 1 provides a map of the distribution of *P. marinus* in ENW including past (up to fall 2019), new published (fall 2019 to date) and original (this work) records.

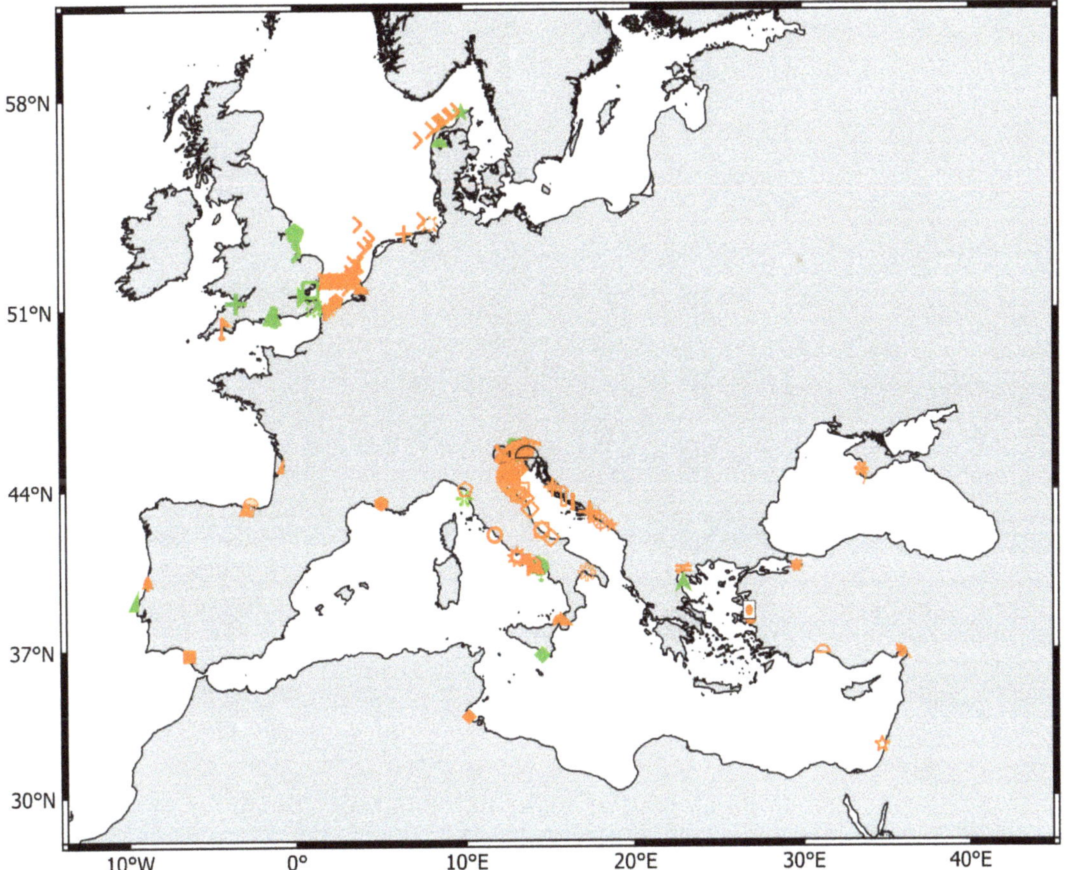

Figure 1. *Cont.*

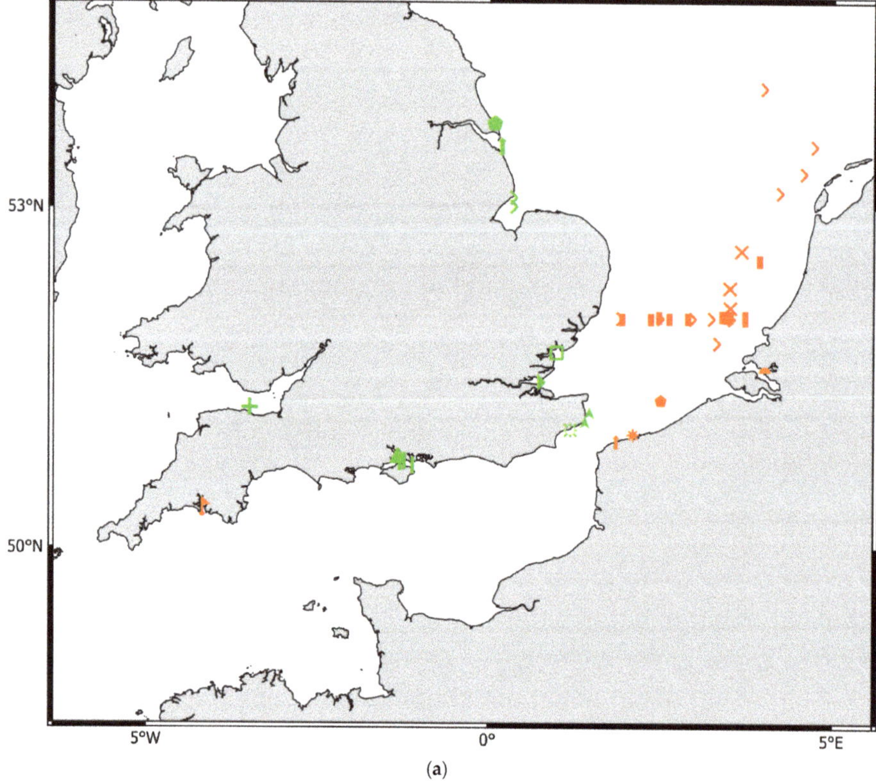

(a)

Figure 1. *Cont.*

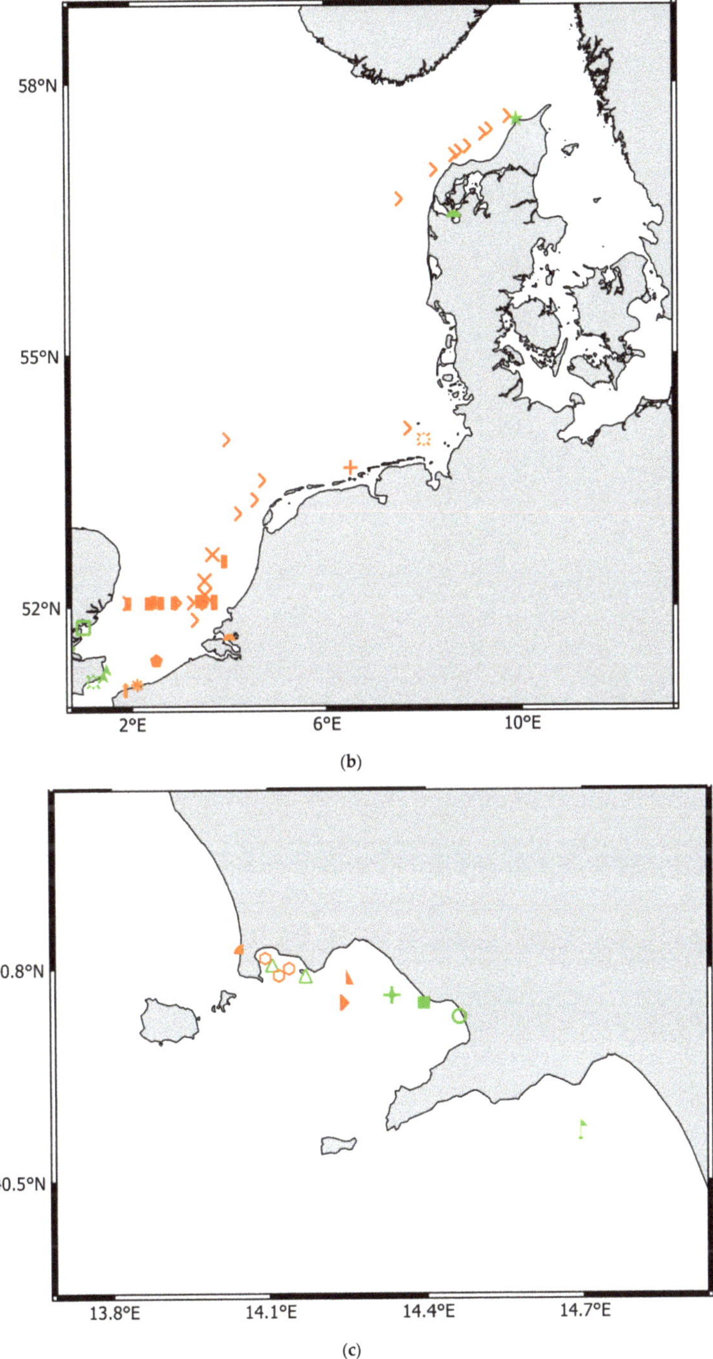

(b)

(c)

Figure 1. *Cont.*

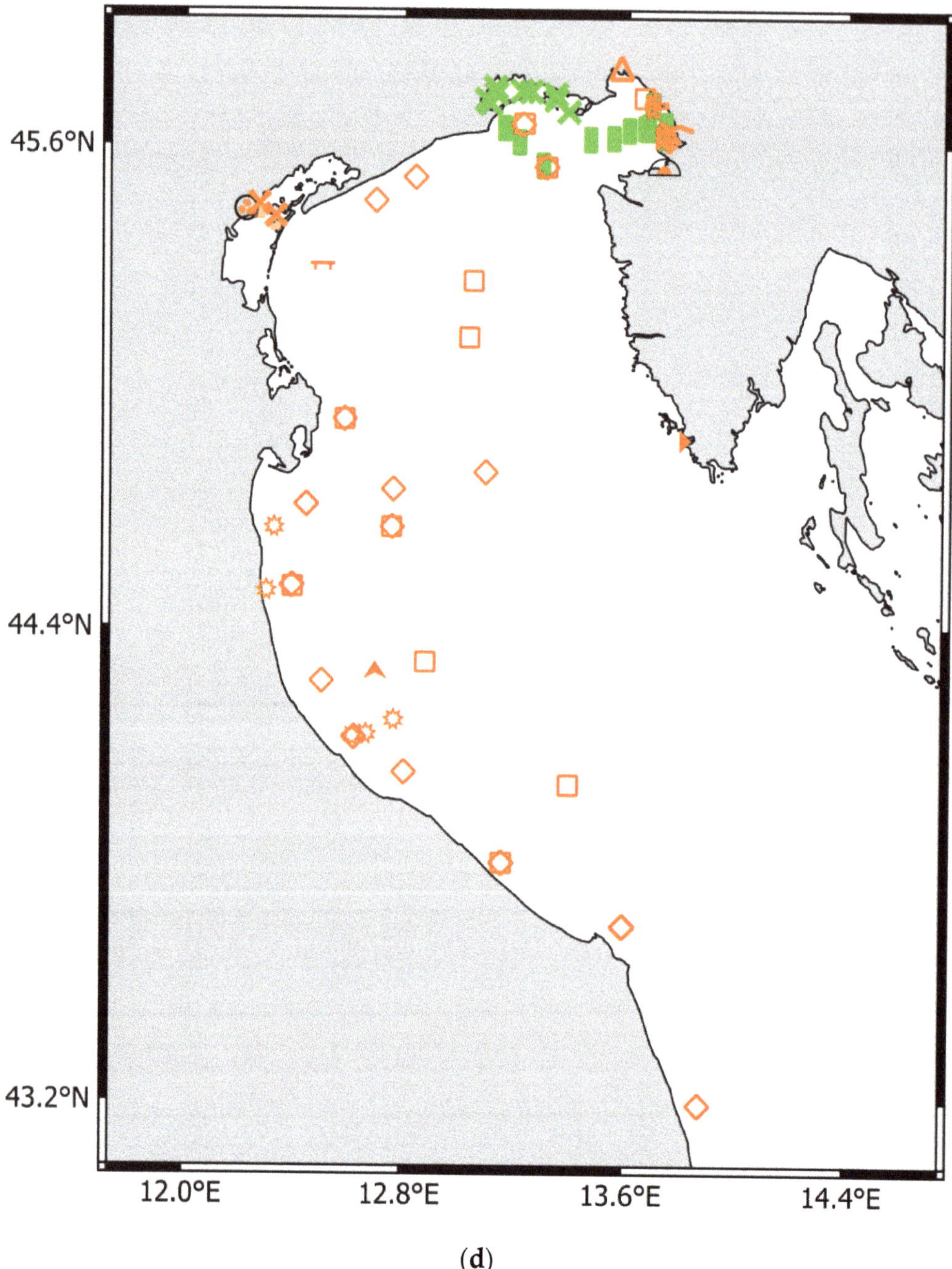

(**d**)

Figure 1. *Cont.*

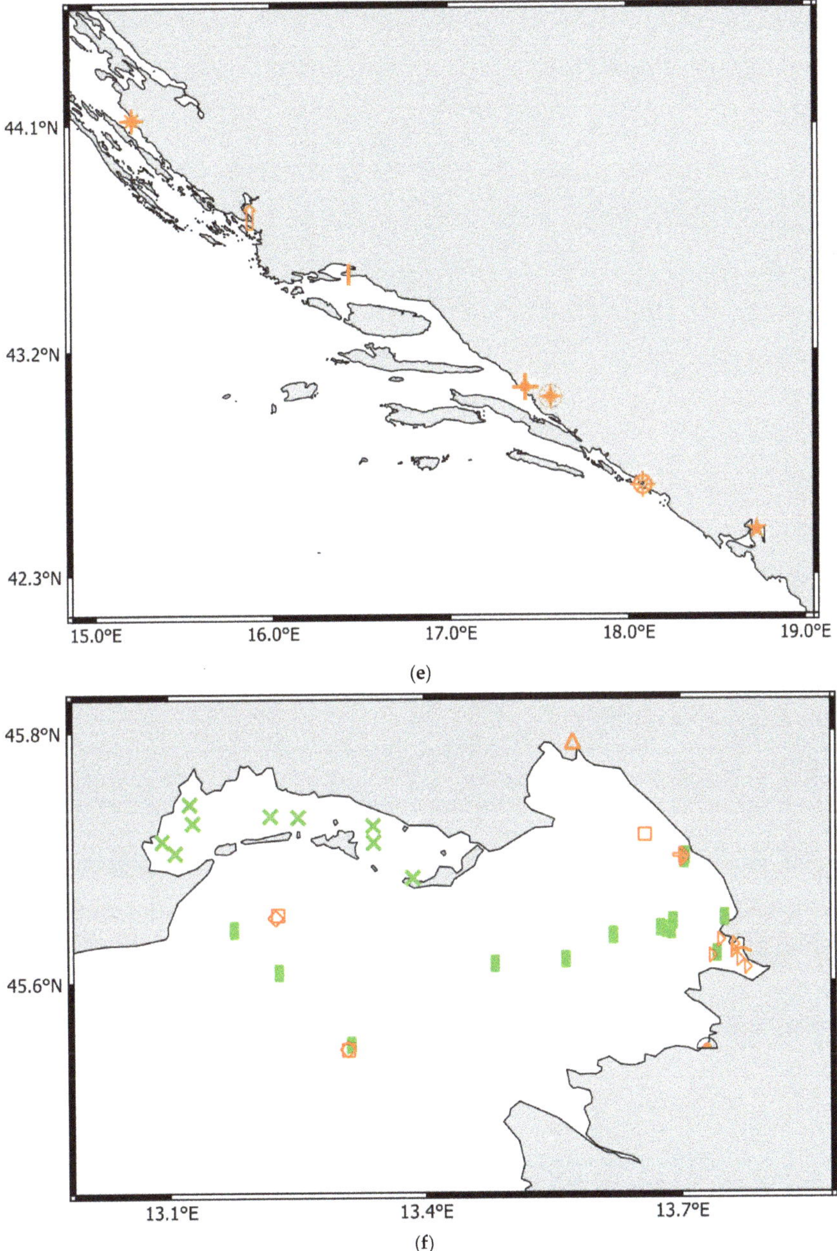

Figure 1. *Pseudodiaptomus marinus* distribution in European and neighbouring waters. The data include published reports (orange symbols) and original records (green symbols). To improve the readability of the map in more densely crowded areas, insets are provided for the following sectors: (**a**) English Channel and southeastern North Sea; (**b**) southern North Sea; (**c**) Gulf of Naples; (**d**) western and northern Adriatic Sea; (**e**) eastern Adriatic Sea; (**f**) Gulf of Trieste. The full bibliographic references of the literature cited are provided in Supplementary Material S1. Details on the geographic coordinates of each site are provided in the Supplementary Material S2.

3. Results

3.1. *Pseudodiaptomus marinus* Records from Published Literature

- *Greater North Sea*

The presence of *P. marinus* was recently recorded in the Eastern Scheldt estuary (the Netherlands; southern North Sea) in September 2018, at a temperature of ~21 °C and a salinity of ~31 [30]. The occurrence of copepodites suggests active reproduction in the estuary. The arrival of this NIS may have been possible through a connection with the adjacent Western Scheldt, where *P. marinus* has been occasionally found since 2011, although other vectors of introduction could have worked as the estuary is subject to intense human activities including shellfish culture ([25] and references therein).

- *Western Mediterranean Sea*

In the Gulf of Naples (Italy; central Tyrrhenian Sea), *P. marinus* was collected for the first time in offshore net samples (0–50 m depth layer) in December 2013 and April 2014 [22]. At the Long-Term Ecological Research site LTER_EU_IT_061 (also labelled as LTER-MC), this NIS was first found only in July 2014 [21], but eDNA metabarcoding (V4-18S rRNA) samples revealed the presence of this species since July 2011 [31]. Afterwards, this NIS was only seldom found in this area, mostly as copepodites.

The Gulf of Pozzuoli, in the northern part of the Gulf of Naples, is characterised by depths between 60 and 110 m and includes an industrial district affecting the surrounding environment [32]. Individuals of *P. marinus* were found in May and July 2019 [32]. Specimens were found with vertical tows in the 0–15 m layer with abundances of 5.6 (males) and 18.7 (copepodites) ind. m^{-3} and in the 0–50 m layer with a copepodite abundance of 3.4 ind. m^{-3}. The supposed vector of introduction was mariculture, as samples were collected close to mussel farms.

- *Adriatic Sea*

P. marinus was found during autumn and winter (November to January) in the period 2015–2017 along the coasts of Emilia Romagna (Italy; northern Adriatic Sea) [33]. This NIS was found at low abundance (maximum abundance: 17.1 ind. m^{-3}), at temperatures between 6.8 and 15.4 °C and salinities between 33.3 and 39.9, over a depth range of 8–25 m [33].

In the Croatian part of the eastern Adriatic, *P. marinus* was first detected in 2015 in the Šibenik Bay located in the Krka River estuary. Specimens of *P. marinus* were found in vertical net hauls collected after sunset in the central part of the bay (depth 36 m), which is consistent with the observed vertical distribution pattern of this species in its native environment [34]. Subsequent monitoring (2016–2022) confirmed the establishment of the *P. marinus* population at anchoring sites in the Šibenik port (depth 6 m) regardless of season, with all life stages present. In quick succession, additional occurrences of *P. marinus* were recorded in other estuarine areas of the eastern Adriatic closely associated with shipping activities: in 2018 at a fixed station in the port of Ploče (delta of the Neretva River, depth ca. 10 m) [22] and in 2020 in the northern (cargo) port of Split located in the eastern part of Kaštela Bay (18 m) [35]. In 2021, evidence of the dispersal of *P. marinus* from the original site in the north port of Split to the wider Kaštela Bay area was found. Apparently, all cases of *P. marinus* presence in the eastern Adriatic share some common features: locations in or near harbours, nocturnal presence in the water column, evidence of established populations and no signs of changes in the local zooplankton community. Considering the detection sites, ballast water is likely to be the main vector of introduction, followed by secondary dispersal to nearby areas.

Recently, genetic sequences irrevocably assigned to *P. marinus* were determined in four other Croatian ports using DNA metabarcoding (V4-18S rRNA) [36]. The results of this study confirmed previous detections in the ports of Šibenik, Ploče, and the northern port of Split and revealed new distributions of *P. marinus* spanning from the northern to the southern Croatian Adriatic coast (ports of Pula, Zadar and Dubrovnik and the city port of Split).

The Neretva River is the largest watercourse on the eastern coast of the Adriatic Sea, and the Neretva Delta is one of the most important and fertile agricultural areas in Croatia. Following the first record of *P. marinus* in the delta [22], a further one was tallied in August 2021 during a one-time check of the zooplankton composition in the upper reaches of the Neretva River at the station Opuzen, with two separate vertical hauls from the bottom to the surface using a modified Nansen net with a 125 μm mesh size [37]. Three individuals (two males and one female) were recorded at this time. Additional detections of this copepod were noted during the monitoring of zooplankton at the same station from May 2022 to January 2023: two males and one copepodite on July 28th; two copepodites, one male and one female on August 17th; one male in October; one male on November 3rd. The entire area is subject to constant stress from humans and fluctuations in water balance, as well as the recent introduction of several NIS [38,39].

- *Aegean-Levantine Sea*

P. marinus was reported in Thessaloniki Bay (Greece) in the North Aegean Sea in August, September and October 2021 [40] during monitoring samplings at a shore-based fixed station on the urban sea front. This shallow bay has restricted water circulation. Salinity and nutrient inputs are variable due to the inflows of rivers around the bay and of the Black Sea [41,42]. Additionally, this bay is undergoing anthropogenic eutrophication pressure due to urban and industrial activities. The sampling station is next to the port of Thessaloniki, one of the main Mediterranean ports with a high traffic load and commercial maritime transport [43]; thus, ship ballast waters can be considered the main vector for the introduction of *P. marinus* into the bay [40].

P. marinus was recorded for the first time in İzmir Bay, Aegean Sea, in November 2015, at a station 29 m deep [44]. İzmir Bay is located on the Turkish coast in the eastern part of the Aegean Sea. Several monitoring studies showed that this NIS easily inhabited the İzmir Bay ecosystem and distributed throughout the bay, with preference for the more productive inner and middle parts of İzmir Bay [44]. The trophic structure of the bay is gradually changing from hypertrophic in the inner to oligotrophic in the outer region, with chlorophyll-*a* values of 1.0–25.4 and 0.1–2.6 μg L^{-1} in the middle-inner and outer bay regions, respectively [45]. The inner bay is shallow and heavily influenced by anthropogenic pressures and river inputs. The TCDD İzmir, Alsancak international port, in the inner bay plays an important role in the transport of several alien species, likely including *P. marinus* [44].

In March 2022, *P. marinus* was also found in Yenifoça Bay (Turkey), adjacent to İzmir Bay, where it was collected at only one station 46 m deep [46,47].

A recent finding of *P. marinus* in the southeastern Levantine Sea was reported at the Hadera meteo-marine station (Israel) 26 m deep [48]. Monthly samplings were performed in the framework of the Israeli National Monitoring Programme between September 2019 and December 2021 using vertical hauls of WP2 net (mesh size: 200 μm; diameter: 57 cm), in which *P. marinus* was identified since February 2020. The initial identification was made using DNA metabarcoding (COI mtRNA and 18S V9 rRNA). Following the initial indication, *P. marinus* specimens were found in the corresponding preserved samples. During the sampling period, the annual mean of the water column integrated values of temperature, salinity and chlorophyll-*a* fluctuated between 16.2 and 32.4 °C, 38.3 and 40.4 and 0.05 and 0.92 μg L^{-1}, respectively [48].

In the coastal waters of İskenderun Bay (Turkey; northeastern Levantine Sea), only one female individual of *P. marinus* was found for the first time in March 2022, in the framework of the Integrated Marine Pollution Monitoring Program [49,50]. The Levantine Sea is ultra-oligotrophic, but in İskenderun Bay, productivity is two to four times higher and chlorophyll-*a* is 0.11–2.86 μg L^{-1} [51]. *P. marinus* could have been transported by the longshore current from the Israeli coast into İskenderun Bay or through ship ballast waters.

İskenderun Bay has many national and international harbours and ports that facilitate NIS introductions [52,53].

- *Black Sea*

The first occurrence of *P. marinus* in the Marmara Sea was recorded in August 2020 [54]. Only two males and three copepodite specimens were found near the offshore waters of Hereke city in İzmit Bay (Turkey), over a water column depth of 106 m. İzmit Bay is an extension of the Marmara Sea at its northeastern-most part. It has a two-layered water system: the upper layer is less saline and originates from the Black Sea, whereas the lower layer originates from the highly saline Mediterranean Sea. The highest chlorophyll-*a* values in the Marmara Sea are generally recorded in the east, including in İzmit Bay. The average seasonal chlorophyll-*a* values range from 1.5 to 9.6 mg L^{-1}, with the highest value of 18 mg L^{-1} measured in the middle of the bay [55]. İzmit Bay is highly eutrophic and suffers from intense anthropogenic, industrial and maritime transport pressures. Its ecological status is classified as "bad" in the inner part and "poor" in its outer region [56]. This bay has an important geographic position, and several large and small ports are located in İzmit bay [54]. *P. marinus* could have been introduced to İzmit Bay by shipping or currents originating in the Aegean Sea. The Marmara Sea ecosystem has favourable conditions for the numerical growth of *P. marinus* populations. This species was found just before the massive mucilage event in April 2021 [57]. Indeed, mucilage events lead to covering the bottom and the destruction of benthic habitats [58]. Since *P. marinus* is a hyperbenthic species [21,22], this phenomenon could affect its persistence in the Marmara Sea. Future monitoring programs can test this hypothesis.

3.2. New Records of *Pseudodiaptomus marinus* in ENW

- *Bay of Biscay and the Iberian Coast*

During the zooplankton monitoring campaigns carried out near the Peniche peninsula (western coast of Portugal), at a sheltered coastal area (ca. 15 m depth), a few *P. marinus* specimens (two ovigerous females, one non-ovigerous female and four copepodites) were observed in November and December 2020. Samples were collected during the day by vertical tow from just above the seabed, approximately 10 m deep using a WP2 net (mesh size: 200 μm; mouth diameter: 57 cm) and preserved in a 4% buffered formaldehyde seawater solution.

This region is characterised by a strongly seasonal upwelling regime, especially during the spring–summer months [59]. Since 1981, it was designated as a natural reserve, along with the Berlengas archipelago, and in 1998, it was declared a Marine Protected Area and in 2011, it was declared a Biosphere Reserve World Heritage by UNESCO. The region is also characterised by two important geomorphological structures, Cape Carvoeiro and the Nazare Canyon, with a significant influence on the physical environment and the ecological features of the region. These structures interact with the circulation associated with coastal upwelling to intensify primary production in the ecosystem. The western coast of Portugal is an important traffic maritime route, so it can be inferred that shipping played a role in the introduction of *P. marinus* into the area.

- *Celtic Seas*

In 2022, the UK government commissioned, for the first time, a routine monitoring programme of zooplankton sampling at inshore sites around England, using morning–early evening standard vertical hauls through a WP2 net (mesh size: 200 μm; mouth diameter: 57 cm). The programme started in August 2022 and covered shallow (depth range 10–30 m) inshore sites. In the Celtic Seas region, *P. marinus* was found for the first time along the western coast of England in the Bristol Channel as a rare species in September 2022. Considering the strong anthropic pressure of the system, ballast waters can be assumed as the primary vector of introduction.

- *Greater North Sea*

In the framework of the same UK government monitoring programme discussed above, *P. marinus* was recorded in numerous sites along English east and south coasts. In August 2022, this species was reported for the first time in Southampton, Isle of Wight, mouth of River Blackwater and South Kent. In September 2022, the NIS was recorded in East Kent and East Anglia, while in October, it also appeared in the mouth of the Thames River and in Southeast Yorkshire. In November 2022, *P. marinus* was also spotted in the mouth of the Humber River. Samples contained males, gravid females and juvenile copepodite stages. Most often, the NIS occurred as a rare species, but on occasions it was the dominant or second component of the samples, scoring a maximum abundance of 730 ind. m^{-3}. Samples with the highest abundance were located near significant sources of freshwater input and large ports. Ballast waters can be assumed as the primary vector of introduction, with coastal circulation acting as a possible principal secondary spreading mechanism.

P. marinus was also found as part of a multispecies culture in an artificial lagoon used for the production of flatfish for restocking in the western part of the Limfjord (Denmark) on 15 September 2021. The temperature was 18 °C and the salinity was 29 at the time of the sampling. *P. marinus* was probably introduced to the lagoons by the inlet of water with a North Sea origin. The lagoon was kept for the next year's production, but *P. marinus* apparently did not survive during winter.

Two specimens of *P. marinus* (a female and a copepodite) were found offshore Hirtshals (Denmark), at an 18 m deep site, on 11 November 2022, in the frame of a national NOVANA programme with monthly samplings. The sampling location, on the mid-eastern North Sea coast, is characterized by a permanently thermally mixed water column. *P. marinus* occurred at salinities from 31.0 to 32.5, at temperatures from 12.2 to 12.8 °C, in fully oxygenated waters. The coastline has great importance as a nursery and spawning ground for fish, shrimp and other larger crustaceans.

- *Western Mediterranean Sea*

P. marinus was recorded for the first time in the pelagic waters off the Tuscany coast, 12 nautical miles from Leghorn harbour (Italy; southern Ligurian Sea), in November 2020, at a station 100 m deep. Only one female and two copepodite specimens were found, collected with night-time horizontal surface sampling (0–5 m). Subsequent monitoring in the same area, conducted seasonally, resulted in the collection of three ovigerous female specimens the following autumn, in November 2021, by night-time vertical sampling from 50 m depth to the surface. The zooplankton samples were taken with a modified WP2 net (mesh size: 300 μm; mouth diameter: 60 cm). The investigated sector, located off the Italian coast, on the border between the northern Tyrrhenian and the Ligurian Seas, is characterized by the large extension of the continental shelf and the limited depth (100 m), even at a considerable distance from the coast (18 miles) [60]. The coastal area of the sampling sites is strongly influenced by anthropogenic pressure, with both tourist and commercial activities. The Leghorn harbour is one of the main commercial ports in the western Mediterranean Sea; consequently, the introduction of *P. marinus* into the area was likely mediated by ballast waters.

In the Gulf of Pozzuoli (Italy; central Tyrrhenian Sea), several *P. marinus* individuals (females, males and copepodites) were found on 19 July 2018 in the framework of the ABBaCo Project. The population abundance was 22.2 ind. m^{-3} at a coastal station situated near Nisida island mussel farms (at the boundary between the Gulf of Pozzuoli and the Gulf of Naples), while the NIS ranked as a rare species at a site inside the Gulf of Pozzuoli; its first record in this area was dated to 2018 rather than 2019 [32].

In the innermost southeasternmost part of the Gulf of Naples (Italy; central Tyrrhenian Sea), *P. marinus* was found as a rare species (one female) in July 2020 in the framework of the NEREA project. The sampling area is affected by the presence of the Sarno River, which carries high concentrations of inorganic and organic pollutants and is considered one of the most polluted rivers in Europe [61]. It is to be noted that in previous samplings in the same area (2002–2005 and 2007–2009), this species was not recorded. At present, it is difficult to formulate hypotheses on the introduction pathway of *P. marinus* in this area, but a realistic possibility is the arrival through secondary spread (water currents, attachment to hull fouling) or from neighbouring introduced areas (e.g., LTER_EU_IT_061).

In April 2021, ten adult females and one male were also found off Torre del Greco, a locality positioned south-east from LTER_EU_IT_061 station in the Gulf of Naples. Subsequently, in June 2021, during the sampling activities of the PO FEAMP project, one female was found as a rare species (0.14 ind. m^{-3}) at a station between Torre del Greco and Torre Annunziata (further south-east from Torre del Greco) in the 0–25 m layer. As for the Gulf of Pozzuoli, in these two cases, the samples were also collected near mussel farms, supporting the introduction through shellfish culture.

The Gulf of Salerno (Italy; central-southern Tyrrhenian Sea) presents an average depth similar to that of the Gulf of Naples (260 and 170 m, respectively), but it is more open to oligotrophic waters from the Tyrrhenian Sea while the human impact is less pronounced [62]. One *P. marinus* female was found in September 2019 in the framework of a monitoring program in the 0–50 m layer at an offshore station during the discharge of sediments dredged from the nearby port of Salerno, which may be assumed as the origin point. Samples collected in the port by vertical hauls did not reveal any presence of *P. marinus*. It is thus reasonable to assume that the species was collected from the bottom, where it stays during the morning, in a layer not sampled by the plankton net.

- *Ionian Sea and the Central Mediterranean Sea*

Several specimens (ovigerous female, male and copepodites) of *P. marinus* were collected in front of Marina di Ragusa (Sicily, Italy), a coastal area facing the Malta Channel. Specimens were collected in November 2022 by daytime vertical and horizontal hauls carried out near the coast, over a water column that ranged from 3 to 9.5 m in depth. In this area, *P. marinus* reaches a maximum abundance of 3.5 ind. m^{-3} at a depth of 4.5 m. The Malta–Sicily Channel is part of the Sicily Channel system, where waters and thermohaline properties between the eastern and western Mediterranean basins mix together. Topographically, the extensive continental shelf between the Sicilian coast and the island of Malta is characterized by a plateau with an average depth of 150 m [63]. The monitored stretch of coastline is characterized by a high intensity of industrial settlements, mainly in the field of hydrocarbon supply and processing. For this reason, it can be hypothesized that the introduction of *P. marinus* was mediated by commercial maritime traffic, particularly through ballast water from ships.

- *Adriatic Sea*

In the Gulf of Trieste (Italy; northern Adriatic Sea), *P. marinus* has been recorded since 2009, when it was found for the first time in the harbour of Monfalcone [20], probably introduced by ballast waters or as a consequence of aquaculture activities. Since then, it was observed in several samples collected at sampling stations in the central part of the gulf, near the harbour of Trieste and at the coastal LTER_EU_IT_056 station [22]. As in samples collected from January 2006 to December 2017 with WP2 vertical nets (mesh size: 200 μm; mouth diameter: 57 cm) [22], *P. marinus* was also present year-round during January 2018–December 2022, mainly as copepodites that were frequently found in summer and autumn. The abundance of *P. marinus* in this last period never reached the maximal values previously observed (172.6 ind. m^{-3}), with the highest values of 15.4 and 15.1 ind. m^{-3} in September 2020 and July 2021, respectively.

Recently, *P. marinus* has also been spreading in the neighbouring Marano and Grado Lagoon, which, together with the Venice Lagoon, is one of the two most important coastal systems in northeastern Italy, located between the Friuli Venezia Giulia lowlands and the northern Adriatic Sea. The Marano and Grado Lagoon is a shallow basin (average depth 1 m) bounded by the Tagliamento River to the west and by the Isonzo River to the east. It extends parallel to the northernmost coast of the Adriatic Sea for a length of about 32 km and covers an area of 160 km^2. This basin plays an important role for fishery, fish and shellfish farming, and it has been designated as a site of the Natura 2000 network. Moreover, it is a Special Area of Conservation (SAC—IT3320037) and a Special Protection Area (SPA—IT3320037). The northernmost part of the basin hosts an industrial area along with the harbour of Porto Nogaro, an international commercial port with a total cargo handled from 2012 to 2022 ranging from around 900,000 to 1,500,000 t y^{-1} [64] accessible through a navigable canal crossing the lagoon.

Zooplankton surveys in the Marano and Grado Lagoon began in March 2019 and were conducted monthly from January 2020 to November 2021, interrupted only in March and April 2020 due to the COVID-19 pandemic. Samplings were performed during the daytime with a Bongo net equipped with a 200 μm net, and floats were attached to support the net collectors. The net was towed at low speed (<1 m s^{-1}) for about five minutes, allowing an average of 6,000 L of water to be filtered per tow. *P. marinus* was found in 49 of the 164 samples analysed, in a wide range of temperature (5.0–30.0 °C) and salinity (4.7–35.4) values. Although *P. marinus* was not always observed at all sites, it was present in almost every sampling, with abundances ranging from 0.1 to 19.8 ind. m^{-3} in February 2021 and July 2019, respectively. In addition to net zooplankton surveys, in spring and autumn 2021, water samples (5 L at each station) for eDNA analyses were collected at 16 stations and filtered through 1.2 μm PES membrane filters (PALL Laboratory, Port Washington, NY, USA). eDNA metabarcoding targeting the mitochondrial cytochrome-c-oxidase I gene (COI) using mlCOIintF and jgHCO2198 primers [65,66] allowed for the first time the detection of *P. marinus* in the lagoon (one site on 27 September 2021) with this approach. The introduction of *P. marinus* in the lagoon was probably associated with the extensive aquaculture activities in the area (as hypothesised also to explain its presence in the artificial channel near the harbour of Monfalcone [20]), but arrival by ballast waters cannot be ruled out.

- *Aegean-Levantine Sea*

Thessaloniki Bay (Greece) is the upper part of the Thermaikos Gulf, which is a marine ecosystem with high complexity that can be divided into three parts: Thessaloniki Bay, the inner Thermaikos Gulf and the outer Thermaikos Gulf. The latter is connected to the Aegean Sea with water exchange taking place and is also influenced by the inflow of the Black Sea waters [67]. Following the first record in Thessaloniki Bay in 2021 [40], an analysis of earlier collected samples from the outer Thermaikos Gulf in 2016 recorded the presence of *P. marinus*. One adult female, one adult male and two copepodites were identified. The adult female was found in a surface sample collected at 3 m depth, while the male and copepodite specimens were found in samples collected from deeper layers (45.5 and 64.5 m, respectively) with temperature ranging from 14.7 to 27.7 °C. These specimens collected in 2016 could have arrived in the Thermaikos Gulf either through ballast waters or through the coastal circulation of waters from the Aegean and the Black Sea, where *P. marinus* was recorded for the first time in 2016 [68]. These new findings bring forward five years the arrival of *P. marinus* in the basin, from 2021 [40] to 2016.

Table 1 summarises the temperature and salinity values (or ranges, depending on the site) associated with the *P. marinus* records in the survey literature and in the original data presented here.

Table 1. Temperature (°C) and salinity values (or ranges, depending on data availability) associated with the occurrence of *Pseudodiaptomus marinus* in European and neighbouring waters, as derived from the literature survey and the original data presented in this work.

Site	Temperature	Salinity
Bay of Biscay and the Iberian Coast		
Peniche (PT)	15.5–17.1 °C	35.0–35.1
Celtic Seas		
Bristol Channel (UK)	n.a.	n.a.
Greater North Sea		
East Anglia (UK)	n.a.	n.a.
East Kent (UK)	n.a.	n.a.
Isle of Wight (UK)	n.a.	n.a.
River Blackwater (UK)	n.a.	n.a.
River Humber (UK)	n.a.	n.a.
River Thames (UK)	n.a.	n.a.
Southampton (UK)	n.a.	n.a.
South Kent (UK)	n.a.	n.a.
Southeast Yorkshire (UK)	n.a.	n.a.
The Eastern Scheldt estuary (NL) [30]	~21 °C	~31
Limfjord (DK)	18 °C	29
Hirtshals (DK)	12.2–12.8 °C	31.0–32.5
Western Mediterranean Sea		
Leghorn (IT)—Nov 2020 §1	17.7 °C	38.3
—Nov 2021 §2	17.3 °C	38.1
Gulf of Naples—Sarno River (IT) §3	15.3 °C	37.6
Gulf of Naples—Torre del Greco (IT) §4	14.9 °C	37.9
Gulf of Naples—Between Torre del Greco and Torre Annunziata (IT)	17.4–28.1 °C	36.9–37.9
Gulf of Pozzuoli (IT)	15.2–27.0 °C	37.6–38.0
Gulf of Pozzuoli (IT) [32]	17.4–27.3 °C	37.5–37.7
Gulf of Salerno (IT) §5	16.6 °C	38.5
Ionian Sea and the Central Mediterranean Sea		
Marina di Ragusa (IT) §6	17.3 °C	38.7
Adriatic Sea		
Emilia Romagna coasts (IT) [33]	6.8–15.4 °C	33.3–39.9
Gulf of Trieste (IT)	9.6–25.9 °C	34.3–38.5
Marano and Grado Lagoon (IT)	5.0–30.0 °C	4.7–35.4
Croatian ports (HR) [35,36]	n.a.	n.a.
Neretva River (HR) [37]	10.6–25.5 °C	0.0–38.4
Aegean-Levantine Sea		
Thessaloniki Bay (GR) [40]	17.2–31.0 °C	34.8–38.5
Thermaikos Gulf (GR)	14.7–27.7 °C	34.9–38.5
İskenderun Bay (TR) [49,50]	15.2–16.1 °C	38.5–38.8
İzmir Bay (TR) # [44]	19 °C	38.9
Yenifoça Bay (TR) [46,47]	14.6–15.1 °C	39.2
Hadera monitoring station (IL) [48]	16.2–32.4 °C	38.3–40.4
Black Sea		
İzmit Bay (TR) [54]	15.6–23.2 °C	29.5–38.7

Vertically integrated values: §1 0–5 m; §2 0–50 m; §3 0–10 m; §4 0–90 m; §5 0–50 m; §6 0–4.5 m. # surface values.

4. Discussion

According to a recent census [69], 874 NIS have been recorded in strictly European marine waters as of 2020, at a rate of 21 new introductions per year over the 2012–2017 period. Their distribution is uneven, with the highest number of aliens found in Italy, France, Spain and Greece, probably due to several factors including increased monitoring efforts and the density of gateways and pathways [29]. Among the species more capable of establishing in a highly diversified range of environments is the calanoid copepod *Pseudodiaptomus marinus*. In a few years after its first spotting in the Adriatic Sea in

2007 [20], this species has recorded a fast spreading in European and neighbouring waters, with a process that is still ongoing. As reviewed in the present contribution, compared to a previous snapshot [22], the distribution of *P. marinus* has further expanded in different sectors of ENW, likely due to either new introductions or secondary spreading from already introduced environments. The presence of *P. marinus* is now verified in eight out of the ten MSFD subregions, now including the Celtic Seas compared to [22]. This spread is particularly impressive considering that in [29] the occurrence of *P. marinus* was verified only in four subregions (Western Mediterranean, Ionian Sea and the Central Mediterranean Sea, Adriatic Sea, Greater North Sea). The continuous monitoring of NIS occurrence is fundamental to assess the real-time evolution of its spreading [70]. Such activity would also crucially benefit from an increase in the use of molecular tools [3,29]. Indeed, metabarcoding studies on plankton communities are contributing to the early detection of non-indigenous and even rare species [71]. With specific reference to *P. marinus* from ENW, over the time window investigated in the present work, new sequences have been made available from the area of interest [31,48,54,72], adding to previous molecular studies [25,73–77]. The increase in new validated reference sequences for this species (e.g., [48,72]) is fundamental for a proper molecular identification of *P. marinus* using metabarcoding approaches (eDNA and/or organismal DNA) (e.g., [31,48]), as well as in the investigation of the genetic connections of geographically distant populations (e.g., [76]).

As discussed in [22], *P. marinus* can exploit different vectors of introduction in new environments, e.g., ballast waters and aquaculture/mariculture, configuring as a polyvectic species [78]. At a regional scale, the secondary arrival of this NIS can be favoured by the local current regimes [22]. For example, the spread of *P. marinus* in the North Sea might have been supported by the mostly cyclonic circulation of the basin [79]. The first observations of this NIS in Calais harbour (France) in 2010 [26] were followed by the first occurrences of *P. marinus* in the Continuous Plankton Recorder samples in 2011 [80] and then further east in subsequent years ([22,30,75,81,82], present study). Recently, it was suggested that species inhabiting the Adriatic Sea could be transported to the Turkish Levantine coast by the Bimodal Oscillation System (BiOS) [83]. This process leads to switching the circulation patterns of the North Ionian Gyre (NIG) between cyclonic and anticyclonic on decadal intervals [84]. BiOS can affect the thermohaline properties in the southern Adriatic Sea and also affect the Levantine Surface Water (LSW) and Levantine Intermediate Water (LIW) via the Ionian Jet current [85]. Therefore, it is hypothesised that *P. marinus* could have arrived in the Köprüçay estuary by the BiOS mechanism or even more likely by ship ballast waters. More research on this topic is needed, likely including the use of molecular tools to reconstruct the phylogeographic connections of the two populations. The original data presented here also identify another possible introduction pathway, i.e., the discharge of sediments dredged in areas where the copepod has already established, as proposed for the Gulf of Salerno (see Section 3.2).

The data presented in this work (both from the literature and original) confirm the ability of *P. marinus* to establish in a variety of environments, as discussed in previous reviews [21,22,26]. This species can establish in coastal areas as well as in estuaries and coastal lagoons, in a wide range of temperatures and salinities. Species of the genus *Pseudodiaptomus* are considered rare, if not completely absent, below the 10 m isobath [86]. The data included here verify the ability of *P. marinus* to establish at sites where the bottom depth is deeper than this threshold, as shown also in [22]. Currently available information also suggests no latitudinal preference in this NIS.

Compared to distributions obtained by model simulations accounting for the species net reproductive rate as a function of water temperature [87,88], *P. marinus* has clearly established in theoretically non-invadable regions, confirming its settlement in areas potentially unsuitable (southern North Sea, Levantine Basin, Black Sea) ([22,75,81,82], present study). It is worth underlining that in [21], the absence (as of 2015) of *P. marinus* in the eastern Mediterranean Sea was supposed to be related to a hypothetical haline restriction of this copepod to salinities < 38.5. Subsequent salinity tolerance experiments [23], however, reported an upper limit to values up to 44, and the records of *P. marinus* in the Levantine Basin [48,49] validate its ability to spread outside its theoretical ecological and geographical boundaries, pointing to species-specific physiological traits supporting a plastic adaptation to a wide range of abiotic conditions [22,23].

The present distribution of *P. marinus* may actually be underestimated. Owing to its epibenthic behaviour, the chances of collecting enough specimens during day-time samplings are reduced, especially considering that vertical tows never allow sampling near the bottom. To date, the introduction of *P. marinus* has not had negative impacts on the pelagic communities of the receiving environments, with the only exception of the Agua Hedionda Lagoon (Agua Hedionda Lagoon, CA, USA) [86], although no information is presently available for the benthic ecosystems where this NIS lives during the morning [22]. In light of this, night-time samplings could be of great help to more efficiently reveal the real occurrence of this species.

As pointed out by [3], the majority of zooplankton invasions are reported for species with documented severe ecological impacts in urbanised and commercially important areas, likely more intensively monitored by different research institutions. Efforts should be devoted to the collection of samples from regions with limited coverage, also resorting to the possibility of using metabarcoding approaches to detect NIS, including *P. marinus* that typically occurs in few numbers and is therefore often rare in zooplankton samples.

The reports available allow us to make tentative hypotheses on the future distribution of *P. marinus*. It may be very likely that this NIS will appear along the North African coasts (where it has been reported so far only in the Gulf of Gabès [22]) and at more sites in the Black Sea, in the Aegean Sea, along the Atlantic coastline and in the English Channel, where this species is already introduced. The occurrence of *P. marinus* along the Danish coast in the southern North Sea suggests a potential future introduction, as well as in the Baltic Sea, a basin where biological invasions have already occurred [89], also including zooplanktonic organisms [90]. The temperature and salinity conditions of the Baltic Sea may be compliant with the tolerance limits of *P. marinus* [23], although to date no record of this species has occurred.

Over the last twenty years, an increase in non-indigenous zooplankton organisms has been evidenced, owing to an increase in awareness and in the number of invaders [3]. Based on its great spreading capacities, *P. marinus* is a target species to improve our knowledge of the mechanisms favouring the introduction of NIS in receiving environments worldwide.

Supplementary Materials: The following supporting information can be downloaded at: https://www.mdpi.com/article/10.3390/jmse11061238/s1, literature survey of *P. marinus* distribution in ENW; published and original *P. marinus* distribution data in ENW.

Author Contributions: Conceptualization, all authors; methodology, all authors; data curation, all authors; writing—original draft preparation, all authors; writing—review and editing, all authors; visualization, all authors. All authors have read and agreed to the published version of the manuscript.

Funding: M.U., E.B., I.D.C., A.G. and V.T. acknowledge the support of NBFC to Stazione Zoologica Anton Dohrn and OGS, funded by the Italian Ministry of University and Research, PNRR, Missione 4 Componente 2, "Dalla ricerca all'impresa", Investimento 1.4, Project CN00000033 (CUP: CN00000033 and F83B22000050001). JV was partly supported by the Project PO FEAMP 2014/2020 (Misura 2.51), funded by Regione Campania (Italy). O.V. acknowledges the support of the Croatian Ministry of Economy and Sustainable Development for funding the monitoring of *P. marinus* in the framework of MSFD Descriptor 2. VT, A.d.O. and A.G. acknowledge ARPA-FVG for funding the monitoring of *P. marinus* in the framework of the Italian MSFD Descriptor 1 and 2. Activities in the Lagoon of Marano and Grado (Italy) were partially funded by Regione Autonoma Friuli Venezia Giulia with the projects NOCE di MARE (legge regionale 30 marzo 2018, n°14, art 2, commi 51-55 e legge regionale 30 dicembre 2020, n°26, art. 4, commi 33-34, CUP F96C18000240002 and CUP: D29J21000900002) and Project ARGOS—ShARed GOvernance of Sustainable fisheries and aquaculture activities as leverage to protect marine resources in the Adriatic Sea (Interreg V-A Italy-Croatia CBC Programme 2014-2020, Project ID 10255153). SCM acknowledges MARE (UIDB/04292/2020 +UIDP/04292/2020 + LA/P/0069/2020) through FCT/MEC national funds, and the cofounding by the FEDER, within the PT2020 Partnership Agreement and Compete 2020. M.W. acknowledges support from Defra for funding the collection of inshore zooplankton through the Marine Natural Capital Ecosystem Assessment Programme. T.G.-H., A.R.M. and X.V. acknowledge the support of the National Monitoring Programme of the Israeli Mediterranean Sea by the IOLR. M.B. thanks Poliservizi for providing water column data for the Ragusa site.

Institutional Review Board Statement: This study did not require ethical approval.

Data Availability Statement: The data presented in this study are available as Supplementary Material.

Acknowledgments: The authors thank the WGEUROBUS and WGIMT of the International Council for the Exploration of the Sea (ICES) for facilitating this research. The authors also thank Okko Outinen and ICES WGBOSV for insightful discussion. O.V. thanks colleagues A. Marasović and D. Udovičić for the field sampling of *P. marinus* and CTD data measurements and T. Damjanović for assistance with the analyses of the samples. A.d.O., A.G., E.B. and V.T. thank colleagues of OGS and ARPA FVG for the help in sampling at the LTER site and in the Lagoon of Marano and Grado, respectively.

Conflicts of Interest: The authors declare no conflict of interest.

References

1. Geburzi, J.C.; McCarthy, M.L. How do they do it?—Understanding the success of marine invasive species. In *YOUMARES 8—Oceans Across Boundaries: Learning from Each Other*; Jungblut, S., Liebich, V., Bode, M., Eds.; Springer: Cham, Switzerland, 2018; pp. 109–124.
2. Simberloff, D. Non-native invasive species and novel ecosystems. *F1000Prime Rep.* **2015**, *7*, 47. [CrossRef]
3. Dexter, E.; Bollens, S.M. Zooplankton invasions in the early 21st century: A global survey of recent studies and recommendations for future research. *Hydrobiologia* **2020**, *847*, 309–319. [CrossRef] [PubMed]
4. Lee, C.E. Evolutionary mechanisms of habitat invasions, using the copepod *Eurytemora affinis* as a model system. *Evol. Appl.* **2016**, *9*, 248–270. [CrossRef] [PubMed]
5. Gollasch, S.; Lenz, J.; Dammer, M.; Andres, H.-G. Survival of tropical ballast water organisms during a cruise from the Indian Ocean to the North Sea. *J. Plankton Res.* **2000**, *22*, 923–937. [CrossRef]
6. Choi, K.H.; Kimmerer, W.; Smith, G.; Ruiz, G.M.; Lion, K. Post-exchange zooplankton in ballast water of ships entering the San Francisco Estuary. *J. Plankton Res.* **2005**, *27*, 707–714. [CrossRef]
7. Cabrini, M.; Cerino, F.; de Olazabal, A.; Di Poi, E.; Fabbro, C.; Fornasaro, D.; Goruppi, A.; Flander-Putrle, V.; France, J.; Gollasch, S.; et al. Potential transfer of aquatic organisms via ballast water with a particular focus on harmful and non-indigenous species: A survey from Adriatic ports. *Mar. Poll. Bull.* **2019**, *147*, 16–35. [CrossRef] [PubMed]
8. Velasquez, X.; Morov, A.R.; Terbıyık Kurt, T.; Meron, D.; Guy-Haim, T. Two-way bioinvasion: Tracking the neritic non-native cyclopoid copepods *Dioithona oculata* and *Oithona davisae* (Oithonidae) in the Eastern Mediterranean Sea. *Mediterr. Mar. Sci.* **2021**, *22*, 586–602. [CrossRef]
9. Zagami, G.; Brugnano, C.; Granata, A.; Guglielmo, L.; Minutoli, R.; Aloise, A. Biogeographical distribution and ecology of the planktonic copepod *Oithona davisae*: Rapid invasion in Lakes Faro and Ganzirri (Central Mediterranean Sea). In *Trends in Copepod Studies—Distribution, Biology and Ecology*; Uttieri, M., Ed.; Nova Science Publishers, Inc.: New York, NY, USA, 2018; pp. 59–82.
10. Feis, M.E.; Goedknegt, M.A.; Arzul, I.; Chenuil, A.; den Boon, O.; Gottschalck, L.; Kondo, Y.; Ohtsuka, S.; Shama, L.N.S.; Thieltges, D.W.; et al. Global invasion genetics of two parasitic copepods infecting marine bivalves. *Sci. Rep.* **2019**, *9*, 12730. [CrossRef]
11. Terbıyık Kurt, T.; Beşiktepe, Ş. First distribution record of the invasive copepod *Oithona davisae* Ferrari and Orsi, 1984, in the coastal waters of the Aegean Sea. *Mar. Ecol.* **2019**, *40*, e12548. [CrossRef]
12. Pansera, M.; Camatti, E.; Schroeder, A.; Zagami, G.; Bergamasco, A. The non-indigenous *Oithona davisae* in a Mediterranean transitional environment: Coexistence patterns with competing species. *Sci. Rep.* **2021**, *11*, 8341. [CrossRef]
13. Dragičević, B.; Anadoli, O.; Angel, D.; Benabdi, M.; Bitar, G.; Castriota, L.; Crocetta, F.; Deidun, A.; Dulčić, J.; Edelist, D.; et al. New Mediterranean biodiversity records (December 2019). *Mediterr. Mar. Sci.* **2019**, *20*, 645–656. [CrossRef]
14. Bollens, S.M.; Breckenridge, J.K.; Cordell, J.R.; Rollwagen-Bollens, G.; Kalata, O. Invasive copepods in the Lower Columbia River Estuary: Seasonal abundance, co-occurrence and potential competition with native copepods. *Aquat. Invasions* **2012**, *7*, 101–109. [CrossRef]
15. Bouley, P.; Kimmerer, W.J. Ecology of a highly abundant, introduced cyclopoid copepod in a temperate estuary. *Mar. Ecol. Prog. Ser.* **2006**, *324*, 219–228. [CrossRef]
16. Barroeta, Z.; Villate, F.; Uriarte, I.; Iriarte, A. Impact of colonizer copepods on zooplankton structure and diversity in contrasting estuaries. *Estuaries Coasts* **2022**, *45*, 2592–2609. [CrossRef]
17. Camatti, E.; Pansera, M.; Bergamasco, A. The copepod *Acartia tonsa* Dana in a microtidal Mediterranean lagoon: History of a successful invasion. *Water* **2019**, *11*, 1200. [CrossRef]

18. Cordell, J.R.; Rasmussen, M.; Bollens, S.M. Biology of the introduced copepod *Pseudodiaptomus inopinus* in a northeastern Pacific estuary. *Mar. Ecol. Prog. Ser.* **2007**, *333*, 213–227. [CrossRef]
19. Adams, J.B.; Bollens, S.M.; Bishop, J.G. Predation on the invasive copepod, *Pseudodiaptomus forbesi*, and native zooplankton in the lower Columbia River: An experimental approach to quantify differences in prey-specific feeding rates. *PLoS ONE* **2015**, *10*, e0144095. [CrossRef] [PubMed]
20. de Olazabal, A.; Tirelli, V. First record of the egg-carrying calanoid copepod *Pseudodiaptomus marinus* in the Adriatic Sea. *Mar. Biodivers. Rec.* **2011**, *4*, e85. [CrossRef]
21. Sabia, L.; Zagami, G.; Mazzocchi, M.G.; Zambianchi, E.; Uttieri, M. Spreading factors of a globally invading coastal copepod. *Mediterr. Mar. Sci.* **2015**, *16*, 460–471. [CrossRef]
22. Uttieri, M.; Aguzzi, L.; Aiese Cigliano, R.; Amato, A.; Bojanić, N.; Brunetta, M.; Camatti, E.; Carotenuto, Y.; Damjanović, T.; Delpy, F.; et al. WGEUROBUS—Working Group "Towards a EURopean OBservatory of the non-indigenous calanoid copepod *Pseudodiaptomus marinus*". *Biol. Invasions* **2020**, *22*, 885–906. [CrossRef]
23. Svetlichny, L.; Hubareva, E.; Khanaychenko, A.; Uttieri, M. Response to salinity and temperature changes in the alien Asian copepod *Pseudodiaptomus marinus* introduced in the Black Sea. *J. Exp. Zool. A* **2019**, *331*, 416–426. [CrossRef] [PubMed]
24. Sabia, L.; Uttieri, M.; Schmitt, F.G.; Zagami, G.; Zambianchi, E.; Souissi, S. *Pseudodiaptomus marinus* Sato, 1913, a new invasive copepod in Lake Faro (Sicily): Observations on the swimming behaviour and the sex-dependent responses to food. *Zool. Stud.* **2014**, *53*, 49. [CrossRef]
25. Sabia, L.; Di Capua, I.; Percopo, I.; Uttieri, M.; Amato, A. ITS2 in calanoid copepods: Reconstructing phylogenetic relationships and identifying a newly introduced species in the Mediterranean. *Eur. Zool. J.* **2017**, *84*, 104–115. [CrossRef]
26. Brylinski, J.M.; Antajan, E.; Raud, T.; Vincent, D. First record of the Asian copepod *Pseudodiaptomus marinus* Sato, 1913 (Copepoda: Calanoida: Pseudodiaptomidae) in the southern bight of the North Sea along the coast of France. *Aquat. Invasions* **2012**, *7*, 577–584. [CrossRef]
27. European Environment Agency. Marine Regions and Subregions. Available online: https://www.eea.europa.eu/data-and-maps/figures/marine-regions-and-subregions (accessed on 19 April 2023).
28. Galanidi, M.; Zenetos, A. Data-driven recommendations for establishing threshold values for the NIS trend indicator in the Mediterranean Sea. *Diversity* **2022**, *14*, 57. [CrossRef]
29. Tsiamis, K.; Palialexis, A.; Stefanova, K.; Gladan, Ž.N.; Skejić, S.; Despalatović, M.; Cvitković, I.; Dragičević, B.; Dulčić, J.; Vidjak, O.; et al. Non-indigenous species refined national baseline inventories: A synthesis in the context of the European Union's Marine Strategy Framework Directive. *Mar. Pollut. Bull.* **2019**, *145*, 429–435. [CrossRef]
30. Horn, H.G.; van Rijswijk, P.; Soetaert, K.; van Oevelen, D. Drivers of spatial and temporal micro- and mesozooplankton dynamics in an estuary under strong anthropogenic influences (the Eastern Scheldt, Netherlands). *J. Sea Res.* **2023**, *192*, 102357. [CrossRef]
31. Di Capua, I.; Piredda, R.; Mazzocchi, M.G.; Zingone, A. Metazoan diversity and seasonality through eDNA metabarcoding at a Mediterranean long-term ecological research site. *ICES J. Mar. Sci.* **2021**, *78*, 3303–3316. [CrossRef]
32. Margiotta, F.; Balestra, C.; Buondonno, A.; Casotti, R.; D'Ambra, I.; Di Capua, I.; Gallia, R.; Mazzocchi, M.G.; Merquiol, L.; Pepi, M.; et al. Do plankton reflect the environmental quality status? The case of a post-industrial Mediterranean Bay. *Mar. Environ. Res.* **2020**, *160*, 104980. [CrossRef]
33. Fiori, E.; Benzi, M.; Ferrari, C.R.; Mazziotti, C. Zooplankton community structure before and after *Mnemiopsis leidyi* arrival. *J. Plankton Res.* **2019**, *41*, 803–820. [CrossRef]
34. Itoh, H.; Tcachibana, A.; Nomura, H.; Tanaka, Y.; Furota, T.; Ishimaru, T. Vertical distribution of planktonic copepods in Tokyo Bay in summer. *Plankton Benthos Res.* **2011**, *6*, 129–134. [CrossRef]
35. Vidjak, O.; Damjanović, T.; Rožić, S.; Šegvić Bubić, T.; Bojanić, N.; Hrabar, J.; Arapov, J.; Skračić, M. Spreading of the non-indigenous Indo-Pacific copepod *Pseudodiaptomus marinus* Sato, 1913 in eastern Adriatic coastal and transitional waters. In *11th International Conference on Biological Invasions. The Human Role in Biological Invasions—A Case of Dr Jekyll and Mr Hyde?* Jelaska, S.D., Ed.; Croatian Ecological Society: Zagreb, Croatia, 2020; p. 95.
36. Lin, Y.; Vidjak, O.; Ezgeta-Balić, D.; Bojanić Varezić, D.; Šegvić-Bubić, T.; Stagličić, N.; Zhan, A.; Briski, E. Plankton diversity in Anthropocene: Shipping vs. aquaculture along the eastern Adriatic coast assessed through DNA metabarcoding. *Sci. Total Environ.* **2022**, *807*, 151043. [CrossRef]
37. Lučić, D.; Onofri, I.; Garić, R.; Violić, I.; Vranješ, M.; Gangai Zovko, B.; Jurinović, J.; Njire, J.; Hure, M. Ingression of the hydromedusa *Neotima lucullana* (Delle Chiaje, 1822) into the ecosystem of the Neretva river estuary (south-eastern Adriatic, Croatia). *Acta Adriat.* **2022**, *63*, 165–174. [CrossRef]
38. Glamuzina, B.; Tutman, P.; Glamuzina, L.; Vidović, Z.; Simonović, P.; Vilizzi, L. Quantifying current and future risks of invasiveness of non-native aquatic species in highly urbanised estuarine ecosystems—A case study of the River Neretva Estuary (Eastern Adriatic Sea: Croatia and Bosnia–Herzegovina). *Fish. Manag. Ecol.* **2021**, *28*, 138–146. [CrossRef]
39. Njire, J.; Bojanić, N.; Lučić, D.; Violić, I. First record of the alien tintinnid ciliate *Rhizodomus tagatzi* Strelkow and Wirketis 1950 in the Adriatic Sea. *Water* **2023**, *15*, 1821. [CrossRef]
40. Kourkoutmani, P.; Michaloudi, E. First record of the calanoid copepod *Pseudodiaptomus marinus* Sato, 1913 in the North Aegean Sea, in Thessaloniki Bay, Greece. *BioInvasions Rec.* **2022**, *11*, 738–746. [CrossRef]
41. Krestenitis, Y.N.; Kombiadou, K.D.; Androulidakis, Y.S. Interannual variability of the physical characteristics of North Thermaikos Gulf (NW Aegean Sea). *J. Mar. Syst.* **2012**, *96–97*, 132–151. [CrossRef]

42. Hyder, P.; Simpson, J.H.; Christopoulos, S.; Krestenitis, Y. The seasonal cycles of stratification and circulation in the Thermaikos Gulf Region of Freshwater Influence (ROFI), north-west Aegean. *Cont. Shelf Res.* **2002**, *22*, 2573–2597. [CrossRef]
43. Angelidis, A. *Fulvia fragilis* (Forsskal in Niebuhr, 1775) (Bivalvia: Cardiidae), first record of an alien mollusk in the Gulf of Thessaloniki (Inner Thermaikos Gulf, North Aegean Sea, Greece). *J. Biol. Res.* **2013**, *20*, 228–232.
44. Besiktepe, S.; Terbıyık Kurt, T.; Gubanova, A. Mesozooplankton composition and distribution in İzmir Bay, Aegean Sea: With special emphasis on copepods. *Reg. Stud. Mar. Sci.* **2022**, *55*, 102567. [CrossRef]
45. Alyuruk, H.; Kontas, A. Seasonal variations and distributions of dissolved free and total carbohydrates at the İzmir Bay, Aegean Sea. *Acta Oceanol. Sin.* **2018**, *37*, 6–14. [CrossRef]
46. Terbıyık Kurt, T.; Beşiktepe, S.; Velasquez, X.; Guy-Haim, T. *New Record of Pseudodiaptomus marinus in the Yenifoça Bay (Aegean Sea)*; Department of Marine Biology, Faculty of Fisheries, Çukurova University: Adana, Turkey, *in prep*.
47. TÜBİTAK-MRC and İzmir Metropolitan Municipality. *Monitoring of İzmir Bay Water Quality and Terrestrial Inputs and Developing Recommendations for the Pollution Prevention (İZİZ)*; TÜBİTAK-MAM ve İzmir Metropolitan Municipality: Kocaeli, Turkey, 2023; p. 522T203.
48. Guy-Haim, T.; Velasquez, X.; Terbiyik-Kurt, T.; Di Capua, I.; Mazzocchi, M.G.; Morov, A.R. A new record of the rapidly spreading calanoid copepod *Pseudodiaptomus marinus* (Sato, 1913) in the Levantine Sea using multi-marker metabarcoding. *BioInvasions Rec.* **2022**, *11*, 964–976. [CrossRef]
49. Terbıyık Kurt, T.; Beşiktepe, Ş.; Velasquez, X.; Guy-Haim, T. *Recent Introduction of Non-Indigeneous Copepods Species Pseudodiaptomus marinus in the İskenderun Bay (Northeastern Mediterranean)*; Department of Marine Biology, Faculty of Fisheries, Çukurova University: Adana, Turkey, *in prep*.
50. Ministry of Environment Urbanization and Climate Change TÜBITAK-MRC. *Integrated Marine Pollution Monitoring 2020–2022 Program: 2022, Mediterranean Sea Report*; Ministry of Environment Urbanization and Climate Change TÜBITAK-MRC: Kocaeli, Turkey, 2023.
51. Polat, S.; Uysal, Z. Abundance and biomass of picoplanktonic *Synechococcus* (Cyanobacteria) in a coastal ecosystem of the northeastern Mediterranean, the Bay of İskenderun. *Mar. Biol. Res.* **2009**, *5*, 363–373. [CrossRef]
52. Terbıyık Kurt, T. Contribution and acclimatization of the swarming tropical copepod *Dioithona oculata* (Farran, 1913) in a Mediterranean coastal ecosystem. *Turk. J. Zool.* **2018**, *42*, 567–577. [CrossRef]
53. Terbiyik, T.; Cevik, C.; Toklu-Alicli, B.; Sarihan, E. First record of *Ferosagitta galerita* (Dallot, 1971) [Chaetognatha] in the Mediterranean Sea. *J. Plankton Res.* **2007**, *29*, 721–726. [CrossRef]
54. Tiralongo, F.; Akyol, O.; Al Mabruk, S.A.A.; Battaglia, P.; Beton, D.; Bitls, B.; Borg, J.A.; Bouchoucha, M.; Çinar, M.E.; Crocetta, F.; et al. New alien Mediterranean biodiversity records (August 2022). *Mediterr. Mar. Sci.* **2022**, *23*, 725–747. [CrossRef]
55. Ergül, H.A. Evaluation of seasonal physicochemical conditions and chlorophyll-a concentrations in Izmit Bay, Marmara Sea. *J. Black Sea Mediterr. Environ.* **2016**, *22*, 201–217.
56. Tan, I.; Beken, Ç.P.; Öncel, S. Pressure-impact analysis of the coastal waters of Marmara Sea. *Fresenius Environ. Bull.* **2017**, *26*, 2689–2699.
57. Ergul, H.A.; Balkis-Ozdelice, N.; Koral, M.; Aksan, S.; Durmus, T.; Kaya, M.; Kayal, M.; Ekmekci, F.; Canli, O. The early stage of mucilage formation in the Marmara Sea during spring 2021. *J. Black Sea Mediterr. Environ.* **2021**, *27*, 232–257.
58. Karadurmuş, U.; Sarı, M. Marine mucilage in the Sea of Marmara and its effects on the marine ecosystem: Mass deaths. *Turk. J. Zool.* **2022**, *46*, 93–102. [CrossRef]
59. Leitão, F.; Baptista, V.; Vieira, V.; Silva, P.L.; Relvas, P.; Teodósio, M.A. A 60-year time series analyses of the upwelling along the Portuguese coast. *Water* **2019**, *11*, 1285. [CrossRef]
60. Battuello, M.; Brizio, P.; Mussat Sartor, R.; Nurra, N.; Pessani, D.; Abete, M.C.; Squadrone, S. Zooplankton from a North Western Mediterranean area as a model of metal transfer in a marine environment. *Ecol. Indic.* **2016**, *66*, 440–451. [CrossRef]
61. Donnarumma, L.; Sandulli, R.; Appolloni, L.; Ferrigno, F.; Rendina, F.; Di Stefano, F.; Russo, G.F. Bathymetrical and temporal variations in soft-bottom molluscan assemblages in the coastal area facing the Sarno River mouth (Mediterranean Sea, Gulf of Naples). *Ecol. Quest.* **2020**, *31*, 53–65. [CrossRef]
62. Marino, D.; Modigh, M.; Zingone, A. General features of phytoplankton communities and primary production in the Gulf of Naples and adjacent waters. In *Marine Phytoplankton Productivity. Lecture Notes on Coastal and Estuarine Studies*; Holm-Hansen, O., Bolis, L., Gilles, R., Eds.; Springer: Berlin/Heidelberg, Germany, 1984; pp. 89–100.
63. Reyes Suarez, N.C.; Cook, M.S.; Gačić, M.; Paduan, J.D.; Drago, A.; Cardin, V. Sea surface circulation structures in the Malta-Sicily Channel from remote sensing data. *Water* **2019**, *11*, 1589. [CrossRef]
64. Consorzio di Sviluppo Economico del Friuli. Porto Nogaro. Available online: https://www.cosef.fvg.it/zona-industriale-aussa-corno/porto-nogaro.html (accessed on 19 April 2023).
65. Leray, M.; Yang, J.Y.; Meyer, C.P.; Mills, S.C.; Agudelo, N.; Ranwez, V.; Boehm, J.T.; Machida, R.J. A new versatile primer set targeting a short fragment of the mitochondrial COI region for metabarcoding metazoan diversity: Application for characterizing coral reef fish gut contents. *Front. Zool.* **2013**, *10*, 34. [CrossRef]
66. Geller, J.; Meyer, C.; Parker, M.; Hawk, H. Redesign of PCR primers for mitochondrial cytochrome c oxidase subunit I for marine invertebrates and application in all-taxa biotic surveys. *Mol. Ecol. Resour.* **2013**, *13*, 851–861. [CrossRef]

67. Androulidakis, Y.; Kolovoyiannis, V.; Makris, C.; Krestenitis, Y.; Baltikas, V.; Stefanidou, N.; Chatziantoniou, A.; Topouzelis, K.; Moustaka-Gouni, M. Effects of ocean circulation on the eutrophication of a Mediterranean gulf with river inlets: The Northern Thermaikos Gulf. *Cont. Shelf Res.* **2021**, *221*, 104416. [CrossRef]
68. Garbazey, O.A.; Popova, E.V.; Gubanova, A.D.; Altukov, D.A. First record of the occurrence of *Pseudodiaptomus marinus* (Copepoda: Calanoida: Pseudodiaptomidae) in the Black Sea (Sevastopol Bay). *Mar. Biol. J.* **2016**, *1*, 78–80. [CrossRef]
69. Zenetos, A.; Tsiamis, K.; Galanidi, M.; Carvalho, N.; Bartilotti, C.; Canning-Clode, J.; Castriota, L.; Chainho, P.; Comas-González, R.; Costa, A.C.; et al. Status and trends in the rate of introduction of marine non-indigenous species in European seas. *Diversity* **2022**, *14*, 1077. [CrossRef]
70. Zenetos, A.; Gofas, S.; Verlaque, M.; Cinar, M.E.; García Raso, J.E.; Bianchi, C.N.; Morri, C.; Azzurro, E.; Bilecenoglu, M.; Froglia, C.; et al. Alien species in the Mediterranean Sea by 2010. A contribution to the application of European Union's Marine Strategy Framework Directive (MSFD). Part I. Spatial distribution. *Mediterr. Mar. Sci.* **2010**, *11*, 381–493. [CrossRef]
71. Brown, E.A.; Chain, F.J.J.; Zhan, A.; MacIsaac, H.J.; Cristescu, M.E. Early detection of aquatic invaders using metabarcoding reveals a high number of non-indigenous species in Canadian ports. *Divers. Distrib.* **2016**, *22*, 1045–1059. [CrossRef]
72. Di Capua, I.; D'Angiolo, R.; Piredda, R.; Minucci, C.; Boero, F.; Uttieri, M.; Carotenuto, Y. From phenotypes to genotypes and back: Toward an integrated evaluation of biodiversity in calanoid copepods. *Front. Mar. Sci.* **2022**, *9*, 833089. [CrossRef]
73. Albaina, A.; Uriarte, I.; Aguirre, M.; Abad, D.; Iriarte, A.; Villate, F.; Estonba, A. Insights on the origin of invasive copepods colonizing Basque estuaries; a DNA barcoding approach. *Mar. Biodivers. Rec.* **2016**, *9*, 51. [CrossRef]
74. Abad, D.; Albaina, A.; Aguirre, M.; Laza-Martínez, A.; Uriarte, I.; Iriarte, A.; Villate, F.; Estonba, A. Is metabarcoding suitable for estuarine plankton monitoring? A comparative study with microscopy. *Mar. Biol.* **2016**, *163*, 149. [CrossRef]
75. Günther, B.; Knebelsberger, T.; Neumann, H.; Laakmann, S.; Martinez Arbizu, P. Metabarcoding of marine environmental DNA based on mitochondrial and nuclear genes. *Sci. Rep.* **2018**, *8*, 14822. [CrossRef]
76. Ohtsuka, S.; Shimono, T.; Hanyuda, T.; Shang, X.; Huang, C.; Soh, H.Y.; Kimmerer, W.; Kawai, H.; Itoh, H.; Ishimaru, T.; et al. Possible origins of planktonic copepods, *Pseudodiaptomus marinus* (Crustacea: Copepoda; Calanoida), introduced from East Asia to the San Francisco Estuary based on a molecular analysis. *Aquat. Invasions* **2018**, *13*, 221–230. [CrossRef]
77. Stefanni, S.; Stanković, D.; Borme, D.; de Olazabal, A.; Juretić, T.; Pallavicini, A.; Tirelli, V. Multi-marker metabarcoding approach to study mesozooplankton at basin scale. *Sci. Rep.* **2018**, *8*, 12085. [CrossRef]
78. Carlton, J.T.; Ruiz, G.M. Vector science and integrated vector management in bioinvasion ecology: Conceptual framework. In *Invasive Alien Species. A New Synthesis*; Mooney, H.A., Mack, R.N., McNeely, J.A., Neville, L.E., Schei, P.J., Waage, J.K., Eds.; Island Press: Washington, DC, USA, 2005; pp. 36–58.
79. Sündermann, J.; Pohlmann, T. A brief analysis of North Sea physics. *Oceanologia* **2011**, *53*, 663–689. [CrossRef]
80. Jha, U.; Jetter, A.; Lindley, J.A.; Postel, L.; Wootton, M. Extension and distribution of *Pseudodiaptomus marinus*, an introduced copepod, in the North Sea. *Mar. Biodivers. Rec.* **2013**, *6*, e53. [CrossRef]
81. Wootton, M.; Fischer, A.C.; Ostle, C.; Skinner, J.; Stevens, D.P.; Johns, D.G. Using the Continuous Plankton Recorder to study the distribution and ecology of marine pelagic copepods. In *Trends in Copepod Studies—Distribution, Biology and Ecology*; Uttieri, M., Ed.; Nova Science Publishers, Inc.: New York, NY, USA, 2018; pp. 13–42.
82. Deschutter, Y.; Vergara, G.; Mortelmans, J.; Deneudt, K.; De Schamphelaere, K.; De Troch, M. Distribution of the invasive calanoid copepod *Pseudodiaptomus marinus* (Sato, 1913) in the Belgian part of the North Sea. *BioInvasions Rec.* **2018**, *7*, 33–41. [CrossRef]
83. Mutlu, E.; Özvarol, Y. Recent record of *Oceania armata* and near-past records of other gelatinous organisms in the Turkish waters presumably derived by basin-scale current. *COMU J. Mar. Sci. Fish.* **2022**, *5*, 48–55. [CrossRef]
84. Civitarese, G.; Gačić, M.; Lipizer, M.; Eusebi Borzelli, G.L. On the impact of the Bimodal Oscillating System (BiOS) on the biogeochemistry and biology of the Adriatic and Ionian Seas (Eastern Mediterranean). *Biogeosciences* **2010**, *7*, 3987–3997. [CrossRef]
85. Ozer, T.; Gertman, I.; Kress, N.; Silverman, J.; Herut, B. Interannual thermohaline (1979–2014) and nutrient (2002–2014) dynamics in the Levantine surface and intermediate water masses, SE Mediterranean Sea. *Glob. Planet. Chang.* **2017**, *151*, 60–67. [CrossRef]
86. Fleminger, A.; Hendrix Kramer, S. Recent introduction of an Asian estuarine copepod, *Pseudodiaptomus marinus* (Copepoda: Calanoida), into southern California embayments. *Mar. Biol.* **1988**, *98*, 535–541. [CrossRef]
87. Rajakaruna, H.; Lewis, M. Temperature cycles affect colonization potential of calanoid copepods. *J. Theor. Biol.* **2017**, *419*, 77–89. [CrossRef]
88. Rajakaruna, H.; Strasser, C.; Lewis, M. Identifying non-invasible habitats for marine copepods using temperature-dependent R_0. *Biol. Invasions* **2012**, *14*, 633–647. [CrossRef]
89. Ojaveer, H.; Jaanus, A.; Mackenzie, B.R.; Martin, G.; Olenin, S.; Radziejewska, T.; Telesh, I.; Zettler, M.L.; Zaiko, A. Status of biodiversity in the Baltic Sea. *PLoS ONE* **2010**, *5*, e12467. [CrossRef]
90. Leppäkoski, E.; Gollasch, S.; Gruszka, P.; Ojaveer, H.; Olenin, S.; Panov, V. The Baltic—A sea of invaders. *Can. J. Fish. Aquat. Sci.* **2002**, *59*, 1175–1188. [CrossRef]

MDPI
St. Alban-Anlage 66
4052 Basel
Switzerland
www.mdpi.com

Journal of Marine Science and Engineering Editorial Office
E-mail: jmse@mdpi.com
www.mdpi.com/journal/jmse

www.ingramcontent.com/pod-product-compliance
Lightning Source LLC
LaVergne TN
LVHW071002170726
843515LV00006B/1053

* 9 7 8 3 0 3 6 5 9 8 7 9 6 *